Empirical Evaluation Techniques
in Computer Vision

Empirical Evaluation Techniques
in Computer Vision

Kevin W. Bowyer
P. Jonathon Phillips

IEEE
COMPUTER
SOCIETY

Los Alamitos, California

Washington ● Brussels ● Tokyo

Library of Congress Cataloging-in-Publication Data

Bowyer, Kevin, 1955–
    Empirical evaluation techniques in computer vision / Kevin Bowyer, P. Jonathon Phillips.
        p.   cm.
    Includes bibliographical references.
    ISBN 0-8186-8401-1
    1. Computer vision.   2. Computer algorithms.   I. Phillips, P Jonathon.   II. Title.
TA1634.B69   1998
006.3 ' 7—dc21   98-18854
                                                                                               CIP

IEEE Computer Society Press Order Number BP08401
Library of Congress Number 98-18854
ISBN 0-8186-8401-1

Additional copies can be ordered from

IEEE Computer Society Press	IEEE Service Center	IEEE Computer Society	IEEE Computer Society
Customer Service Center	445 Hoes Lane	13, avenue de l'Aquilon	Ooshima Building
10662 Los Vaqueros Circle	P.O. Box 1331	B-1200 Brussels	2-19-1 Minami-Aoyama
P.O. Box 3014	Piscataway, NJ 08855-1331	BELGIUM	Minato-ku, Tokyo 107
Los Alamitos, CA 90720-1314	Tel: (732) 981-1393	Tel: +32-2-770-2198	JAPAN
Tel: (714) 821-8380	Fax: (732) 981-9667	Fax: +32-2-770-8505	Tel: +81-3-3408-3118
Fax: (714) 821-4641	mis.custserv@computer.org	euro.ofc@computer.org	Fax: +81-3-3408-3553
Email: cs.books@computer.org			tokyo.ofc@computer.org

Publisher: Matt Loeb
Developmental/Acquisitions Editor: Cheryl Baltes
Advertising/Promotions: Tom Fink
Production Editor: Lisa O'Conner
Cover Design: Alex Torres. The fingerprint image is from "Fingerprint Image Enhancement:
Algorithm and Performance Evaluation" by L. Hong, Y. Wan, and A. Jain, pp. 117–134.

Printed in the United States of America by BookCrafters

IEEE
COMPUTER
SOCIETY

IEEE

Contents

Overview of Work in Empirical Evaluation of Computer Vision Algorithms

Kevin W. Bowyer[1] and P. Jonathon Phillips[2]

[1]Department of Computer Science and Engineering
University of South Florida
Tampa, Florida 33620-5359
kwb@csee.usf.edu

[2]National Institute of Standards and Technology
Gaithersburg, MD 20899, USA
jonathon@nist.gov

1: Introduction

Computer vision emerged as subfields in computer science and electrical engineering in the 1960s. (The landmark paper of Roberts on object recognition was published in 1965 [29].) Two main motivations for research in computer vision are to develop algorithms to solve vision problems, and to understand and model the human visual system. It turns out that finding satisfactory answers to either motivation is significantly harder than common wisdom initially assumed.

Research in computer vision has continued actively to the current time. Most of the research in the computer vision and pattern recognition community is focused on developing solutions to vision problems. This chapter will address issues from that point of view. There are many researchers who are interested in the human visual system, with collaborations among researchers in computer vision, neuroscience and psycho-physics. Issues of empirical evaluation are important in this area too, but we not directly address this type of research in this chapter.

With three decades of research behind current efforts, and the availability of powerful and inexpensive computers, there is a common belief that computer vision is poised to deliver reliable solutions. Unfortunately, for most applications there are no methods available to test whether computer vision algorithms can live up to their claims. Nor is there any way to measure performance among algorithms, or to reliably determine the state-of-the-art of solutions to a particular problem.

In the absence of accepted methods of empirical evaluation of algorithm performance, advances in computer vision algorithms must naturally come to be judged by other criteria. Such criteria might include considerations such as conceptual elegance, the sophistication of the mathematical methods used and the computational complexity of the algorithm. Unfortunately, conceptual elegance and sophistication of the mathematics are not necessarily correlated in a positive way with performance of an algorithm in application. If the use of more sophisticated mathematics requires more specific assumptions about the application,

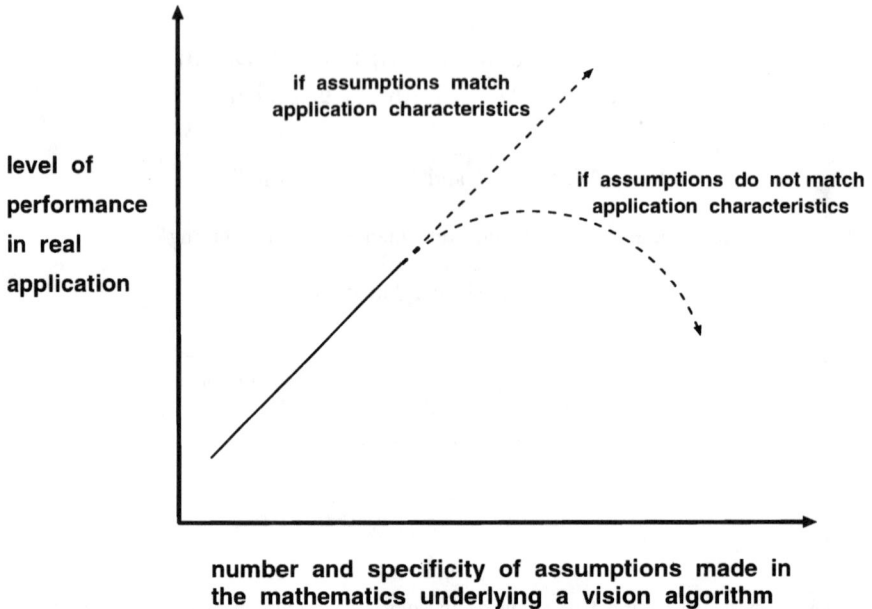

level of
performance
in real
application

**if assumptions match
application characteristics**

**if assumptions do not match
application characteristics**

**number and specificity of assumptions made in
the mathematics underlying a vision algorithm**

Figure 1. Performance as a function of mathematical sophistication.

and these assumptions are not satisfied by the application, performance could even degrade. Conceptually, the situation might be as depicted in Figure 1.

In the computer vision literature, various articles and discussions have argued for methods that would allow comparative assessments of algorithms [10, 12, 16, 20, 25]. The benefits of such methods would include (1) placing computer vision on a solid experimental and scientific ground, (2) assisting in developing engineering solutions to practical problems, (3) allowing accurate assessment of the state of the art, and (4) providing convincing evidence to potential users that computer vision research has indeed found a practical solution to their problems.

Despite these well-founded arguments, the computer vision community for the most part has not yet heeded the call. This has started to change in the last few years. There have been a number of workshops, conferences, and special issues of journals on the topic of empirical evaluation, including papers in this volume from the IEEE Workshop on Empirical Evaluation of Computer Vision Algorithms.

2   Approaches to empirical evaluation

We divide evaluation work into three basic categories. As is the risk with any classification, the categories will not necessarily be clean divisions. Evaluation work could fit into more than one category, or not neatly fit into any category. Despite this risk, we believe that the categories provide insights useful for the developing field of empirical evaluation of computer vision algorithms.

The first category is evaluations that are independently administered. In the prototypical independent evaluation, one group collects a set of images, designs the evaluation protocol, provides images to the testees, and evaluates the test results. This method allows for a high degree of standardization in the evaluation, since all algorithms are tested on the same images and scored by the same method. Thus, independent evaluations usually allow for a direct comparison between competing approaches to a problem. The competing approaches are usually state-of-the-art algorithms, and the individual competitors are often the original developers of the algorithms. Independent evaluation by a non-competitor gives a greater sense of impartiality and objectivity to the results. The major drawback to this form of evaluation is the level of ongoing effort required by the group administering the evaluation. Ideally, the evaluation mechanism needs to evolve and be refined over time.

The second category is evaluations of a set of classification algorithms by one group. The group wanting to do the evaluation will often not be able to get access to original implementations of all of the algorithms of interest, and so will have to implement some of the algorithms based on information in the literature. This introduces the possibility that the version of the algorithm evaluated will not be identical to that used by the original developers of the algorithm. However, implementation and evaluation of a set algorithms by one group can at least establish performance for baseline algorithms. When a new algorithm is first developed, it may not be sufficiently refined to be able to compete against the start-of-the-art algorithms. Comparing against a baseline allows for an initial assessment. Comparing between state of the art and baseline algorithms provides a measure of how much the additional performance costs. The cost could be a function of computational cost or complexity, reliability of the algorithm, or development effort.

An important feature of either of the first two categories is that ground truth is in theory not fuzzy and can be determined accurately. Classification problems often exhibit this property. For example, the identity of a person in a face image is not fuzzy, and the particular character that is written in a certain location is known. As long as provision for recording ground truth is made at data collection time, it should be possible to get reliable and accurate ground truth. However, in practice things are sometimes not so simple. Further discussion of ground truthing techniques is beyond the scope of this overview, but it is a problem touched on in several of the chapters.

The third category is problems where the ground truth is not self evident and a major component of the evaluation process to develop a method of obtaining the ground truth. The classic example here is developing methods of evaluating edge detectors. The question of what "should be" marked as an edge in real images is often problematic. Heath et al [13] managed to avoid a direct answer to the question and still conduct a careful comparative evaluation of five edge detectors. They used human observers to rate the quality of an edge image in the context of object recognition, and then performed a statistical analysis of the resulting quality ratings. Those without experience in conducting human observer experiments may find it difficult to conduct this method of evaluation. Also, it is conceivable that human perception of edge quality may use properties of edges which are different than those needed for machine vision tasks.

3: Tour of Papers in This Volume

The workshop had two invited speakers, Henry Baird and Michael Fitzpatrick. Baird spoke on the impact of standard databases, benchmarks, and generative models on document image analysis research.

The work described in "A blinded evaluation and comparison of image registration methods" by Fitzpatrick and West is an example of an independently administered evaluation procedure. They assessed the performance of multimodal registration algorithms for magnetic resonance imagery (MRI) and position emission tomography (PET), and PET and computer tomograghy (CT). They acquired the images, generated the ground truth, designed the testing protocol, provided the data to the groups participating in the test, and scored the results. (Additional references to their work are West et al. [37, 38, 39].)

The work presented in "A Benchmark for Graphics Recognition Systems" by Chhabra and Phillips is an independent testing procedure for algorithms that automatically analyze engineering drawings. They describe how the synthetic images and the corresponding ground truth are generated, the testing protocol, and the results. Evaluation of document image analysis algorithms is an important topic which has received substantial attention in recent years.

The paper "Performance evaluation of clustering algorithms for scalable image retrieval" by Abdel-Mottaleb, Krishnamachari, and Mankovich provides baseline performance for color-based retrieval of images. They compare performance of two clustering algorithms and three image-to-image metric-based algorithms on a database of 2,000 images.

In face recognition, principal component analysis (PCA) techniques form the basis for a number of face recognition algorithms, and provide a baseline against which other face recognition algorithms are compared. Unfortunately, there are numerous implementations of PCA algorithms. "Analysis of PCA-based face recognition algorithms" by Moon and Phillips presents an empirical evaluation of different implementations of PCA-based algorithms on the standard FERET database of facial images. Also, they look at determining what are statistically significant differences in performance between two implementations of a PCA algorithm. This is done by generating multiple (resampling) test sets from a larger database of facial images.

"Performance assessment by resampling: rigid motion estimators" by Matei, Meer, and Tyler examines resampling techniques for evaluating algorithms. They develop a set of computational tools for applying the bootstrap, jackknife, and empirical influence function to computer vision problems. Their approach is demonstrated on rigid motion estimators.

Algorithm performance can be affected by the quality of the image that goes into an algorithm. Image quality is obviously affected by noise in the image acquisition system. "Sensor errors and the uncertainties in stereo reconstruction" by Kamberova investigates the effects of sensor noise in three-dimensional reconstruction from stereo images. Based on the noise model, the author developed a radiometric correction procedure to reduce the effects of the noise.

"Fingerprint image enhancement: algorithm and performance evaluation" by Hong, Wan, and Jain presents an image enhancement algorithm for fingerprints. The effects of the enhancement are assessed by measuring the performance of a fingerprint matcher on the original and enhanced fingerprints.

In some applications it is not feasible to collect enough real data to design an algorithm. In which case, it is necessary to generate synthetic data and design an algorithm from the synthetic data. "Empirical evaluation of laser radar recognition algorithm using synthetic and real data" by Der and Qinfen discusses the process of designing and evaluating a laser radar autonomous target recognizer. The algorithm is designed from synthetic data generated by a model of a laser radar sensor. They describe the method for validating the model, developing the algorithm from the synthetic data, and validating the algorithm on real data.

The availability of a standard database for a problem makes it possible for researchers to report results on a common database of images and provides a source of images to researchers who do not each individually have the time and money to collect them. "A WWW-accessible database for 3D vision research" by Flynn and Campbell describes a database of three-dimensional models, range images, and code to support research and evaluation of model based recognition and range image analysis algorithms. The database can be browsed and retrieved on the world wide web.

Because numerous algorithms being developed are visual functions that humans can perform, one can gain insight into the visual problem by understanding how humans accomplish the task. "Shape of motion and the perception of human gaits" by Boyd and Little is a case study of the interaction of algorithm development and psycho-physics for the design of an algorithm to understand the human gait.

"Empirical evaluation of automatically extracted road axes" by Wiedemann, Heipke and Mayer looks at the problem of evaluating algorithms that attempt to extract roads in aerial images. Road extraction is a classic problem in aerial image interpretation. Wiedemann et al have manually specified the ground truth for three aerial images, and have an automated method to evaluate an extracted road map in comparison to the manually specified map. Their evaluation includes several quality metrics for the results, such as completeness of the extracted road map, location error, number of gaps per kilometer and mean gap length.

Three chapters in this volume address methods of evaluating edges. "Analytical and empirical performance evaluation of subpixel line and edge detection" by Steger presents a method of analyzing the localization quality of subpixel line and edge detectors. His work includes analytical modeling, simulation experiments and experiments with real data. The nature of his task requires relatively simple images and careful consideration of the details of the experimental setup and image acquisition. This work is most relevant to industrial inspection tasks, as opposed to more general object recognition tasks.

"Objective evaluation of edge detectors using a formally defined framework" by Dougherty and Bowyer presents a pixel-level method of evaluating edge detectors. Their method uses a set of real images for which they have marked ground truth in terms of true edges and no-edge regions. The results of an edge detector are matched to the ground truth to give a (true positive, false positive) count. A given edge detector is adaptively tuned to give its best result in terms of a receiver operating characteristic curve (ROC) for the given set of images. They use four groups of ten images, each representing a different application domain. They present a performance analysis of six edge detectors, including accuracy, ease of tuning parameters and execution time.

"Evaluation of edge detection algorithms using a structure from motion task" by Shin, Goldgof, and Bowyer presents a "higher level" or more "tasked based" evaluation method than that of Dougherty and Bowyer. They have created image sequences for which they have independently measured the 3-D structure and motion. They use the quality of results produced by the line-based structure-from-motion routine of Taylor and Kriegman [32] as the basis of their evaluation. In this context, a better edge detector is one which results in more accurate SFM results. Interestingly, Shin et al reach overall conclusions in their comparison of edge detectors which are similar to the conclusions reached by Dougherty and Bowyer.

4: Other Evaluation Work

This section provides pointers to other recent work in evaluation methods. Some of these works have received substantial attention, and some address areas not covered by the papers in this volume. This section is not meant to be an exhaustive survey. There are undoubtedly many relevant papers which we do not mention. However, we hope that this section will help to give a feel for the current state of the art in empirical evaluation of computer vision algorithms.

Barron et al [1] implemented a number of algorithms for calculating optic flow and studied the performance of these algorithms. However, their work used primarily synthetic images. Use of synthetic images obviously makes the construction of ground truth much simpler, but also introduces the question of how well results obtained on synthetic data extrapolate to real data.

Bolles et al [3] attempted to lead an empirical evaluation of algorithms for shape-from-stereo algorithms. This work was done in the style of distributing data for the evaluation to various groups for them to work on and then report results back. However, only three of five groups were able to complete enough of the evaluation to report results.

Zhang et al [44] report on an empirical evaluation of shape-from-shading algorithms. In this work, all the algorithms were implemented by the group doing the evaluation. This obviously reduces problems arising due to incompatibility of different implementations.

Hoover et al [15] conducted a detailed evaluation of four different range image segmentation algorithms. This comparison used forty images from each of two different types of range sensor. Groups from the University of South Florida, Washington State University, Edinburgh University, and the University of Bern collaborated to compare their respective algorithms. Materials for the comparison were made available on a web site for future researchers to be able to replicate/extend the comparison. The publicly-available materials include both the training and test sets of images, and software tools to automatically compare a machine segmentation to ground truth and score the result.

In the area of face recognition, the FERET evaluation procedures for automatic face recognition algorithms are well known. There have been three FERET tests, the August 1994 and March 1995 tests [23, 24], and another in September 1996 [21, 22, 28]. The FERET evaluation procedures tested the ability of algorithms to recognize faces from still images where the face occupied most of image. The difficulty of the tests increased over time. For the FERET tests, a database of face images was collected. A portion of the images was distributed to researchers for algorithm development and tuning. A second portion was sequestered for testing, which allowed algorithms to be evaluated on images that the researchers had not seen before.

The National Institute of Standards and Technology (NIST) has sponsored two evaluation conferences for optical character recognition (OCR) systems. At the evaluation conference, the results of an independent test are reported and the systems tested are described. For the evaluation, there is a dataset for development and training, and one for testing. A summary, overview, and discussion of the results of these two conferences can be found in Wilson [41] and Wilson et al. [42]. The reports of the evaluation conferences can be found Wilkinson et al. [40] and Geist et al. [11].

A number of researchers have evaluated OCR algorithms. Wilson et al. [2, 43] report performance on a number of classifiers, including nearest-neighbor, multi-layer perceptron, and radial basis function classifiers. The test was performed on NIST datasets. Ho and

Baird [14] describe a method for generating synthetic data and report performance of nearest neighbor, decision tree, and distribution map classifiers on the synthetic data. One advantage of synthetic data is that very large datasets can be generated without the cost that would be incurred for real data.

Automatic target recognition (ATR) is a vision problem of interest to the military. An ATR algorithm detects and identifies a class of objects (usually military) in different forms of imagery. Prior to fielding ATR algorithms, the military is evaluating their performance. The US Army has procedures for performing independent of evaluations on infrared ATRs [27] and the US Air Force has procedures for synthetic aperture radar [19, 30]. Li [17] reports results on a number of infrared ATR algorithms that they implemented.

Borra and Sarkar conducted a performance evaluation of edge-based perceptual grouping algorithms [4]. The performance measures used in this work are related to the speed and the accuracy of both constrained-search-based and indexing-based object recognition strategies.

In 1996, a Workshop on Performance Characteristics of Vision Algorithms was held in conjunction with the European Conference on Computer Vision [6]. Two of the organizers of this workshop, Christensen and Förstner, then edited a special issue of the journal *Machine Vision and Applications* containing extended versions of selected papers from the workshop [5]. Their overview to the special issue summarizes a number of philosophical and methodological points related to performance evaluation.

The *MV&A* special issue contains ten papers selected from the workshop. Courtney et al [7] discuss issues of algorithmic modeling. They contrast analytical and numerical approaches, in the context of a feature detection algorithm, a stereo matching algorithm and an object recognition algorithm. Ramesh and Haralick [26] present a method of analyzing the performance of edge finding, linking and gap-filling. They give both analytical predictions and empirical results for simulated data. Wenyin and Dori [36] outline a protocol for evaluating line detection algorithms in document image analysis. They give experimental results from the analysis of synthetic images. Sheinvald and Kiryati [31] analyze the performance of the "subspace-based line detection" ("SLIDE") algorithm relative to the Hough transform. They conclude that the SLIDE algorithm is less robust than the Hough transform. Vanrell et al [34] use a multi-dimensional scaling approach to analyze a particular texture perception algorithm. They use images from the well-known Brodatz texture album in their work. Eggert et al [9] compare the performance of four well-known algorithms for estimating 3-D rigid body transformations. They reach the conclusion that "... for real-world noise levels, there is no difference in the robustness of the final solutions." Madsen [18] compares the performance of two pose estimation techniques. He looks at performance both for "unstable viewpoints" and "robust viewpoints." Venetianer et al [35] investigate the performance of vision-based localization in robotic systems. Their work has the goal of being able to make statements about the performance of autonomous agents. Torr and Zisserman [33] look at an algorithm for estimating the "fundamental matrix" which represents the epipolar geometry between two images. They consider various levels of lossy JPEG compression and the effect that this has on performance. Cozzi et al [8] analyze the performance of phase-based algorithms for estimating stereo disparity. They hope to be able to estimate the expected accuracy of phase-based stereo analyzers in the face of common sources of error.

5   Summary

Empirical evaluation of algorithms is slowly emerging as a serious subfield in computer vision. This is demonstrated by the papers in this volume, recent evaluation workshops, and special issues and sections in computer vision journals. This is laying the foundation to develop accepted practices and methods for evaluating algorithms.

Evaluating algorithms lets researchers know the strengths and weaknesses of a particular approach and identifies aspects of a problem where further research is needed. From a practical point of view, successful evaluations help convince potential users that an algorithm has matured to the point that it can be successfully fielded. This helps to alleviate the situation where a solution is oversold, and the resulting disappointment when the algorithm is fielded.

To help empirical evaluations to become an accepted and expected part of the community we feel that the following steps need to be addressed. For well-established problems, there needs to be standard databases, evaluation protocols, and scoring methods available to researchers. Also, when papers in these areas are submitted for publication, journal editors and referees need to insist that authors evaluate their algorithms. Finally, just as journal reviewers and editors need to insist on appropriate evaluation methods, funding agencies need to give greater attention to these issues. While too great of a rigidity in what constitutes appropriate evaluation may stifle creative new work, the current near-total lack of emphasis of evaluation methods by funding agencies has its own dangers. The FERET program in face recognition algorithms is perhaps the best current example of evaluation methods being enforced by the funding agency.

References

[1] J. L. Barron, D. J. Fleet, and S. S. Beauchemin. Systems and experiment: Performance of optical flow techniques. _International Journal of Computer Vision_, 12(1):43–77, 1994.

[2] J. L. Blue, G. T. Candela, P. J. Grother, R. Chellappa, and C. L. Wilson. Evaluation of pattern classifiers for fingerprint and OCR applications. _Pattern Recognition_, 27(4):485–501, 1994.

[3] R. C. Bolles, H. H. Baker, and M. J. Hannah. The JISCT stereo evaluation. In _DARPA Image Understanding Workshop_, pages 263–274, 1998.

[4] S. Borra and S. Sarkar. A framework for performance characterization of intermediate level grouping modules. _IEEE Trans. PAMI_, 19(11):1306–1312, 1997.

[5] H. Christensen and W. Foerstner. Special issue on performance evaluation. _Machine Vision and Applications_, 9(5), 1997.

[6] H. Christensen, W. Foerstner, and C.B. Madsen. _Workshop on performance characterization of vision algorithms._ http://www.vision.auc.dk/hic/perf-proc.html, 1996.

[7] P. Courtney, N. Thacker, and A.F. Clark. Algorithmic modelling for performance evaluation. _Machine Vision and Applications_, 9(5):219–228, 1997.

[8] A. Cozzi, B. Crespi, F. Valentinotti, and F. Worgotter. Performance of phase-based algorithms for disparity estimation. _Machine Vision and Applications_, 9(5):334–340, 1997.

[9] D.W. Eggert, A. Lorusso, and R.B. Fisher. Estimating 3-d rigid body transformations. *Machine Vision and Applications*, 9(5):272–290, 1997.

[10] O. Firschein, M. Fischler, and T. Kanade. Creating benchmarking problems in machine vision: scientific challenge problems. In *DARPA Image Understanding Workshop*, pages 177–182, 1993.

[11] J. Geist, R. A. Wilkinson, S. Janet, P. J. Gother, B. Hammond, N. J. Larsen, R. M. Klear, M. J. Matsko, C. J. C. Burges, R. Creecy, J. J. Hull, T. P. Vogl, and C. L. Wilson. The second census optical character recognition systems conference. Technical Report NISTIR 5452, National Institute of Standards and Technology, 1994.

[12] R. Haralick. Computer vision theory: the lack thereof. *Computer Vision, Graphics, and Image Processing*, 36:372–386, 1986.

[13] M. Heath, S. Sarkar, T. Sanocki, and K. Bowyer. A robust visual method for assessing the relative performance of edge detection algorithms. *IEEE Trans. PAMI*, 19(12):1338–1359, 1997.

[14] T. K. Ho and H. S. Baird. Large-scale simulation studies in image pattern recognition. *IEEE Trans. PAMI*, 19(10):1067–1079, 1997.

[15] A. Hoover, G. Jean-Baptiste, X. Jiang, P. Flynn, H. Bunke, D. Goldgof, K. Bowyer, D. Eggert, A. Fitzgibbon, and R. Fisher. An experimental comparison of range image segmentation algorithms. *IEEE Trans. PAMI*, 18(7):673–689, 1996.

[16] R. Jain and T. Binford. Ignorance, myopia, and naivete in computer vision systems. *CVGIP: Image Understanding*, 53(1):112–117, 1991.

[17] B. Li, Q. Zheng, S. Z. Der, and R. Chellappa. FLIR-ATR techniques. *Automatic target Recognition VIII*, Proceedings of SPIE Vol. 3371, (in press 1998).

[18] C.B. Madsen. A comparative study of the robustness of two pose estimation techniques. *Machine Vision and Applications*, 9(5):291–303, 1997.

[19] J. C. Mossing and T. D. Ross. MSTAR evaluation breaks new ground: methodology, results, infrastructure, and data analysis. *Algorithms for synthetic aperture radar imagery V*, Proceedings of SPIE Vol. 3370, (in press 1998).

[20] T. Pavlidis. Why progress in machine vision is so slow. *Pattern Recognition Letters*, 13:221–225, 1992.

[21] P. J. Phillips, H. Moon, P. Rauss, and S. Rizvi. The FERET evaluation methodology for face-recognition algorithms. In *Proceedings Computer Vision and Pattern Recognition 97*, pages 137–143, 1997.

[22] P. J. Phillips, H. Moon, S. Rizvi, and P. Rauss. The FERET evaluation. In H. Wechsler, P. J. Phillips, V. Bruce, F. Fogelman-Soule, and T. S. Huang, editors, *Face Recognition: From theory to applications*. Springer Verlag, Berlin, (to appear 1998).

[23] P. J. Phillips, P. Rauss, and S. Der. FERET (face recognition technology) recognition algorithm development and test report. Technical Report ARL-TR-995, U.S. Army Research Laboratory, 1996.

[24] P. J. Phillips, H. Wechsler, J. Huang, and P. Rauss. The FERET database and evaluation procedure for face-recognition algorithms. *Image and Vision Computing Journal*, in press, 1998.

[25] K. Price. Anything you can do, I can do better (no you can't). *Computer Vision, Graphics, and Image Processing*, 36:387–391, 1986.

[26] V. Ramesh and R.M. Haralick. Random perturbation models for boundary extraction sequence. *Machine Vision and Applications*, 9(5):229–239, 1997.

[27] J. A. Ratches, C. P. Walters, R. G. Buser, and B. D. Guenther. Aided and automatic target recognition based upon sensory inputs from image forming systems. *IEEE Trans. PAMI*, 19(9):1004–1019, 1997.

[28] S. Rizvi, P. J. Phillips, and H. Moon. A verification protocol and statistical performance analysis for face recognition algorithms. In *Proceedings Computer Vision and Pattern Recognition 98*, (to appear) 1998.

[29] L. G. Roberts. Machine perception of three-dimensional solids. In J.T. Tippett et al., editor, *Optical and Electro-Optical Information Processing*, pages 159–197. MIT Press, Cambridge Massachusetts, 1965.

[30] T. D. Ross, S. W. Worrell, V. J. Velten, J. C. Mossing, and M. L. Bryant. Standard SAR ATR evaluation experiments using the MSTAR public release data set. *Algorithms for synthetic aperture radar imagery V*, Proceedings of SPIE Vol. 3370, (in press 1998).

[31] J. Sheinvald and N. Kiryati. On the magic of slide. *Machine Vision and Applications*, 9(5):251–261, 1997.

[32] C. Taylor and D. Kriegman. Structure and motion from line segments in multiple images. *IEEE Trans. PAMI*, 17:1021–1032, 1995.

[33] P.H.S. Torr and A. Zisserman. Performance characterization of fundamental matrix estimation under image degradation. *Machine Vision and Applications*, 9(5):321–333, 1997.

[34] M. Vanrell, Jvitrià, and X. Roca. A multidimensional scaling approach to explore the behavior of a texture perception algorithm. *Machine Vision and Applications*, 9(5):262–271, 1997.

[35] P.L. Venetianer, E.W. Large, and R. Bajcsy. A methodology for evaluation of task performance in robotic systems. *Machine Vision and Applications*, 9(5):304–320, 1997.

[36] L. Wenyin and D. Dori. A protocol for performance evaluation of line detection algorithms. *Machine Vision and Applications*, 9(5):240–250, 1997.

[37] J. West, J. M. Fitzpatrick, M. Y. Wang, B. M. Dawant, C. R. Maurer, Jr., R. M. Kessler, R. J. Maciunas, C. Barillot, D. Lemoine, A. Collignon, F. Maes, P. Suetens, D. Vandermeulen, P. A. van den Elsen, P. F. Hemler, S. Napel, T. S. Sumanaweera, B. Harkness, D. L. G. Hill, C. Studholme, G. Malandain, X. Pennec, M. E. Noz, G. Q. Maguire, Jr., M. Pollack, C. A. Pelizzari, R. A. Robb, D. Hanson, and R. P. Woods. Comparison and evaluation of retrospective intermodality image registration techniques. *Medical Imaging 1996: Image Processing*, Proc. SPIE 2710:332–347, 1996.

[38] J. West, J. M. Fitzpatrick, M. Y. Wang, B. M. Dawant, C. R. Maurer, Jr., R. M. Kessler, and R. J. Maciunas. Retrospective intermodality registration techniques: Surface-based versus volume-based. In J. Troccaz, E. Grimson, and R. Mösges, editors, *CVRMed-MRCAS '97*, pages 151–160. Springer-Verlag, Berlin, 1997.

[39] J. West, J. M. Fitzpatrick, M. Y. Wang, B. M. Dawant, C. R. Maurer, Jr., R. M. Kessler, R. J. Maciunas, C. Barillot, D. Lemoine, A. Collignon, F. Maes, P. Suetens, D. Vandermeulen, P. A. van den Elsen, S. Napel, T. S. Sumanaweera, B. Harkness, P. F. Hemler, D. L. G. Hill, D. J. Hawkes, C. Studholme, J. B. A. Maintz, M. A.

Viergever, G. Malandain, X. Pennec, M. E. Noz, G. Q. Maguire, Jr., M. Pollack, C. A. Pelizzari, R. A. Robb, D. Hanson, and R. P. Woods. Comparison and evaluation of retrospective intermodality image registration techniques. *J. Comput. Assist. Tomogr.*, 21:554–566, 1997.

[40] R. A. Wilkinson, J. Geist, S. Janet, P. J. Gother, C. J. C. Burges, R. Creecy, B. Hammond, J. J. Hull, N. J. Larsen, T. P. Vogl, and C. L. Wilson. The first optical character recognition systems conference. Technical Report NISTIR 4912, National Institute of Standards and Technology, 1992.

[41] C. L. Wilson. Effectiveness of feature and classifier algorithms in character recognition systems. Technical Report NISTIR 4995, National Institute of Standards and Technology, 1992.

[42] C. L. Wilson, J. Geist, M. D. Garris, and R. Chellappa. Design, integration, and evaluation of form-based handprint and ocr systems. Technical Report NISTIR 5932, National Institute of Standards and Technology, 1996.

[43] C. L. Wilson, P. J. Grother, and C. S. Barnes. Binary decision clustering for neural-network-based optical character recognition. *Pattern Recognition*, 29(3):425–437, 1996.

[44] R. Zhang, P. Tsai, J. Cryer, and M. Shah. Analysis of shape from shading techniques. In *Computer Vision and Pattern Recognition*, pages 377–384, 1994.

A Blinded Evaluation and Comparison of Image Registration Methods

J. Michael Fitzpatrick*   Jay B. West*

1: Introduction

During the period from March, 1995 through February, 1998 a project supported by the National Institutes of Health [1] was conducted at Vanderbilt University. The goal of the project was to perform an evaluation of a set of methods for registering medical images of the human head. The project was the first of its kind in medical imaging and has not of this writing been duplicated. The challenge was to provide a means for carrying out a blinded study of image processing algorithms on a common data set when the algorithms were being applied in laboratories that were scattered over the world. The term "blinded" here means that the people who were applying the registration methods were ignorant of the correct answers. By keeping the study blind and using a common data set we feel that we heightened the credibility of the evaluation considerably.

To carry out a study of this nature we had to solve many problems, some of which were specific to the particular problem of registration (e.g., how to describe a registering transformation), some of which were specific to image processing (e.g., how to describe an image format), and some of which would apply to any study involving the blinded application of algorithms at remote sites that were to be evaluated at some central site (e.g., how to make sure that communication errors did not confuse the evaluation results). One problem that is universal to any such study is this: How are the correct answers determined? This problem can be a difficult one in general, but in our case it had been solved before this project was undertaken by the development of a highly accurate registration system at Vanderbilt. That system gets its accuracy by taking advantage of special preparation of the patients whose images were being used for the study.

This project encountered many problems, but they were all solvable, given enough attention and given sufficient cooperation among the many groups involved. After the first twelve groups' registrations had been evaluated we declared that the first study was completed. The results of the study were published immediately after its completion [6]; instead of concentrating here on the results, the emphasis will instead be on the procedures and potential pitfalls inherent in performing such a comparison study, in the belief that this is equally applicable to similar evaluation studies in any field.

Maintaining blindedness was a difficult challenge but we feel that the increased credibility is worth the effort.

*Department of Computer Science, Vanderbilt University, Nashville, TN

[1] The images and the standard transformations were provided as part of the project "Evaluation of Retrospective Image Registration", National Institutes of Health, Project Number 1 R01 NS33926-01, Principal Investigator, J. Michael Fitzpatrick, Vanderbilt University, Nashville, TN.

1.1: The Registration Problem

The problem of image registration is as follows: given two different representations of the same object (*e.g.* Magnetic Resonance (MR) and Computerized Tomography (CT) images of a patient's head), to find a transformation which, when applied to one image, will align (or *register*) points in that image with corresponding points in the other. In this study, we considered three-dimensional MR, CT and Positron Emission Tomography (PET) images of patients who were due to undergo neurological surgery. Each image volume consisted of a stack of two-dimensional "slices". There were two tasks evaluated: the registration of CT images to the corresponding MR images, and the registration of PET to MR. The MR images could be subdivided into submodalities – Proton Density (PD) weighted, T1 weighted and T2 weighted, both with and without a geometrical distortion correction algorithm [3, 5] having been applied. As the human head behaves approximately as a rigid object, we evaluated only rigid-body registration methods, *i.e.* those that produced a transformation consisting solely of a rotation and translation. As we shall see later, this greatly simplified the evaluation step of this study.

1.2: The gold standard

In order to compare a set of registration techniques, it is necessary to have a single technique that is generally agreed to have superior accuracy to those in the evaluation set, and to use the results generated by this technique as the "ground truth" for comparison purposes. The Acustar^{TM} 1 Surgical Navigation System was used as the gold standard in this study. This system was developed at Vanderbilt in collaboration with Johnson & Johnson Professional, Inc. and relies on markers that are rigidly attached to the skull in order to derive the registration transform. The markers (called *fiducial* markers) are designed to show up brightly in the images, and may be replaced by devices called *localization caps* in the operating room in order to allow the surgeon to navigate interactively during surgery. Systems such as this one are labeled *prospective*: they require preparatory steps to be taken before imaging in order to perform the registration. The preparatory step in this case is the drilling of holes into the outer table of the patient's skull. This step is highly invasive and is considered a disadvantage of using this registration method. However, it is generally agreed that prospective techniques such as this one give more accurate results than *retrospective* techniques (*i.e.* those that do not require any special preparation to be performed before the images are acquired). Given that all the methods to be evaluated were retrospective in nature, we felt it appropriate to use as the gold standard the results generated by the Acustar system.

1.3: The techniques to be evaluated

Researchers at twelve separate sites performed a total of sixteen distinct retrospective registration techniques on the image data we provided. It is possible (with a few exceptions) to categorize these methods into two groups: volume-based and surface-based. The volume-based methods make use of a correlation between the intensity values of pixels in either image that correspond to the same anatomical location; the label stems from the fact that all the information in each image volume is used to compute the registration transform. The surface-based methods consist of a segmentation step that derives a representation of surfaces in each image that are either corresponding or closely fitting (*e.g.* the skull in

CT and the brain surface in MR) and a registration step that finds the transformation minimizing some distance measure between the two surfaces.

2: Data preparation

The gold standard transformations were based on the positions of the fiducial markers in the images: the correct transformation is defined as the one that minimizes the root-mean-square (RMS) distance between corresponding fiducial points after registration. We used the closed-form solution developed by Arun et al. [1] to find this optimal mapping between the two images. All the registrations to be evaluated were retrospective in nature, and therefore were supposed to be able to register the images without recourse to the fiducial markers. It was therefore necessary for us to remove all trace of these markers from the images, while attempting to reconstruct the removed regions to make them appear as close as possible to the natural background pattern. Also present in the image was a stereotactic frame. This is a cage-like structure rigidly attached to the patient's head that was used for intrasurgical navigation before the advent of the marker-based system. The frame was present in addition to the markers in the images used in the study as these images were acquired during a testing phase of the fiducial markers, where the frame was used as a backup measure to the markers. Clearly the frame also had to be removed from the images in order to provide a fair test for the techniques being evaluated.

In order to prepare the images for use in this study, we developed a technique that we dubbed *air brushing*. This was performed by manually outlining the regions containing the structures to be removed, followed by approximately reconstructing the image background in each missing region R. In MR, where the background consists of unstructured noise, pixels at random positions between the edge of R and the lateral image boundary were sampled and placed in R. In CT, as the outer regions of the image are comprised mainly of reconstruction artifacts that take the form of quasi-radial stripes, the approach taken was to interpolate these stripes within the removed region. This was done at a given point P in region R by the following method (Fig. 1):

1. Calculate the radial distances e and i of the point P, respectively, from the external and internal boundaries of the region.

2. Identify points E and I at radial distances e and i, respectively, external and internal to the region. If E lies outside the image, set E to be on the border of the image. A similar precaution is taken to insure that I does not lie within the head.

3. Let the intensity at E be I_e and at I be I_i. Assign P's intensity I_p as $I_p = I_e$ with probability ρ, or $I_p = I_i$ with probability $1 - \rho$, where $\rho = i/(i + e)$.

A similar technique was used for the PET images; in this case, however, pixels in region R were set to an intensity value linearly interpolated between the intensity of the internal and external boundaries of R.

This process is illustrated in Figure 2. The top row (Fig. 2a–c) shows slices of original volumes from each of the three modalities. The window and level have been set so that the background artifacts may be seen. The middle row (Fig. 2d–f) shows the same slices after the region R has been outlined and zeroed. This procedure is applied to each slice in the volume. The last row (Fig. 2g–i) shows the slices after reconstruction of the background in the region R. For the MR case, it can be seen that the replaced area is indistinguishable from the rest of the background. In CT and PET, there are slight discontinuities in the direction of the stripes, but the intensity changes smoothly.

3: Communication

The communication phase of this study may be divided into two phases: the transfer of the image data to the remote sites (those performing the registrations) and the transmission of the registration transformations back to the central site (Vanderbilt) where the evaluations were to be performed.

3.1: Transferring the images

When choosing the method by which the image volumes would be distributed to the remote sites there were several options available, including writing the data on to CD-ROMs and sending them by physical mail, and the use of FTP (File Transfer Protocol) to allow electronic transmission of data from the central site to the remote sites. We selected FTP for several reasons:

- It did not require extra hardware at the central site (*e.g.*, a writeable CD-ROM drive) or at the remote sites (*e.g.*, a CD-ROM drive connected to the computer being used to perform the registration algorithm).
- It allowed incremental changes in the image dataset to be made, *e.g.*, if the file containing one of the image volumes was found to be corrupted, the problem could be remedied with little additional work.
- The work contained in transmitting the information was largely distributed amongst the remote sites, rather than being concentrated at the central site.

In addition to the binary data corresponding the images themselves, it was also necessary that the remote sites have access to other pertinent data, *e.g.*, the resolution and physical dimensions of each voxel, and orientation of each image. This data was provided for each image by means of an ASCII header file that corresponded to the Interfile standard [2, 4]. This standard was developed specifically to facilitate the transmission of information describing three-dimensional medical images.

Another concern was the protection of patient confidentiality: the Vanderbilt Institutional Review Board deemed it permissible to distribute the patient image data to qualified researchers, but not to make them generally accessible to anyone who wished to view them. To this end, we created a login and password that allowed FTP access to the images; in order to be given this password, a responsible individual at each remote site was required to submit to the central site a signed document promising that the image data would not be redistributed.

3.2: Submission of the registration transforms

A three-dimensional rigid body transformation may be specified in terms of six parameters (three rotation angles and three translation components). Because of the fact that the amount of data involved in communicating a transformation is relatively small, Internet e-mail seemed the obvious choice as a protocol for transmission of results. There still remained the problem, however, that there are many different ways in which a rigid-body transform may be specified (*e.g.* rotation angles, a rotation matrix and translation vector, quaternions). Each remote site had its own internal representation format for the results; the task of analyzing the results was performed entirely at the central site, and we saw the danger of great confusion if each remote site were to be allowed to submit results in its own format. Hence we insisted on the adoption of a standard submission format that was to be used in all results submissions. This format was as follows: in the "From" volume (e.g., CT in the case of CT-to-MR registration), the positions of eight points are calculated; taking the origin to be the center of the first voxel in the volume (i.e., the top left pixel of slice zero), the x, y, and z coordinates of the centers of the eight corner voxels in the volume were derived. These positions were provided via FTP by Vanderbilt for every CT and PET volume in the form of a partially completed (columns headed "x","y", and "z") "transformation table" (Fig. 4 and Table 1) for each pair of volumes.

After the retrospective registration transformation was determined, the transformed positions of these eight points *relative to the origin of the "To" modality* (Fig. 5) were computed by each site (see Table 1, "new_x","new_y", and "new_z"). As depicted in Fig. 5, the field of view of the two volumes is typically different, so it is important to specify which volume provides the origin relative to which the transformed positions are calculated.

All coordinates were specified to at least four decimal places in units of millimeters. Such high precision insures that any round-off error inherent in converting between a registration transformation and the eight-point sets is negligible. In order to convert back from the transformation table to a rigid body transformation, a singular value decomposition (SVD) algorithm [1] was used, giving the rigid body transformation that minimized the mean square distance between the two sets of points. This was the same algorithm used to register the positions of the fiducial markers in the two image volumes.

Only three points are necessary to uniquely specify such a transformation, but the full set of eight was used for reasons of symmetry, error reduction, and error prevention.

Clearly, this method of data transmission allows only rigid-body transformations to be accurately communicated, since any nonrigid transformation would be approximated by a rigid one. The use of this protocol thus limited the scope of this project to an evaluation restricted to rigid-body transformations. However, by measuring the *fiducial registration error* (FRE), i.e., the distance between the pairs of corresponding points after the application of the SVD algorithm, it is possible to determine whether or not the retrospective registration uses nonrigid deformations, since the FRE in this case would be on the order of millimeters, as opposed to hundredths of a millimeter for a well-specified rigid-body transformation. This feature is facilitated by the use of a larger point set than necessary, and guards against the possibility of a nonrigid transformation being mistakenly supplied and evaluated as if it were rigid.

Each transformation was transmitted to Vanderbilt by e-mailing an ASCII file containing a completed transformation table (see Table 1).

4: Data Analysis

At Vanderbilt, after the transformation tables had been received from each site and the corresponding rigid-body transformations determined, the next step in the evaluation was to perform a comparison between these registrations and the fiducial-based ones. In collaboration with a neurological and a neurosurgical expert, a set of *volumes of interest*

(VOIs) was chosen representing areas of neurological and/or surgical interest. Each VOI was manually segmented within one of the MR image volumes; this procedure was repeated for each patient data set used. The VOIs were stored as sets of x, y, and z voxel coordinates.

An estimate of the accuracy of the retrospective registration at the position of each VOI was computed as follows (Fig. 3). The centroid pixel of the VOI was found, and its position was converted from a voxel index to a millimetric position c in the "To" modality using the known voxel size for the image volume. Let R_g and t_g be the rotation matrix and translation vector, respectively, of the gold-standard rigid-body transformation G, and R_r and t_r be the rotation and translation components of the retrospective transformation R. The point c' in the "From" modality was defined so that c is the mapping of c' under the gold-standard transformation. Thus,

$$c = G(c') = R_g c' + t_g. \tag{1}$$

By inverting equation (1), we obtain

$$c' = G^{-1}(c) = R_g^{-1} c - R_g^{-1} t_g. \tag{2}$$

The point c'' in the "To" modality is defined as the mapping of c' under the retrospective transformation. Thus,

$$c'' = R(c') = R_r c' + t_r. \tag{3}$$

The difference between the registered target position of the retrospective method and that of the gold standard is $d = c'' - c$. We define the *Target Registration Error* (TRE) of the retrospective registration at the anatomical location of the VOI to be the Euclidean distance d between c and c'', i.e., TRE $= d = \|d\|$.

The particular anatomical positions corresponding to the VOIs used in this evaluation are as follows: 1) maximum aperture of fourth ventricle, 2) junction of fourth ventricle with aqueduct, 3) right globe, 4) left globe (not shown), 5) optic chiasm, 6) apex of left Sylvian fissure, 7) apex of right Sylvian fissure, 8) junction of central sulcus with midline, 9) left occipital horn, and 10) right occipital horn.

Just as there are many ways to specify a rigid-body transformation, there are also many formats in which the difference between the retrospective and gold standard registration may be written. We chose to give the TRE in millimeters, as specified at a set of major landmark locations in the brain, because we felt that this would allow physicians reading the results easily to assess how useful the registration techniques would be if they gave the same degree of accuracy when used in clinical practice.

5: Presentation of results

When we finished analyzing the registration transforms that were sent to us, we were left with error estimates for sixteen registration techniques on 76 image matching tasks. For each task, we estimated the error at ten VOIs, thus giving a total of 12,160 numbers. Clearly this was far too much data to present, so we had to devise a method of compiling the data into statistics that gave a good description of the accuracy of each registration method. We chose to present six tables: median, 75th percentile, and maximum values for PET-MR and CT-MR registration. Each table contained a column for each of the remote sites that submitted transforms for that modality pair (some sites performed only CT-MR or PET-MR registration); each row of the table contained statistics for matching PET or CT volumes to a particular MR submodality (PD, T1 and T2, with and without distortion correction having been applied). Each statistic was then compiled from the errors measured at all VOIs for all patient datasets containing that modality pair. A sample table of results is shown in Table 2.

The rationale for our choice of statistics was as follows. We presented the maximum value for each technique to give insight into the stability and reliability of the technique. We used the median rather than mean value to give an estimate of the typical registration accuracy, unweighted by large errors occurring in a small number of instances where the technique performed very poorly. The 75th percentile data was added to give extra insight into reliability, *i.e*, whether a particular method failed only in one instance or for a sizeable percentage of the registration tasks.

The question that everyone wanted our study to answer was "which of the registration techniques tested was the best?". Unfortunately, because of the fact that the number of techniques was large compared to the number of tasks on which they were tested, our data lacked the statistical power to proclaim a "winner". We did notice, however, that the volume-based techniques tended to perform better than the surface-based ones. This led us to perform another analysis of the data. We partitioned the registration methods into two groups: volume-based and surface-based (omitting the few that fell into neither group). We performed the same analysis on the aggregate results of each group. In this case, there was enough statistical power in the data for us to be able to proclaim that on many of the modality pairs, the volume-based methods performed significantly more accurate registrations than the surface-based ones [7]. We were also able to conclude that, for some modality pairs, mainly those including a CT volume, the application of geometrical distortion correction to the MR volume [3, 5] resulted in a significant reduction in registration error.

6: Blindedness

We use the adjective *blinded* to describe the remote sites, to reflect the fact that they remained ignorant of the gold standard solutions until the results had been submitted. The degree to which the gold standard solutions were to be kept secret, however, was a matter that needed detailed consideration. At one extreme was the possibility of making the fiducial-based registrations freely available throughout the study: this approach seemed to have few advantages and would lead many to dismiss the results as unreliable. It would also have been feasible to withhold the gold standard indefinitely from the remote sites, and to withhold the calculated errors of the retrospective transformations until publication of the results, as the evaluation step was carried out entirely at Vanderbilt. This protocol has the advantage that it would allow researchers at the remote sites to find and correct errors in their registration transformations after submission, without suspicion that the gold standard solutions (or the error statistics derived from them) had led them to discover mistakes that in clinical practice would have remained undetected. However, for reasons that will be explained in Section 7, some of the remote sites strongly felt that this approach was unacceptable, and we elected to adopt a compromise solution.

We withheld all results and gold standard transforms until all the remote sites had completed their submissions. At this point, we sent by e-mail to the researchers at each site a table containing the TRE values that we had calculated for their registration method. We also created a separate login and password that gave access to the gold standard information, and distributed this. We advised each site to examine carefully the numbers in that table, paying particular attention to any values that seemed larger than expected.

The blindedness of the study would clearly have been compromised if we had routinely accepted resubmissions of results after the gold standard had been released; it was of concern, however, that the assessment might be inaccurate because of mistakes in, for example, conversions between the internal representation format of remote sites and the format required for submission to the central site. These assessments would not give an accurate view of the clinical performance of a registration technique, as such conversions would not have to be performed for real-world registration tasks. For this reason, we allowed remote sites to review the results and if necessary to send a statement that an error had been made in the submission of a particular subset of the results. We did not allow resubmission of this subset; instead, we omitted the erroneous results from the calculation of the overall error statistics for that registration method and added a footnote describing the omission in the articles containing the results. We made only one exception to the "no resubmission" rule, as described in Section 7. We also allowed remote sites to send a declaration that the registration had failed in particular cases. This declaration would have allowed the omission of those cases from the overall error calculations, and a note would have been added stating how many of the registration tasks were declared to be failures for each method. We received no such declarations, however.

7: Errors

It is natural to expect that, because of the complexity of and number of participants in this study, there would be some degree of error and miscommunication. Rather than striving for the elimination of error, we tried to develop a protocol that would give a good opportunity for such errors, made by both the central and remote sites, to be detected and corrected before the final results were published.

7.1: The practice patient

In addition to the datasets that were used for the study, we provided an additional patient dataset that was similar to the others but different in two respects. First, the gold standard registration transforms for the dataset were immediately made accessible to the remote sites. Second, the images in this dataset had had randomly chosen gross rotations and translations applied to them, so that the image pairs were profoundly misregistered. The images used for the evaluations were only slightly misregistered.

The rationale for providing this "practice" dataset was that the remote sites would be able to perform registrations on this dataset and compare results against the gold standard. Inexplicably large errors would be a warning that the conversion between registration representation formats was possibly being performed incorrectly. The images were significantly misaligned so that such incorrect conversions would show up more clearly.

The practice patient rapidly proved valuable, but in a way that was quite surprising. A researcher at one of the remote sites complained that his registrations were significantly different from the gold standard, but that his results appeared to provide superior registrations on visual inspection. This led to the discovery of an error in the algorithm used at Vanderbilt to compute the gold standard transformations. Although this error was related to the preparatory reformatting steps taken in producing the practice dataset and would therefore not have affected the calculation of error statistics for the evaluations, it was imperative that the mistake be found and corrected lest the other remote sites be confused by the failure of their registration transforms to match the erroneous gold standard results on the practice patient. Because of this error one of the researchers urged us to release the gold standard transformations for all the image pairs before publication of the results, therefore allowing remote sites to check visually the accuracy of the transforms against which theirs were to be compared.

After we corrected the gold standard transformations, several remote sites still reported a large mismatch with their registration solutions. Communication with these sites revealed several conversion errors between transformation formats.

7.2: Large misregistrations

When we had completed the analysis of all the results submitted to us, we saw that, for some sites, the maximum registration errors were extremely large (greater than 40 mm). There were four possible explanations for a large error being observed for a particular matching task and remote site:

1. The gold standard was in error for that transform.

2. The remote site submitting the result had not correctly performed the conversion between transformation formats.

3. The retrospective registration technique in question had failed for that particular matching task.

4. An administrative error had been made by the remote site (*e.g.*, the images matched by the registration transform had been mislabeled).

In all instances, we ruled out the first and second possibilities. For the first possibility, an erroneous gold standard transform would have given consistently large error measures for all remote sites. We did not observe such consistent errors for any particular image pair. For the second possibility, the remote site would have submitted transforms all of which showed large error measures. We did not see this consistency either.

When the remote sites were presented with the results, several of them explained their large maximum errors by describing administrative mistakes that had been made. We considered these explanations and in each case this allowed us to omit the transforms in question from the overall calculation of error statistics (there were three such cases). There were still some large error measurements that remained unexplained, however. As stated in Section 6, we allowed the researchers performing the registrations to declare that their technique had failed for a particular task, but no such declarations were received. It is thus our belief that the large error values were because of registration failures that had remained undiagnosed because of a lack of visual inspection of the image matching. The conclusion of our publication of the results of the study highlighted the importance of using visual inspection to check the results of automated image registration techniques.

Finally, there was one case in which an exception was made to the rule that no registration transforms could be submitted once the gold standard had been released. Shortly before the results were due to be published, and after the gold standard had been made available, we received a statement from one remote site that they had made a mistake converting from their own transform representation to the submission format. Specifically, they claimed that the sign of one of the rotation angles of the transformation had been consistently negated during the conversion. As the orientation of the images being matched were fairly close, the rotation angles were small. The observed registration errors were changed by the sign negation on the order of 0.5 mm, which was about 50% of the median error. As we were able to independently verify that the new transforms were derived from the old ones simply by negating one of the rotation angles, we felt that the blindedness of the study was not compromised by allowing the resubmission of these results. In every case in which mistakes led to omission or changing of data a description of the nature of the mistake was included in our published results along with the evaluations both with and without the alterations.

8: Conclusion

By performing this study, we were able to publish what we believe to be a fair, quantitative comparison of many different retrospective image registration methods. Judging from the reprint requests we have received for the article containing the results [6] and from the number of research groups who have subsequently asked for and been granted access to the database so that their registration techniques may be similarly evaluated (sixteen), we believe our study was successful and hope it will provide encouragement to others to conduct blinded studies via the Internet. It appears to have provided both a service and information that was of great interest to many people active in this field. As a final note, we would point out that this study involved many people in many research groups in many countries. It could not have been successful unless these people felt strongly that it should be successful. As in any project in which the efforts of people are to be evaluated with the results to be published, there is a great potential for misunderstanding, impatience, and disagreement. These problems are especially critical when the people involved are competing in the same field for publications, funding, and peer approval. We would like to add a personal note of gratitude to the people with whom we interacted at the remote sites during this project for their professional approach. We have learned from our experience with this project, and we suspect that it applies to any such project, that while careful preparation and attention to detail are essential on the part of the evaluators, the project would fail without active, supportive cooperation on the part of those who are being evaluated.

References

[1] K. S. Arun, T. S. Huang, and S. D. Blostein. Least-squares fitting of two 3-D point sets. *IEEE Trans. Pattern Anal. Mach. Intell.*, 9:698–700, 1987.

[2] B. S. Baxter, L. E. Hitchner, and G. Q. Maguire, Jr. A standard format for digital image exchange. *American Association of Physicists in Medicine*, 1982.

[3] H. Chang and J. M. Fitzpatrick. A technique for accurate magnetic resonance imaging in the presence of field inhomogeneities. *IEEE Trans. Med. Imaging*, 11:319–329, 1992.

[4] T. D. Cradduck, D. L. Bailey, B. F. Hutton, F. de Conninck, E. Busemann-Sokole, H. Bergmann, and U. Noelpp. A standard protocol for the exchange of nuclear medicine image files. *Nucl. Med. Comm.*, 10:703–713, 1989.

Table 1. Example Transformation Table

Transformation Parameters
Investigator(s): A. B. Cee and D. E. Eff
Site: Extra University, Somewhere, New Country
Method: 1
Date: 23 August 1995
Patient number: 001
From: CT
To: MR-PD

Point	x	y	z	new_x	new_y	new_z
1	0.0000	0.0000	0.0000	-5.7884	-29.5052	-23.9565
2	333.9870	0.0000	0.0000	326.4905	-62.6706	-30.1350
3	0.0000	333.9870	0.0000	27.4767	302.7901	-19.3302
4	333.9870	333.9870	0.0000	359.7436	269.6247	-25.5087
5	0.0000	0.0000	112.0000	-3.8810	-31.2550	88.0136
6	333.9870	0.0000	112.0000	328.3798	-64.4204	81.8351
7	0.0000	333.9870	112.0000	29.3721	301.0404	92.6399
8	333.9870	333.9870	112.0000	361.6509	267.8750	86.4614

(All distances are in millimeters.)

[5] C. R. Maurer, Jr., G. B. Aboutanos, B. M. Dawant, S. Gadamsetty, R. A. Margolin, R. J. Maciunas, and J. M. Fitzpatrick. Effect of geometrical distortion correction in MR on image registration accuracy. *J. Comput. Assist. Tomogr.*, 20:666–679, 1996.

[6] J. West, J. M. Fitzpatrick, M. Y. Wang, B. M. Dawant, C. R. Maurer, Jr., R. M. Kessler, R. J. Maciunas, C. Barillot, D. Lemoine, A. Collignon, F. Maes, P. Suetens, D. Vandermeulen, P. A. van den Elsen, S. Napel, T. S. Sumanaweera, B. Harkness, P. F. Hemler, D. L. G. Hill, D. J. Hawkes, C. Studholme, J. B. A. Maintz, M. A. Viergever, G. Malandain, X. Pennec, M. E. Noz, G. Q. Maguire, Jr., M. Pollack, C. A. Pelizzari, R. A. Robb, D. Hanson, and R. P. Woods. Comparison and evaluation of retrospective intermodality brain image registration techniques. *J. Comput. Assist. Tomogr.*, 21:554–566, 1997.

[7] J. West, J. M. Fitzpatrick, M. Y. Wang, B. M. Dawant, C. R. Maurer, Jr., R. M. Kessler, and R. J. Maciunas. Retrospective intermodality registration techniques: Surface-based versus volume-based. In J. Troccaz, E. Grimson, and R. Mösges, editors, *CVRMed-MRCAS '97*, pages 151–160. Springer-Verlag, Berlin, 1997.

Table 2. Median Errors for CT-to-MR Registration

MR Modality	Technique														N
	BA	CO	EL	HA	HE	HI†	MAI	MAL	NO†	PE	RO1	RO2	RO3	RO4	
T1	1.6	1.5	1.6	3.4	1.4	1.2	5.1	4.3	3.3	2.7	4.2	5.2	5.7	5.4	7
PD	1.9	1.5	2.0	3.1	2.4	1.9	4.1	4.0	7.8	1.9	4.5	5.5	4.9	4.8	7
T2	2.5	1.5	1.6	4.2	4.7	1.5	3.9	5.0	3.9	2.5	4.5	4.5	5.4	4.7	7
T1 rect.	1.4*	0.7	0.9	3.3	1.0	0.7	4.9	5.4	3.4	2.2	5.9	5.9	6.3	5.9	6
PD rect.	1.7*	0.8	1.1	3.0	1.7	0.7	3.0	4.0	4.6	2.1	5.9	5.7	5.5*	5.5*	7
T2 rect.	2.1*	0.8	1.6	3.5	1.6	0.8	4.3	5.3	4.2	2.9	5.5	5.3	5.3	5.3	7

The label "rect." indicates that the MR image was corrected for geometrical distortion before registration. See text for technique abbreviations. *One patient omitted. †Non-rigid transformations. ‡Results resubmitted after gold standard released. All errors are in units of mm.

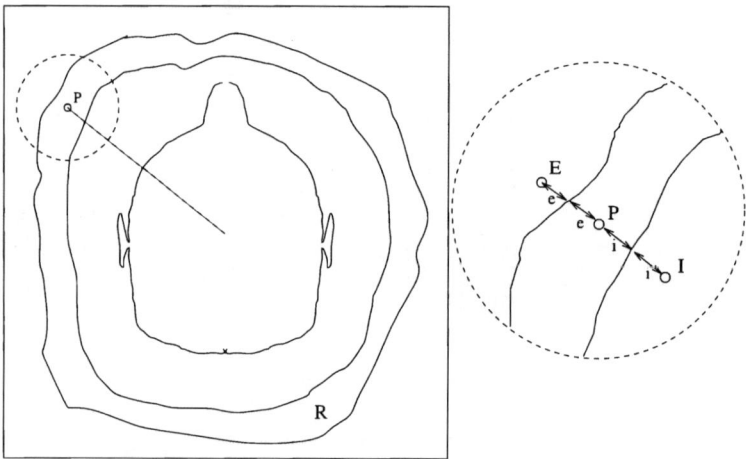

Figure 1. Calculation of points for interpolation of background patterns in CT and PET images. See text for an explanation.

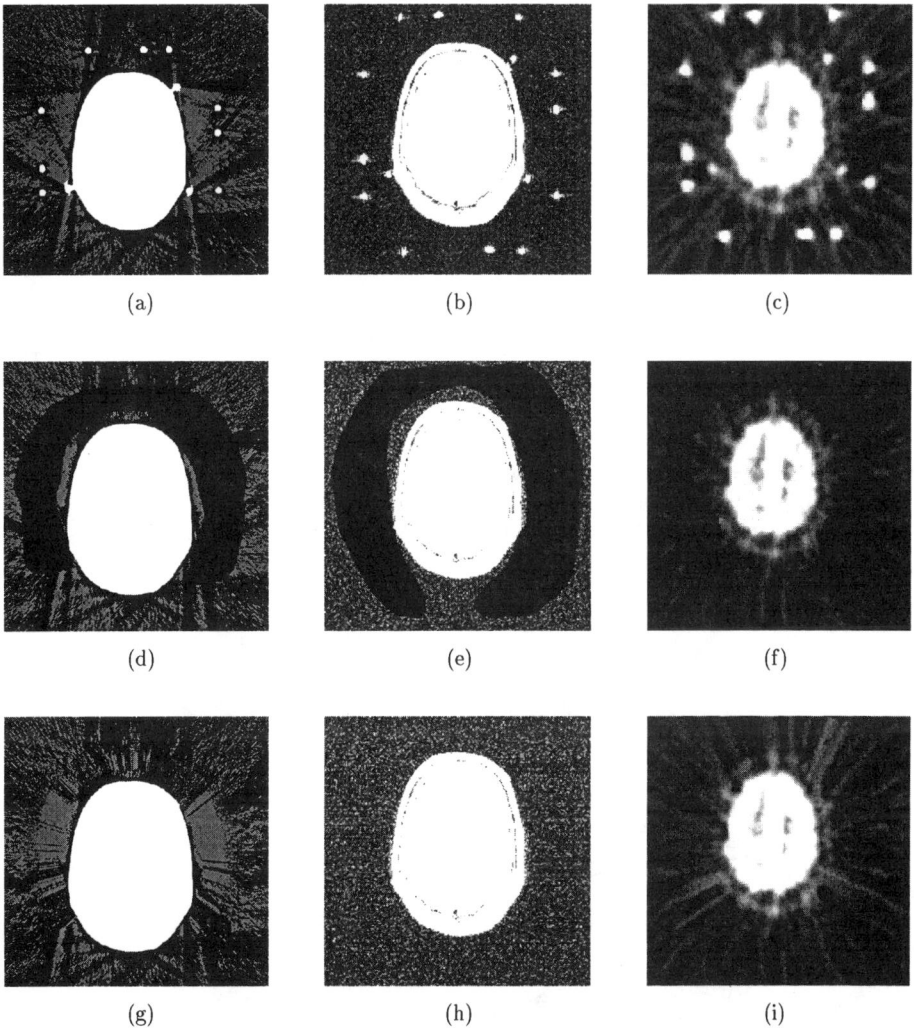

Figure 2. Removal of fiducial markers and stereotactic frame. The top row (*a*–*c*) shows sample original image slices from CT (*a*), MR (*b*), and PET (*c*). The stereotactic frame (bright circular spots towards the edge of the image) and fiducial markers (three bright spots near the head) are clearly visible in all three modalities. The window and level have been set to show the background artifacts. The middle row (*d*–*f*) shows the same image slices after the region R has been outlined and zeroed. All trace of the stereotactic frame and fiducial markers has been removed. The bottom row (*g*–*i*) shows the image slices after reconstruction of the background in the region R. For MR, the replaced area is indistinguishable from the rest of the background. For CT and PET, there are slight discontinuities in the direction of the stripes, but the intensity changes relatively smoothly.

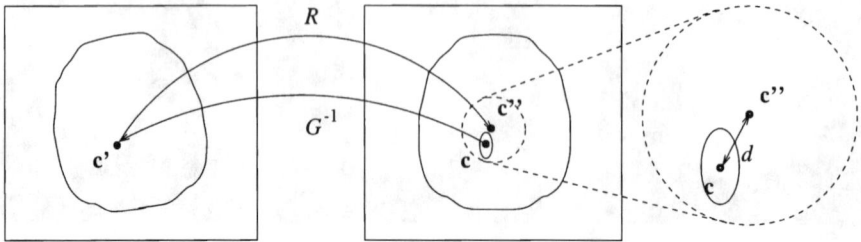

Figure 3. Calculation of the accuracy of a retrospective registration at a VOI. A VOI (represented here by an ellipse) is defined in the "To" modality (right image). The centroid voxel of the VOI is found and converted from a voxel index to a millimetric position c using the known voxel size for the image volume. The inverse of the gold-standard rigid-body transformation G is applied to the point c, giving point $c' = G^{-1}(c)$ in the "From" modality (left image). Then, the retrospective transformation R is applied to c', giving point c''. The registration error of the retrospective transformation at the centroid of the VOI is taken to be the Euclidean distance d between the points c and $c'' = R(G^{-1}(c))$.

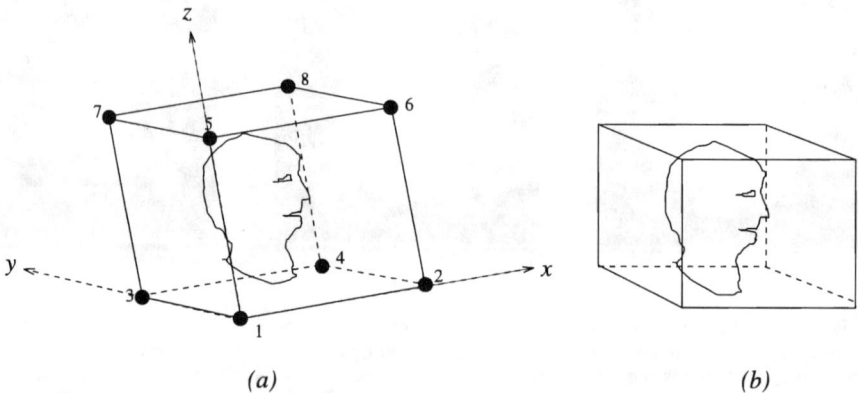

Figure 4. Image volumes prior to registration: (a) "From" volume and (b) "To" volume. The points 1 to 8 are defined as the corners of the volume, relative to the axes shown, and form the left-hand side of the registration table shown in Table 1.

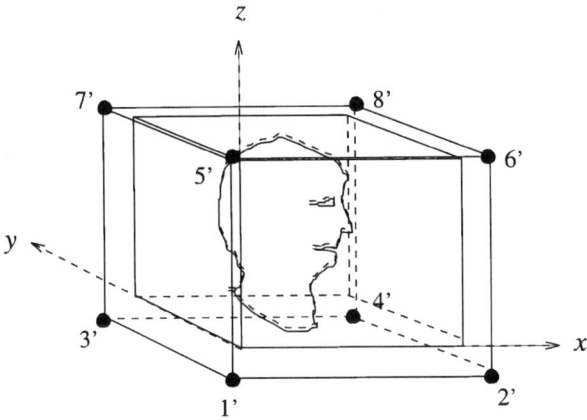

Figure 5. Image volumes after registration, so that the heads in the two volumes are almost perfectly aligned with each other. The points 1' to 8' are used to specify the transformation and form the right-hand side of the transformation table.

A Benchmark for Graphics Recognition Systems

Atul K. Chhabra[1] and Ihsin T. Phillips[2]

[1] Bell Atlantic Network Systems, Advanced Technology
500 Westchester Avenue, White Plains, NY 10604, USA
atul@basit.com

[2] Department of Computer Science/Software Engineering
Seattle University, Seattle, Washington 98122, USA
yun@seattleu.edu

Abstract

This paper presents a benchmark for evaluating graphics recognition systems. The benchmark is designed for evaluating the performance of graphics recognition systems on images that contain straight lines (solid or dashed), circles (solid or dashed), circular arcs (solid or dashed), and text blocks. This benchmark enables an objective assessment and comparison of vectorization software. It uses practical performance evaluation methods that can be applied to complete vectorization systems. At a recent graphics recognition contest, three systems were evaluated using this benchmark. The results of the contest are summarized in this paper. We hope that this benchmark will help assess the state of the art in graphics recognition, and highlight the strengths and weaknesses of current vectorization technology and evaluation methods.

1: Introduction

Performance evaluation and benchmarking have been gaining acceptance in all areas of computer vision. An overview of this area is available at [1]. However, the performance evaluation of graphics recognition is still a very young field. Although graphics recognition systems have existed for several years (e.g., [2, 3, 4, 5]), there has been no way to compare their performance quantitatively (some qualitative and subjective comparisons do exist, e.g., [6]). Objective and quantitative methods for evaluation of graphics recognition systems have been proposed very recently [7, 8, 9, 10].

Kong et al. [7] propose a quantitative method for evaluating the recognition of dashed lines. Hori and Doermann [8] propose a quantitative performance measurement methodology for task-specific raster to vector conversion. Wenyin and Dori [9] present a protocol for evaluating the recognition of straight and circular lines. These methods do not lend themselves to an objective evaluation of a complete graphics recognition system. Some of the methods evaluate low level algorithms and not complete systems. Other methods, although quite comprehensive, use performance metrics that favor good recognition algorithms over good recognition systems. For instance, often the most important requirement for a graphics recognition system is that the manual post-processing or clean-up work be minimized. This goal is often at odds with, or not necessarily related to, performance measures designed for evaluating recognition algorithms. Moreover, all of these methods lack a systematic way of generating benchmark training and testing data and ground truth.

We present a benchmark designed for evaluating the performance of graphics recognition systems on images of engineering drawings. The benchmark allows images that contain straight lines (solid or dashed), circles (solid or dashed), circular arcs (solid or dashed), and text blocks. Extending the benchmark for a new type of graphical entity is straightforward. We need to provide the evaluator a list of parameters for the new entity along with the entity matching criteria. We also need to extend the vector file format to allow the new type of entity, and the image rendering software to rasterize this type of entity. This benchmark enables a scientific comparison of vectorization software. It uses practical performance evaluation methods that can be applied to complete vectorization systems. The benchmark consists of a data set for training (tuning) and testing graphics recognition systems, data format specifications for the image and vector files, data conversion utilities, and performance metrics. We hope that this benchmark will help assess the state of the art in graphics recognition and highlight the strengths and weaknesses of current vectorization technology and evaluation methods.

This benchmark was used in the second international graphics recognition contest held recently in Nancy, France [11, 12]. Three systems were evaluated using this benchmark. The results of the contest are summarized towards the end of this paper. A detailed report on the contest and the methods for evaluating a match between a ground truth entity and a recognized entity are available in [12] and [13] respectively.

This paper is organized as follows. In Section 2, the selection of the set of drawings is discussed. The approach adopted for generating synthetic data is also introduced in Section 2. The image and vector file formats are specified in Section 3. Utilities for converting data formats, including the tool for rendering vector files into images, are presented in Section 4. In Section 5, we propose various metrics that can be used to measure the performance of a graphics recognition system. In Section 6, we present the results of the graphics recognition contest and analyze the performance of the participating systems. We conclude, in Section 7, with a discussion of the current limitations of the benchmark and ideas for future work.

2: The data set

In order to do a thorough evaluation and comparison of graphics recognition systems, one must use real images of scanned line drawing documents. However, that requires a very large effort to carefully create the ground truth data for each image. Moreover, using real scanned images, it is not possible to control the vector entities in various ways so as to test the breaking points of the systems. Therefore, we decided to use only synthetic images. The obvious advantages of synthetically generated images are that it is easy to create the ground truth files, and also to have complete control over the individual graphical entities and the image rendering process. Although it is always desirable to eventually test with real images, synthetic images offer a good starting point in the evaluation. A lot can be done with synthetic data that cannot be accomplished with real images. Further, synthetic images and the corresponding ground truth files are very inexpensive to generate.

The term 'synthetically generated image' conjures up a picture of some kind of mathematical model for lines and objects that, when rendered with some set of parameters, creates a line drawing of arbitrary objects. However, we chose a very pragmatic approach of using *real* CAD drawings to create synthetic images and ground truth data. We obtained the CAD drawings from the University of Washington's UW-III document image database [14]. The selected CAD drawings are complex, 'real life' archived drawings. Each of these drawings contains over 500 entities (a mix of lines, arcs, circles, and text.)

Image	Text Strings	Continuous Lines	Dashed Lines	Continuous Arcs	Dashed Arcs	Continuous Circles	Dashed Circles
ds29	82	201	31	28	0	14	12
ds30	91	250	46	24	0	19	13
ds31	135	385	19	29	0	41	17
ds32	80	319	82	13	0	16	3
ds35	186	368	0	8	0	77	0
ds36	119	801	0	105	0	3	0
ds09	141	263	13	23	0	7	0
ds07	482	791	1	0	0	86	0

Table 1. Counts of various types of entities in the eight test drawing images

In the process of CAD to image conversion, we converted polylines in the original drawings into their constituent straight line segments and then removed entities other than the types that the benchmark was designed for. Complete counts of each entity type within each test drawing are given in Table 1.

The simplicity afforded by the synthetic image generation process is not at the expense of realism or complexity of the rendered images. Realism in drawing images has two components – the complexity of the vector drawing and the realism of the image degradation. In this benchmark, we focused only on the first component. We used real-life complex CAD drawings as our starting point. We did not add any 'noise' to the images. Instead, we perturbed the CAD elements to help make the resulting images resemble real images. The distortions were simple, such as differing thickness of different lines, arcs, and circles; varying length of dashes and gaps in dashed lines; and the varying orientation, size, and stroke width of text (this causes some text to touch or overlap with graphics – a very real problem in real images). We used clean images of realistic CAD drawings (which are quite complex), degraded by manipulating the vector entities, but not degraded at the image level.

Within the selected CAD drawings and rendered images, there are four kinds of drawings – mechanical, architectural, and two distinct types of utility drawings. The images are carefully partitioned into two sets so that the images in the training set and the testing set have similar characteristics. Four of the training images are shown in Figures 1 through 4. More details on the image rendering process can be found in Section 4.

3: File format specifications

Here we specify the image and vector file formats used in this benchmark. The vector file format applies to both the ground truth file and the recognition result file.

3.1: Image file format

Only bi-level images are used in this benchmark. The images are in TIFF 6.0 CCITT Group 4 format. The software for generating synthetic images was built using several publicly available components [15, 16, 17]. The software, several sample images and the associated ground truth (VEC) files are available at the web site listed in [11].

3.2: Output and ground-truth specification: Vector file format

Most commercial raster to vector conversion systems output the vector results in the DXF format [18]. The DXF format is too complex for the quantitative evaluation of recognition results, especially given the types of graphical entities considered in the benchmark. In order to make the evaluation simple, we specify a simpler vector file format (the VEC format).

VEC files are ASCII files. The first line of a VEC file looks as follows.

```
%VEC-1.0 xsize ysize [dpi]
```

If dpi is present, xsize and ysize are in units of pixels. If dpi is absent, xsize and ysize could be in any units. The x axis points rightwards and the y axis points downwards. The coordinates of the ground truth VEC files are in pixels.

The VEC format has only four graphical primitives – lines, arcs, circles, and text regions. Below are the attributes associated with each of the four primitives in the above order. Each primitive is contained in a single row in the VEC file. All dimensions are interpreted as floating point values.

```
L C|D x1 y1 x2 y2 width
A C|D xcenter ycenter radius start_angle end_angle width
C C|D xcenter ycenter radius width
T x1 y1 x2 y2 orientation fontHeight fontWidthFactor \
    fontStrokeWidth %TEXT
```

The first letter of each row stands for the entity type. The L in the first row indicates a line. C indicates a continuous line. D indicates a dashed line. (x1,y1) and (x2,y2) are the two endpoints of the line. The width of lines and other entities is used only for rendering a VEC file into a raster image; it is not needed for matching recognition results with the ground truth. The A in the second row indicates an arc. The move from start_angle to end_angle is always clockwise. Angles are measured as positive in the clockwise direction. Angle zero points in the positive x direction. The C in the third row indicates a circle. The T in the fourth row indicates text. (x1,y1) and (x2,y2) are two diagonally opposite corners of the bounding box of the text string. The ground truth file always uses (x1,y1) to refer to the insertion point of left justified text. But that is not required in the recognition result. Recognition results may contain any two diagonally opposite corners of the bounding box in any order. Orientation is the angle of the baseline of the text. FontHeight, fontWidthFactor, and fontStrokeWidth are used for rendering text. The width to height proportion of the raw font is changed by fontWidthFactor. None of these three variables is required for evaluating results of recognition. One may output a value -1 for all of these variables to indicate that these values are just place holders. TEXT is a single line text string. The TEXT field may be left blank (the % sign must be included).

AutoCAD uses %%D, %%C, and %%P text strings to represent the special characters °, ϕ, and ±, respectively. In the image rendering program, we re-mapped the symbols rarely used in engineering drawings, !, ?, and $, to the special characters °, ϕ, and ±, respectively. In the VEC files, these symbols (!, ?, and $) should be used to represent the special symbols °, ϕ, and ±.

When evaluating graphics recognition systems using this benchmark, the systems are not required to OCR the text. Only the bounding boxes for text regions are needed by the performance evaluation software. Most commercial graphics recognition systems cannot directly report text bounding boxes. They report the OCR'd text (and the respective text insertion points and text justifications), along with the recognized graphical entities, in a DXF file. We wrote software to convert DXF format into our VEC format. In order to correctly generate the text bounding boxes in the VEC file, the OCR'd text strings in the DXF files are used. No attention is given to the accuracy of the recognized text. Only the size attributes of the text strings are used.

4: Data conversion utilities

In this section, we describe the DXF2VEC and VEC2TIFF conversion tools. DXF2VEC is used for converting the CAD drawings in the data set into our VEC file format. This tool is also used to convert the DXF output of graphics recognition systems to the VEC format. The VEC2TIFF software renders a VEC file into a raster image. This tool uses several parameters that can be set to control the amount of perturbations imposed on the vector entities.

4.1: Converting DXF files to VEC files

The CAD files in the UW-III database are in AutoCAD DWG format. The first filter we wrote was for the conversion of the DWG files into Autodesk's public drawing interchange format, DXF [18]. DWG2DXF is a small script written in AutoCAD's batch scripting language. The resulting DXF files are in AutoCAD units for length and degrees for angles (positive in the counter-clockwise direction).

To convert the DXF drawings into VEC files, we wrote a filter called DXF2VEC in the PERL language (version 5). The units of length in the resulting VEC file are the AutoCAD units. Only the drawing extents are clipped out of the DXF file. The y axis is transformed so that y increases downwards. Angles are measured in degrees, positive angles being clockwise.

The VEC format has only four graphical primitives – lines, arcs, circles, and text regions. At present, DXF2VEC interprets and converts only the following types of AutoCAD graphical objects – lines, polylines, arcs, circles, and text regions. Polylines are broken up into the individual lines and arcs. AutoCAD allows polyline segments to be splines. We do not interpret splines. Only lines and arcs are interpreted. Among the objects that are not interpreted or converted are blocks (compound objects), dimensions, solids, and trace.

You may notice that the example TIFF images contain several dimensioning lines and text. In all of these instances, the dimensioning had been done in AutoCAD by manually drawing lines, arcs, text, arrow-heads etc. To AutoCAD, these appear as individual graphical objects and not as dimensions. Therefore, they end up being interpreted as lines, text, etc. You don't see arrowheads because in these AutoCAD drawings, the arrowheads were entered as solids or blocks which we do not interpret.

If we were to create AutoCAD drawings from scratch where we let AutoCAD do the dimensioning, then the dimensions would appear as 'dimension' entities in AutoCAD rather than lines, solids, and text. There are inumerable parameters associated with dimension entities and writing a script to interpret them is quite a time consuming task. Therefore, we decided not to handle AutoCAD 'dimensions' for the current benchmark. Consequently, the benchmark does not handle arrowheads.

4.2: Rendering VEC files as TIFF images

The VEC2TIFF filter (witten in C) renders the vectors and text in the input VEC file into a bi-level raster image. The image is saved as a TIFF file compressed using the CCITT Group 4 format. During the rendering process, the AutoCAD units of length are converted into pixel units based upon the required arguments `xsize` (in inches) and `resolution` (in dpi). A default line width *minLineWidth* pixels is used for rendering all lines whose scaled width would translate to less than

Parameter	Value Used	Minimum Value	Maximum Value
minLineWidth	3	1	10
maxLineWidth	20	1	32
minFontStrokeWidth	2	1	3
maxFontStrokeWidth	3	3	5
dashLength	30	5	99
dashGap	14	5	50
minLineLengthToRender	15	5	50
minArcRadiusToRender	5	2	50
minCircleRadiusToRender	5	5	50
renderDashedLinesAsContinuous	0	0	1
fixedWidthFont	0	0	1
stdDevOfLineWidth	2.0	0.0	3.0
stdDevOfDashLength	3.0	0.0	10.0
stdDevOfDashGap	2.0	0.0	5.0
stdDevOfFontHeight	2.0	0.0	5.0
stdDevOfTextOrientation	0.4	0.0	2.0

Table 2. The VEC2TIFF rendering parameters, their values used, and valid ranges.

minLineWidth (e.g., the line entity in AutoCAD is a zero width object). Several other parameters are used to control the rendering of the image. The parameters, their valid range of values, and the values used for the benchmark images are given in Table 2. The valid range for each parameter and its default value are hard coded. However, the default values for the parameters can be overridden at run time with values stored in a file.

Figures 1 through 4 show the drawings belonging to the training set as rendered by the VEC2TIFF program.

Some parameters impose a degree of randomness on line width, text height, text orientation, etc. This makes the rendered image more realistic. For instance, the random perturbation of text orientation causes the text to touch or overlap with graphics on several occasions (this is clearly evident in Figures 2 and 3). The random variables are generated using a portable random number generator [17] that generates uniform variates. The uniform variates are then transformed to normal variates. The portable random number generator generates repeatable results on any computer/OS/compiler.

VEC2TIFF also creates an output VEC file. The similarities and differences between the input and output VEC files are highlighted below.

- The output VEC file uses pixel coordinates as opposed to AutoCAD coordinates of the input VEC file.

- Both VEC files have the y coordinate increasing downward.

- Both VEC files use degrees as a measure of angle, clockwise angles being positive. In both VEC files, arcs are always clockwise, i.e., the move from start angle to end angle is always clockwise.

- The TIFF image matches the output VEC file. The variability (randomness) imposed on the input VEC entities is reflected in the output VEC file. Therefore, the output VEC file contains the ground truth data corresponding to the TIFF image.

The VEC2TIFF software makes use of several public domain software packages such as libtiff [16] for saving the image in the TIFF CCITT Group 4 format, the portable random number generator [17] that is platform/OS/compiler independent, and the VOGLE device portable graphics library [15] for rendering text strings into the image.

5: Performance evaluation

The recognition results of a graphics recognition system on a test image are matched by the performance evaluator against the corresponding ground truth for the test image. The matching process results in the following counts – the numbers of one-to-one matches, one-to-many matches, many-to-one matches, as well as the numbers of false-alarms and misses (a complete description of the evaluator is given in [13].) We present some system level metrics that make use of these low level matching counts.

5.1: System performance metrics

Measures of performance for a recognition system can be formulated using a linear combination of some or all of the matching results. Let *one2one* be the count of the one-to-one matches, *one2many* be the count of the ground truth entities having one-to-many (one ground truth to many detected) matches, *many2one* be the count of ground truth entities having many-to-one (many ground truth to one detected) matches, *false_alarms* be the count of the false-alarms (detected entities with no match among the ground truth), *misses* be the count of the misses (ground truth entities with no match among the detected entities), N be the count of the entities in the ground truth file, and M be the count of the entities in the recognition result file.

We propose the following measures of system performance.

$$DetectionRate = w_1 \cdot \frac{one2one}{N} + w_2 \cdot \frac{one2many}{N} + w_3 \cdot \frac{many2one}{N}.$$

DetectionRate is roughly the percentage of ground truth entities detected by the recognition system. Here, w_1 should weigh more than w_2 and w_3 since one should favor one-to-one matches over one-to-many and many-to-one matches (one-to-many and many-to-one matches require a manual clean-up effort during post-processing).

$$MissRate = \frac{misses}{N}.$$

MissRate is the percentage of the ground truth entities which are not detected by the recognition system. Note that *DetectionRate* and *MissRate* may not necessary add up to one, because of the weighting factors involved in the computation of *DetectionRate*.

$$FalseAlarmRate = \frac{false\_alarms}{M}.$$

FalseAlarmRate is the percentage of the detected entities that do not correspond to any ground truth entity.

$$EditingCost = w_4 \cdot false\_alarms + w_5 \cdot misses + w_6 \cdot one2many + w_7 \cdot many2one.$$

EditingCost is an estimated cost for a human post-processing effort to clean-up the recognition results. It should be clear that a higher *EditingCost* requires a higher clean-up effort. Entities missing from the recognition result file need to be added and the false-alarms need to be removed. Moreover, for each one-to-many match, one needs to remove the many (partial matches) from the result file and add the real one to it. And each many-to-one match requires one removal and many additions. Note that w_4 is the factor for one deletion effort and w_5 is the factor for one insertion effort. The two factors should be weighted according to the tool one uses for post-processing. The weights w_6 and w_7 are more complex; they depend on the number and type of partial matches. At the graphics recognition contest, we set w_4 and w_5 to one and w_6 and w_7 to zero.

6: Contest results and performance analysis

The benchmark presented here was used recently in the second international graphics recognition contest [11, 12]. Of the three systems that participated in the contest, two were commercial products and one came from an university. A detailed report on the contest and the methods for evaluating a match between a ground truth entity and a recognized entity are available in [12] and [13] respectively. Here, we only present a summary of the results. The participating systems are referred to as Systems A, B, and C.

Prior to the contest, we provided the participants a set of training images. The participants were told to use these images to determine the optimal operating parameters of their software for each image category. The test images used at the contest closely resemble the training images. The participants provided the contest organizers the fine-tuned parameter sets for each image class, and a copy of their software for evaluation.

The test images in this benchmark consist of four mechanical drawings, one architectural drawing, two utility drawings, and one structural drawing – a total of eight test images. Each of three participating systems was tested on all these eight images. Matching scores were computed for the recognition results.

The performance evaluator uses an acceptance threshold to determine whether a pair of entities matches. The matching criteria defined in [13] is roughly a similarity measurement. When the acceptance threshold is set high, the evaluator accepts only those pairs that are very similar (having high matching scores). Lowering the acceptance threshold causes the evaluator to lower its matching requirement. We expected that with a high acceptance threshold, only highly accurate recognition systems could score well in their performance measurements. However, we were interested in learning the trends of system performance with respect to changes in the acceptance threshold. Our presumption was that for the highly accurate recognition systems, lowering the acceptance threshold would increase their performance a little, not drastically. On the other hand, for the not so accurate systems, the performance measurements would increase greatly when the evaluator's acceptance threshold was set lower. Thus, using various acceptance thresholds in the evaluation would reveal the stability and robustness of a recognition system.

With the above in mind, we evaluated the systems using nine levels of the acceptance threshold (ranging from 0.5 to 0.9, in steps of 0.05). That is, for each recognition result file produced by a system, we obtained nine sets of matching counts using these nine acceptance thresholds.

6.1: Analysis of performance characteristics

We were interested in learning whether the performance characteristics of recognition systems could be observed by varying the evaluator's acceptance threshold. We are happy to report that we did indeed observe some performance characteristics of the participating systems.

To illustrate the trend of change in the performance of the three participating systems with respect to the nine acceptance thresholds, we plot the counts of the false-alarms vs. the misses (for four of the eight test drawings) in Figures 5 through 8. Each of these figures contains three nine-point curves (one curve per system) where the nine points correspond to the nine thresholds used. The first point on each curve corresponds to a threshold of 0.5 and the last point on each curve corresponds to a threshold of 0.9. We observed the following from these plots.

- In general, all three curves in each of the plots show upward trends. That is, as the acceptance threshold was increased, all three systems produced more misses and more false-alarms.

- In general, the first three or four points on most of the curves (corresponding to the threshold values 0.5, 0.55, 0.6, and 0.65) either form a tight cluster, or have equal or higher counts of misses and false-alarms than the counts for the points corresponding to 0.65, 0.7 or 0.75 thresholds. This is visible as odd kinks on some plots, especially for the System B. The interpretation for this trend may be that using an acceptance threshold below 0.65 does not yield a better evaluation for a given system. Or, it may be that the performance measurements produced by the evaluator using thresholds below 0.65 are not reliable (we suspect that with the acceptance threshold set too low, the evaluator may be making matching errors, resulting in more misses and false-alarms.) We are currently investigating this.

- In most of the cases, all the systems produced more false-alarms than misses. This may be partly due to one of the following reasons. (1) At present, the evaluator does not match any dashed entity to any solid entity. So, if a dashed-line in a test image is detected by a vectorization system as several little straight line segments, the evaluator produces counts of one miss (dashed-line) and several false-alarms (little line segments.) (2) When a text string in a test image is not correctly detected as a text region, it is often 'vectorized' into several small lines, arcs, etc. In this case, the evaluator currently produces counts of one miss (the missing text string) and several false-alarms (the little 'vectors').

We also observed some performance characteristics for each of the three systems. For example, we observed that two of the systems (Systems B and C) had very small increases in the counts of misses and false-alarms. If we take the amount of increases, with respect to the increase in the acceptance threshold, as an indicator of the stability of a system, these two systems outperform the third by a significant margin. One of these two systems also produces much fewer misses than the others. For the four mechanical drawings, System C, which was designed specifically for mechanical drawings, produced much fewer false-alarms and fewer misses than the other two systems. It is apparent that customizing a system for a specific type of drawing can lead to a significant improvement in performance.

7: Discussion

The benchmark limits itself to a quantitative evaluation of the automatic vectorization capability of graphics recognition systems. While the benchmark was being developed for the contest, several other constraints were imposed either due to lack of time and resources or in order to keep the evaluation protocol simple. The primary constraints were as follows. (1) The benchmark only includes synthetic bi-level images for both training and testing. (2) The only 'noise' in the images is in the form of thickness of lines, length of dashes and gaps in dashed lines, and the orientation and size of text. No 'image noise' has been added. (3) Up to now, we have only tested at the image resolution of 200 dots per inch. (4) We only test for detection of straight lines, arcs, circles, and text. Evaluation of the detection of polylines, dimensioning, objects, symbols, etc. is not supported yet. (5) Currently, only one kind of dashed line is used. This is the simple dash-dash line. (6) No match is attempted between dashed entities and solid entities.

There are some known shortcomings in our evaluation process which we will address in the near future. If a vectorization system erroneously recognizes a dashed line as a sequence of short continuous lines, then our evaluation method assigns a single miss but a large number of false-alarms (because we do not attempt to match a dashed line with continuous line segments). We need to allow matching of dashed lines with several small line segments, but this should be penalized somewhat due to the fragmentation introduced.

If a text region is not correctly identified, then we assign a single miss accompanied by a large number of false-alarms. This happens because if a text region is not correctly identified, then the vectorization software will invariably try to 'vectorize' the region. The resulting short lines ('vectors') count as false-alarms because we do not attempt to match a text area with any other type of entity. In order to correct the misinterpretation, one only needs to box a text region and mark it as text. This is a very simple post-processing operation. Therefore, this kind of error should not be penalized so heavily.

Gathering data to test and compare graphics recognition systems is very time consuming. This benchmark only uses synthetic images with associated ground truth. Future benchmarks should include synthetic images with image degradation and real images with manually created ground truth. The graphics recognition community needs to collaborate in building a database of images and ground truth files.

The real strengths and weaknesses of a system are revealed by stress testing the system. We can accomplish this by testing the performance of a vectorization system with increasing image degradation and increasing image complexity. This should be attempted in a future contest.

Future contests will hopefully attract participation from many more graphics recognition companies and research groups. Evaluation of a larger number of systems will provide us broader trends and a real assessment of the state of the technology.

References

[1] ECVNet. Benchmarking and Performance Evaluation web site. http://pandora.imag.fr/ECVNet/benchmarking.html.

[2] I/Vector (Vectory) ver. 3.8 Raster to Vector Conversion Software. Graphikon, Berlin, Germany, and IDEAL Scanners & Systems, Rockville, MD, USA. http://www.graphikon.com and http://www.ideal.com.

[3] VPstudio ver. 6 rev. 2 Raster to Vector Conversion Software. Softelec, Munich, Germany, and Austin, TX, USA. http://www.softelec.com and http://www.hybridcad.com.

[4] L. Wenyin and D. Dori. Automated CAD conversion with the machine drawing understanding system. In *Proceedings of the IAPR Workshop on Document Analysis Systems*, pages 241–259, Malvern, PA, USA, October 1996. ftp://ftp.technion.ac.il/pub/supported/ie/dori/MDUS/sunmdus.gz.

[5] L. Wenyin and D. Dori. Generic graphics recognition of engineering drawing objects. In *Proceedings of Second IAPR Workshop on Graphics Recognition*, pages 10–11, Nancy, France, August 1997.

[6] David Byrnes. Raster-to-vector comes of age with AutoCAD Release 14. *CADALYST*, pages 48–70, December 1997.

[7] B. Kong, I. Phillips, R. Haralick, A. Prasad, and R. Kasturi. A benchmark: Performance evaluation of dashed line detection algorithms. In R. Kasturi and K. Tombre, editors, *Graphics Recognition: Methods and Applications, First International Workshop, University Park, PA, USA, August 1995, Selected Papers*, volume 1072 of *Lecture Notes in Computer Science*, pages 270–285. Springer, Berlin, 1996.

[8] O. Hori and S. Doermann. Quantitative measurement of the performance of raster-to-vector conversion algorithms. In R. Kasturi and K. Tombre, editors, *Graphics Recognition: Methods and Applications, First International Workshop, University Park, PA, USA, August 1995, Selected Papers*, volume 1072 of *Lecture Notes in Computer Science*, pages 57–68. Springer, Berlin, 1996.

[9] L. Wenyin and D. Dori. A protocol for performance evaluation of line detection algorithms. *Machine Vision and Applications*, 9(5/6):240–250, 1997. Special Issue on Performance Characterisitics of Vision Algorithms.

[10] L. Wenyin and D. Dori. A protocol for performance evaluation of algorithms for text segmentation from graphics-rich documents. In *Proceedings of Second IAPR Workshop on Graphics Recognition*, pages 317–324, Nancy, France, August 1997.

[11] A. Chhabra and I. Phillips. Web page for the Second International Graphics Recognition Contest – Raster to Vector Conversion. http://graphics.basit.com/iapr-tc10/contest.html.

[12] A. Chhabra and I. Phillips. The second international graphics recognition contest – raster to vector conversion: A report. In K. Tombre and A. K. Chhabra, editors, *Graphics Recognition—Algorithms and Systems, Second International Workshop, Nancy, France, August 1997, Selected Papers*, volume 1389 of *Lecture Notes in Computer Science*, pages 390–410. Springer Verlag, Berlin, 1998.

[13] I. Phillips, J. Liang, A. Chhabra, and R. Haralick. A performance evaluation protocol for graphics recognition systems. In K. Tombre and A. K. Chhabra, editors, *Graphics Recognition—Algorithms and Systems, Second International Workshop, Nancy, France, August 1997, Selected Papers*, volume 1389 of *Lecture Notes in Computer Science*, pages 372–389. Springer Verlag, Berlin, 1998.

[14] I. Phillips. Users' reference manual: CD-ROM, UW-III document image database-III. Technical report.

[15] VOGLE, a public domain device portable graphics library. ftp://munnari.oz.au/pub/graphics/vogle.tar.gz.

[16] Sam Leffler and Silicon Graphics, Inc. TIFF software distribution. ftp://ftp.sgi.com/graphics/tiff/tiff-v3.4beta036-tar.gz.

[17] D. Knuth. The portable random number generator. http://www-cs-faculty.stanford.edu/ knuth/programs.html. Also published in *The Art of Computer Programming, Volume 2/Seminumerical Algorithms*, 3rd edition, section 3.6. Addison-Wesley, Reading, MA, USA, 1997.

[18] Autodesk, Inc. *AutoCAD Release 13 Customization Guide*. 1995.

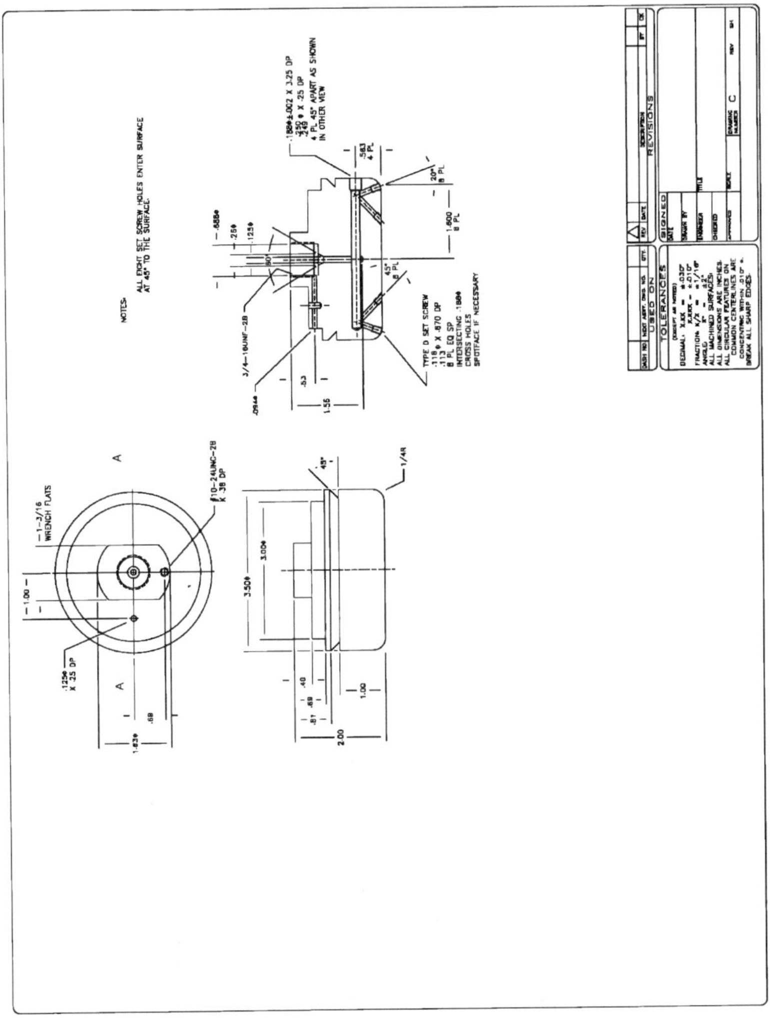

Figure 1. Training image mech.tif (mechanical drawing)

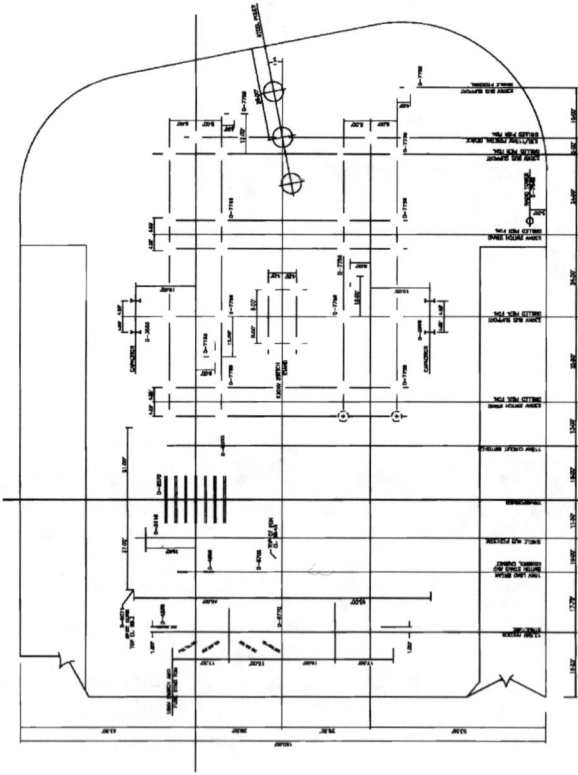

Figure 2. Training image arch.tif (architectural drawing)

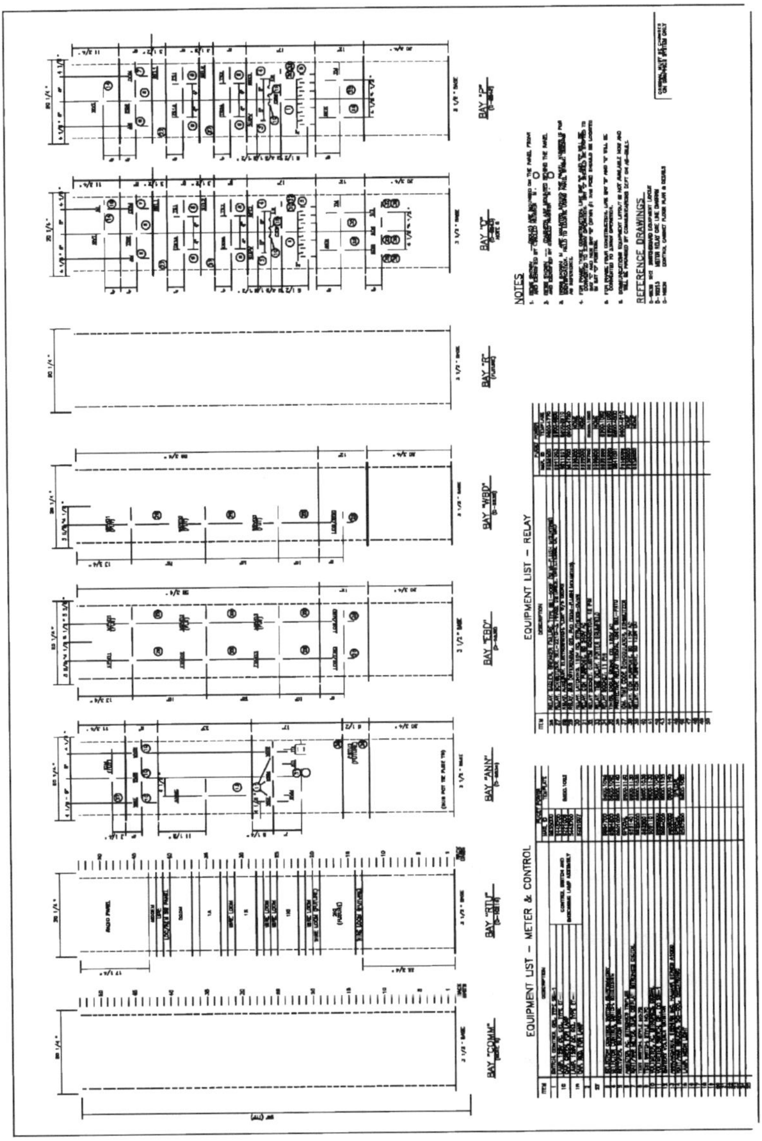

Figure 3. Training image util1.tif (utility drawing, type 1)

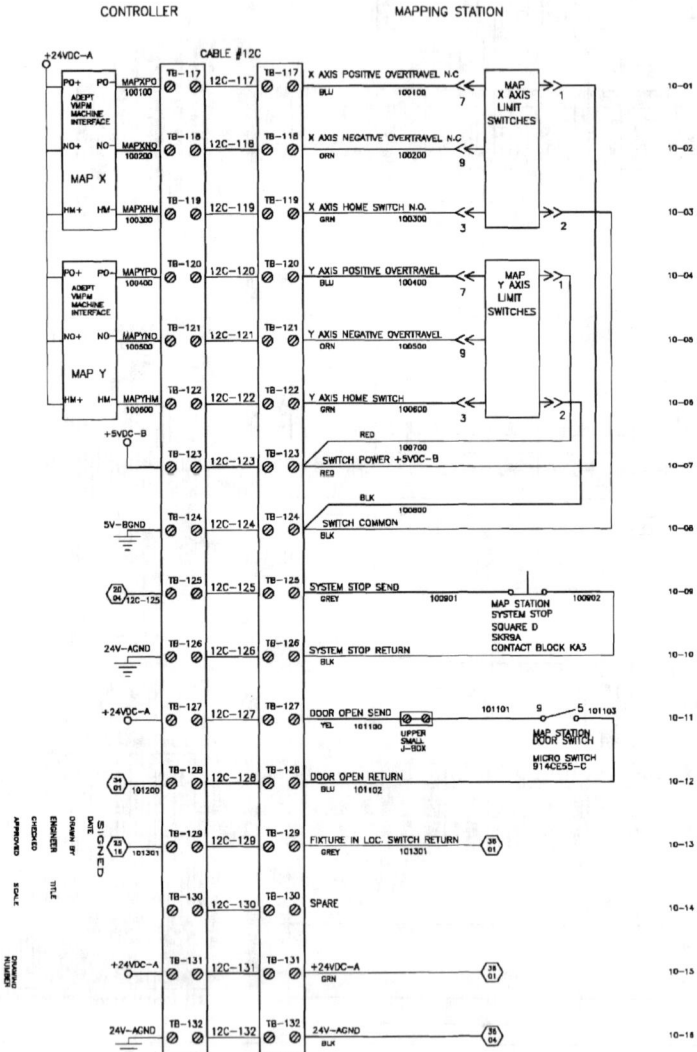

Figure 4. Training image util2.tif (utility drawing, type 2)

ds29.tif: 368 entities

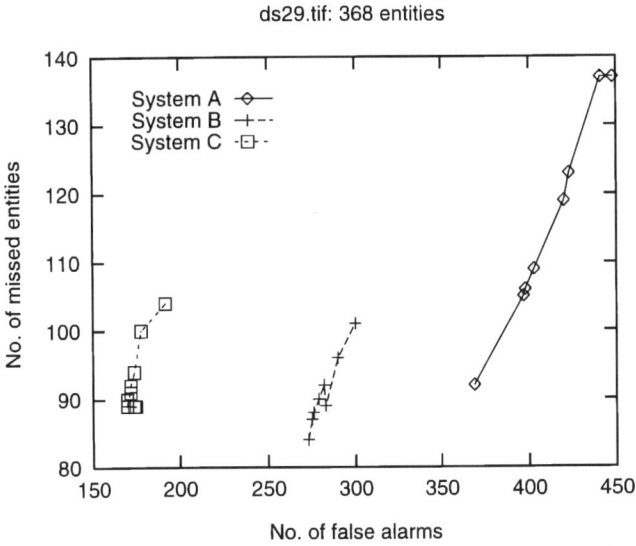

No. of false alarms

Figure 5. Performance curves of the systems for the image ds29.tif (image of a mechanical drawing)

ds09.tif: 447 entities

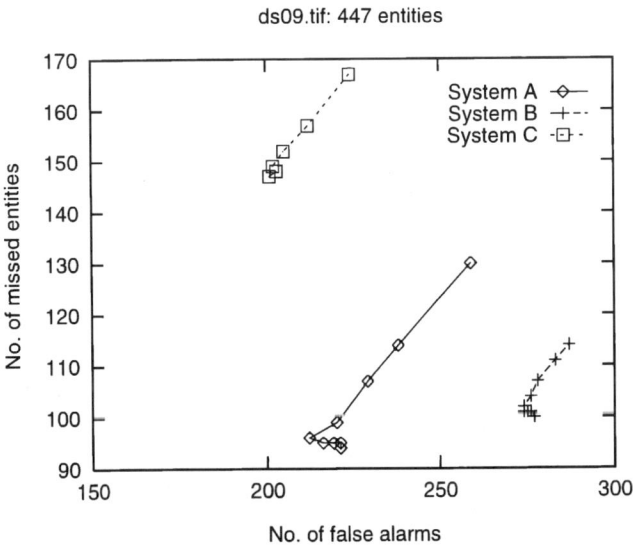

No. of false alarms

Figure 6. Performance curves of the systems for the image ds09.tif (image of an architectural drawing)

Chhabra, Phillips

ds07.tif: 1360 entities

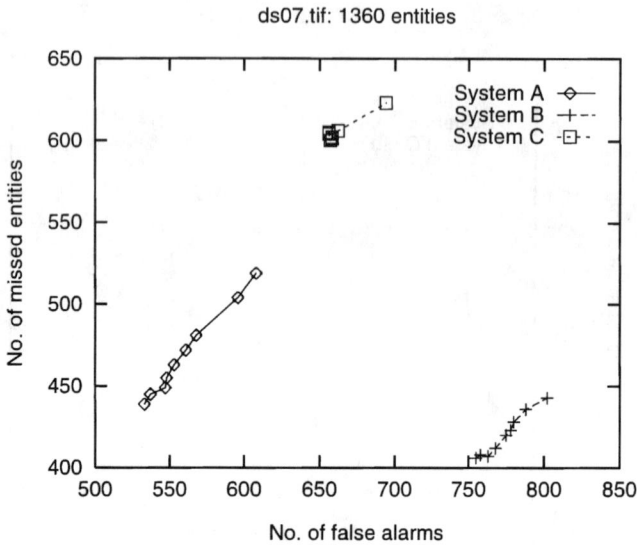

Figure 7. Performance curves of the systems for the image ds07.tif (image of a utility drawing, type 1)

ds35.tif: 639 entities

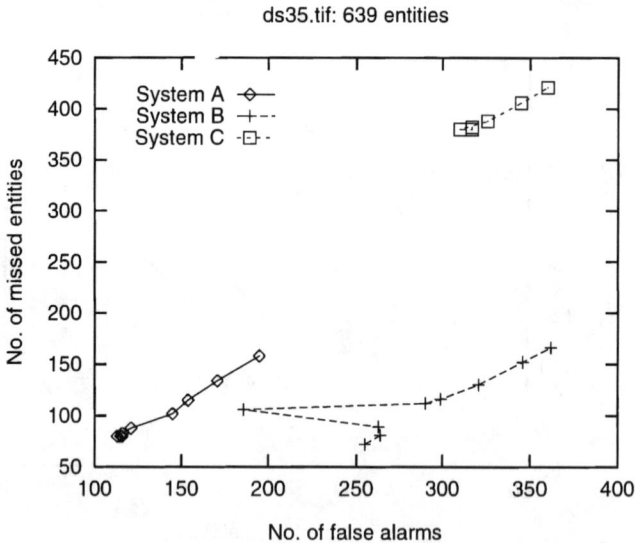

Figure 8. Performance curves of the systems for the image ds35.tif (image of a utility drawing, type 2)

Performance Evaluation of Clustering Algorithms for Scalable Image Retrieval

Mohamed Abdel-Mottaleb
Santhana Krishnamachari
Nicholas J. Mankovich
Philips Research
345 Scarborough Road
Briarcliff Manor, NY 10510
{msa, sgk, njm}@philabs.research.philips.com

Abstract

In this paper we present scalable algorithms for image retrieval based on color. Our solution for scalability is to cluster the images in the database into groups of images with similar color content. At search time the query image is first compared with the pre-computed clusters, and only the closest set of clusters is further examined by comparing the query image to the images in that set. This obviates the need to compare the query image with every image in the database, thus making the search scalable to large databases. We have used the hierarchical clustering and the K-means clustering techniques. Performances of these two clustering algorithms are compared when three similarity measures, the histogram intersection measure, the L_1, and the L_2 measures, are used for image retrieval.

The retrieval accuracy of the clustering algorithms is computed by comparing the results of retrieval with clustering against the results of retrieval without clustering. Our experiments with a database of 2000 color images show that both clustering techniques offer a retrieval accuracy of over 90% with only an average of 300 similarity comparisons (as opposed to 2000 comparisons that are required for retrieval without clustering). Our evaluations show that the hierarchical clustering algorithm outperforms the K-means clustering algorithm for all three similarity measures, although only marginally in some cases.

1: Introduction

Content-based image retrieval has become a prominent research topic in recent years. Research interest in this field has escalated because of the proliferation of video and image data in digital form. The goal in image retrieval is to search through a database to find images that are perceptually similar to a query image. An ideal image retrieval engine is one that can completely comprehend a given image, *i.e.*, to identify the various objects present in the image and their properties. Given the state of the art of research in the image analysis community, such an ideal retrieval system is far from being reality. Moreover retrieval based on human annotation is to no avail, because of the size of the video and image databases and the varying interpretations that different humans can attach to an image. Examples of color content based image retrieval algorithms can be found in [1-5, 7, 8].

In a practical scenario, like the Internet, the number of images can be of the order of millions and is ever growing. Even if the time required to compare two images is very short, the cumulative time needed to compare the query image with all the database images is rather long and is probably longer than the time an average user wants to wait. We solve this problem by grouping or clustering the images according to their similarity beforehand, so that at the time of the query, it is not necessary to perform an exhaustive comparison with all the images in the database. The clustering is performed based on visual features extracted automatically from the images.

Performance evaluation has been a challenging issue in the field of content-based retrieval, primarily because of the difficulty associated with calculating quantitative measures to evaluate the quality of retrieval. The *precision* and *recall* measures have been frequently used by many researchers [5] to evaluate the performance of retrieval algorithms. In this paper we introduce a quantitative method to evaluate the retrieval accuracy of clustering algorithms. Our goal is not to subjectively evaluate the quality of retrieval, but to quantitatively compare the performance of retrieval with and without clustering.

In this paper we present clustering techniques for scalable image retrieval from large databases and evaluate the performances of two clustering techniques: the hierarchical and the K-means. Each clustering technique is applied with three different similarity (or distance) measures: the L_1 norm, the L_2 norm, and the histogram intersection measure. The images in the database are clustered into groups based on local color similarity. During retrieval, the query image is initially compared with the representative of each cluster. Then the query image is compared only with the images in the closest clusters and these images are ranked according to their similarity with the query image. Thus it is not necessary to compare the query image with every image in the database. For each clustering technique and similarity measure, the retrieval accuracy is obtained by comparing the results of retrieval with clustering against the results of retrieval without clustering (comparing the query with all the images in the database) using the *same* similarity measure.

Techniques for fast image retrieval from large databases have been presented in [9-11]. All these techniques require that the similarity (distance) measure used to compare images be a metric, *i.e.*, the similarity measure should satisfy the triangle inequality. However, many similarity measures, like the histogram intersection measure used here, do not satisfy the triangle inequality. The scalable retrieval technique presented here does not require the similarity measure to be a metric and hence is more general.

The rest of the paper is organized as follows. Section 2 carries the details of retrieval based on clustering. Section 3 presents the details of different similarity measures and clustering algorithms that are used. Section 4 presents the experimental set-up and the experimental results. Section 5 presents the conclusions of the performance evaluation and avenues for future work.

2: Image clustering

Searching large databases of images is a challenging task especially for retrieval by content. Most search engines calculate the similarity between the query image and all the images in the database and rank the images by sorting their similarities. One problem with this approach is that it does not scale up for large databases. The retrieval time is the sum of two times: T_{sim} and T_{sort}. T_{sim} is the time to calculate the similarity between the query and every image in the database, and T_{sort} is the time to rank all the images in the database according to their similarity to the query.

$$T_{total} = nT_{1sim} + O(n\log n)$$

where n is the number of images in the database, T_{1sim} is the time to calculate the similarity between two images, and $O(n\log n)$ is the time to sort n elements.

When the images in the database are clustered, the retrieval time is the sum of three times, the time to calculate the similarity between the query and the cluster centers, the time to calculate the similarity between the query and the images in the nearest clusters and the time to rank the images. Therefore the total search time is:

$$T_{cluster} = kT_{1sim} + lT_{1sim} + O(l\log l)$$

Here k is the number of clusters, l is the number of images in the clusters nearest to the query. Since $k \ll n$ and $l \ll n$, $T_{cluster} \ll T_{total}$.

3: Image representation, similarity measures and clustering

3.1: Image representation

Several histogram-based approaches have been proposed for image retrieval by color [1, 3, 7]. These approaches are based on calculating a similarity measure between the color histogram of the query image and the images in the database. The difference between these approaches is mainly in their choice of the color space and the similarity measure. Since these approaches use a single image histogram to calculate similarities, the results are expected to reflect only global similarity. For example, if a user submits a query image with a sky at the top and sand at the bottom, the retrieved results would have a mix of blue and beige, but not necessarily with blue at the top and beige at the bottom. This can be achieved only if the image representation reflects the local color information.

In this paper, we use the scheme that we presented in [2], to allow retrieval based on local color features. Images in the database are divided into rectangular regions. Then every image is represented by the set of normalized histograms corresponding to these rectangular regions. It should be noted here that the choice of the rectangular region size is important. In one extreme, the whole image is considered as a single region which reflects the global color information. As the size of the region becomes smaller, the local variations of color information is captured by the histograms. The size of the region should be small enough to emphasize the local color and large enough to offer a statistically valid histogram. In the experiments, images were divided into 16 rectangular regions.

3.2: Similarity measures

The similarity between two images is measured by calculating the similarity between the histograms of the corresponding rectangular regions. Then a single measure of similarity between the two images is calculated by combining the individual similarities. For performance evaluation, we have used three similarity (distance) measures that are frequently used in the literature, namely, the histogram intersection, the L_1, and the L_2 norms. The L_1 and the L_2 norms are distance measures, whereas the histogram intersection measure is a similarity measure. We converted distance measures to similarity measures by taking the negative values of the distance measures. Given two *normalized* histograms, $P = \{p_1, p_2, ..., p_m\}$, $Q = \{q_1, q_2, ..., q_m\}$, the histogram intersection measure is defined as [1]:

$$H(P, Q) = \sum_{i=1}^{m} \min(p_i, q_i).$$

The L_1 norm is defined as:

$$L_1(P, Q) = \sum_{i=1}^{m} |p_i - q_i|$$

and the L_2 norm is defined as:

$$L_2(P, Q) = \sqrt{\sum_{i=1}^{m} |p_i - q_i|^2}$$

3.3: Clustering

We have used two clustering algorithms, the hierarchical and the K-means clustering algorithms to group the images into clusters based on the color content. Both these clustering algorithms have been frequently used in the pattern recognition literature. Brief details on the implementation of these two clustering algorithms are presented below.

The hierarchical clustering algorithm [6], is implemented as shown below:

> *Let n be the number of images in the database, the similarity between all pairs of images is precomputed.*
> *1. The n images in the database are placed in n distinct clusters indexed by $\{C_1, C_2,..., C_n\}$.*
> *2. Two distinct unmerged clusters C_k and C_l are picked such that their similarity measure is the largest.*
> *3. These two clusters are merged into a new cluster C_{n+1}. At each step two clusters are merged to form a new cluster. Therefore, the number of clusters is reduced by one.*
> *4. Steps 2 and 3 are repeated until the number of unmerged clusters has reduced to a required number n_c or the largest similarity measure between clusters has dropped to some lower threshold.*

The K-means clustering algorithm [6] is implemented as follows:

> *1. The number of clusters n_c is chosen a priori. The n_c centers are chosen by randomly picking n_c images from the database.*
> *2. For each image in the database, the similarity measure between the image and the clusters centers are computed and the image is assigned to the cluster with which it exhibits the largest similarity measure.*
> *3. New cluster centers are computed as the centroids of the clusters.*
> *4. Steps 2 and 3 are repeated until there is no further change in the cluster centers.*

4. Experimental setup

The results presented in this paper are obtained with a database of 2000 images, 200 of which are taken from two collections of COREL Professional Photo CD-ROMs. The rest of the images are obtained from the Department of Water Resources, California. The COREL images are obtained from two CD-ROM collections, the Sampler II - Series 400000 and the Sampler - Series 200000. The list of images that have been used here can be made available to anyone that may be interested in using them by contacting the second author.

The images are of widely varying colors and scene content. The number of clusters n_c is chosen to be 133. The number of clusters, n_c, is chosen such that the average number of images per cluster is 15, i.e., $n_c=2000/15=133$. The number of clusters in both the clustering techniques are the same to ensure fair comparison. The hierarchical and K-means clustering algorithms are applied to the 2000 images in the database using each of the three different similarity measures. For the hierarchical clustering, the number of images in the smallest and the largest cluster is, 2 and 40 with the histogram intersection measure, 2 and 48 with the L_1 measure, and 1 and 63 with the L_2 measure. For the K-means clustering, the corresponding numbers are, 1 and 49 with the histogram intersection measure, 1 and 52 with the L_1 measure, and 1 and 69 with the L_2 measure.

4.1: Retrieval accuracy with clustering

After clustering and selecting the cluster centers, the given query image is first compared with all the cluster centers. The clusters are ranked according to their similarity with the query. Few close clusters from the top of this ranked list are chosen. Then the query image is compared directly with the images in these clusters. Thus the number of comparisons is reduced considerably from comparing the query with all the images in the database. It is shown below that the retrieval accuracy is not compromised in this process. The number of similarity comparisons required depends on the sizes of the clusters and the number of clusters being examined.

We have used a quantitative measure to compare the retrieval results *with* clustering against the retrieval results *without* clustering. A user searching through a large database, is interested in only the top few best matches (say 10 or 20). Hence, if the retrieval with clustering returns the same few best matches as the ones returned by retrieval without clustering, then the retrieval with clustering is very accurate. Assume that the user is interested in only top N best matches and that M is the number of images that are present both in the top N results returned by retrieval with and without clustering. The retrieval accuracy with clustering ψ_i, when the ith image is used as a query is defined as:

$$\psi_i = \frac{M}{N} 100$$

The *average retrieval accuracy* with clustering A_N is obtained by taking the average ψ_i over all the query images.

$$A_N = \frac{1}{n} \sum_{i=1}^{n} \psi_i$$

4.2: Discussion of results

The experimental results from using the two clustering algorithms with the three similarity measures are presented in Figures 1-7. Each of these plots contain eight points, obtained by examining the top 3, 4, 7, 10, 13, 19, 25, and 31 clusters. In all the figures, the leftmost point corresponds to the result obtained by examining 3 clusters and the rightmost point corresponds to the result obtained by examining 31 clusters. The retrieval accuracy increases as the number of examined clusters is increased.

We conducted two sets of experiments. In the first set, we used each of the 2000 images in the database as a query image. For each query, the retrieval accuracy and the number of image similarity comparisons for the eight different cases are calculated. The averages of the retrieval accuracies and the averages number of comparisons are plotted in Figures 1-3. The results are discussed in Section 4.2.1. In the second set of experiment, we used a set of 300 images that are not a part of the database of 2000 images as query images. These 300 images are also obtained from the Department of Water Resources, California. Again, for each query, the retrieval accuracy and the number of image comparisons for the eight different cases are computed. Figures 4-6 carry the results of this experiment. The results of this experiment are discussed in Section 4.2.2. In the case of the K-means clustering, experiments were repeated with different random selections of the initial centers. We found that for different random initializations the variations in the average retrieval accuracy and the number of similarity comparisons were very small.

4.2.1: Results with database images as queries: Figure 1(a) shows the plot of average retrieval accuracy against the number of comparisons for the hierarchical clustering using the three different similarity measures. It is worth mentioning that for each similarity measure, the retrieval results obtained with clustering are compared against the retrieval results obtained without clustering using the same similarity measure. We have not used a common *ground-truth* against which the retrieval results of different similarity measures are compared against. Figure 1(a) shows that for the hierarchical clustering, the histogram intersection measure offers the largest retrieval accuracy for a given number of comparisons, closely followed by the L_1 measure. The retrieval accuracy for the L_2 measure is significantly lower than the rest. Similar results are obtained with the K-means clustering as shown in Figure 1(b).

One inference that can be drawn from Figure 1 is that both clustering algorithms offer a large reduction in the number of comparisons without sacrificing the retrieval accuracy. A retrieval accuracy of over 90% can be obtained for both clustering algorithms with the histogram intersection and the L_1 measure by examining only the top 13 clusters (out of 133 clusters). The average number of comparisons required to examine the top 13 clusters is less than 300, compared to the 2000 comparisons that are required if the retrieval is performed without clustering. We expect that the reduction in the number of comparisons with clustering will be much larger as the size of the database is increased.

Figures 2 shows the retrieval accuracies for the two clustering algorithms with the histogram intersection measures for values of N=10 and 20. As expected the average retrieval accuracy for N=20 is lower than that for N=10, but only a few percentage points. The difference in retrieval accuracies between N=10 and N=20 reduces as more number of clusters are examined, *i. e.*, the gap between the two plots, in Figure 2, narrows as we move towards the right in the abscissa. The plots are very similar for the other two similarity measures and hence are not presented.

Figure 3(a) shows the comparison between the hierarchical clustering and the K-means clustering with the histogram intersection measure. The hierarchical clustering performs better than the K-means clustering, but only very marginally. Figure 3(b) shows the same comparison for the L_1 measure and the conclusion here is the same. For the L_2 measure (not shown) the results are similar.

4.2.2: Results with external images as queries: Figures 1-3 showed the results obtained when the query images are taken from the database. A more thorough evaluation should include queries that are not part of the original database. Figures 4-6 show the results obtained from the second set of experiment, where we used a set of 300 images that are not a part of the original database as queries. The performances reported in Figures 4-6 are very similar to the corresponding results in Figures 1-3 showing that there is no degradation in the retrieval performance when the queries are chosen from outside the database.

Figure 4 shows the retrieval performance for the three different similarity measures. Similar to Figure 1, the histogram intersection measures performs slightly better than the L_1 measure. The L_2 measure exhibits the lowest retrieval accuracy among the three measures. Figure 5 shows the comparison of retrieval accuracies for values of N=10 and 20. The results are similar to Figure 2, with the retrieval accuracy for N=20 slightly lower than N=10 and the difference reduces as the number of clusters examined is increased. Figure 6 shows the comparison of the hierarchical and the K-means clustering algorithms. Again, the results are similar to Figure 3, with the hierarchical clustering performing marginally better than the K-means clustering.

Figure 7 shows the comparison between the two sets of experiments. The average retrieval accuracies obtained with the 2000 images as queries (*internal queries*) and with the 300 images (external to the database) as queries are plotted in Figure 7. It is interesting to see that when less than seven clusters are examined (the third point in each plot), the retrieval accuracy for external queries is less than that of the internal queries. But, as more clusters are examined, the retrieval accuracy for external queries outperformed the retrieval accuracy for internal queries. The results are similar for all three similarity measures (the L_2 measure plots are not shown).

5: Conclusions and future work

The results obtained with the hierarchical and the K-means clustering algorithms show that both clustering algorithms drastically reduce the number of required similarity comparisons without sacrificing the retrieval accuracy. These clustering techniques can be used for scalable image and video retrieval from large databases. The hierarchical clustering algorithm outperforms the K-means clustering algorithm in all cases, even though only marginally in some cases. Out of the three similarity measures that are investigated, the histogram intersection and the L_1 measures perform very similarly, whereas the L_2 measure is significantly worse. We also found that the histogram intersection and the L_1 measures result in more uniform clusters, whereas the L_2 measure results in many very small and very large clusters.

We chose the histogram intersection, the L_1, and the L_2 measures, because they are well known measures. Our goal is to show the effectiveness of clustering for scalable retrieval and to compare the results when using different clustering algorithms. We are presently experimenting with more sophisticated similarity measures that have been used in the content-based retrieval community and these results will be presented in the near future.

In our computations of retrieval accuracy for each similarity measure, we have obtained the retrieval accuracy by comparing the results of retrieval with clustering against the results of retrieval without clustering using the same similarity measure. One direction of future work, is to identify a benchmark against which the performances of different similarity measures can be compared against.

The clustering technique presented here for image retrieval can potentially be used for fast browsing of large image and video databases. From each cluster, one or more representative images can be chosen and thumbnail representations of these images can be used to navigate and browse through the database.

Acknowledgments

We wish to thank Dave Kearney of the California Department of Water Resources and Ginger

Ogle of the University of California, Berkeley, for their assistance in providing us with the Cypress database images.

References

[1] M. J. Swain and D. H. Ballard, "Color Indexing", *Intl. J. of Computer Vision*, 7(1), pp. 11-32, 1991.

[2] M. Abdel-Mottaleb, N. Dimitrova, R. Desai, and J. Martino, "CONIVAS: CONtent-based Image and Video Access System", *Proc. of ACM Intl. Multimedia Conference*, Nov. 1996.

[3] W. Niblack, R. Barber, *et.al.*, "The QBIC Project: Querying Images by Content Using Color, Texture and Shape", In *Storage and Retrieval for Image and Video Databases I*, Vol. 1908, SPIE Proceedings, Feb. 1993.

[4] J. Smith and S.-F. Chang, "A Fully Automated Content-based Image Query System", *Proc. of ACM Intl. Multimedia Conference*, Nov. 1996.

[5] H. Zhang, Y. Gong, C. Y. Low and S. W. Smoliar, "Image Retrieval Based on Color Features: An Evaluation Study", *Proc of SPIE*, Vol 2606, pp. 212-220, 1995.

[6] A. K. Jain and R. C. Dubes, "Algorithms for Clustering Data", Prentice Hall, 1988.

[7] M. Stricker and M. Orengo, "Similarity of Color Images", *SPIE Proceedings*, Vol 2420, pp. 381-392, 1995.

[8] W.Y. Ma and B.S. Manjunath, NeTra: "A Toolbox for Navigating Large Image Databases", pp 568-571, *Proc. of ICIP*, Vol 1, 1997.

[9] J-Y. Chen, C. A. Bouman, and J. P. Allebach, "Fast Image Database Search using Tree-Structure VQ", pp. 827-830, *Proc. of ICIP*, Vol II, 1997.

[10] J. Barros, J. French, W. Martin, P. Kelly, and M. Cannon, "Using the triangle inequality to reduce the number of computations required for similarity-based retrieval", *Proc. of SPIE/IS&T Conf. on Storage and Retrieval for Image and Video Databases IV*, Vol. 2670, 1996.

[11] A. Berman and L. Shapiro, "Efficient image retrieval with multiple distance measures", *Proc. of SPIE/IS&T Conf. on Storage and Retrieval for Image and Video Databases V*, Vol. 3022, 1997.

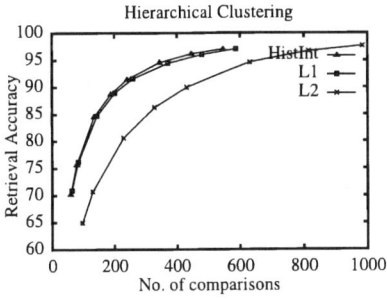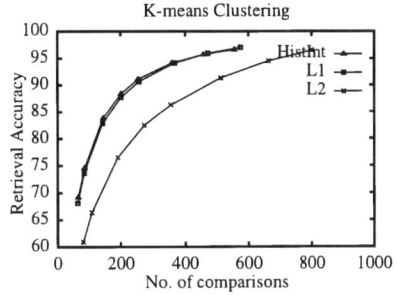

Figure 1: Average retrieval accuracy of (a) the hierarchical clustering and (b) the K-means clustering with N=10. (The eight points in these and the rest of the plots are obtained by examining the top 3, 4, 7, 10, 13, 19, 25, and 31 clusters. The retrieval accuracy and the number of comparisons required increase with the number of clusters examined).

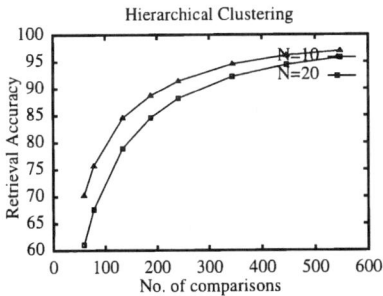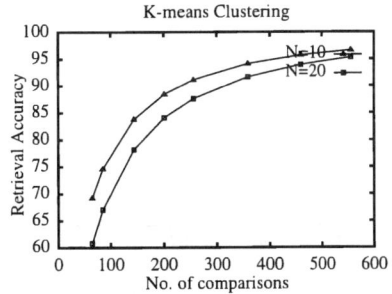

Figure 2: Average retrieval accuracy of (a) the hierarchical clustering and (b) the K-means clustering with the histogram intersection measure for different values of N.

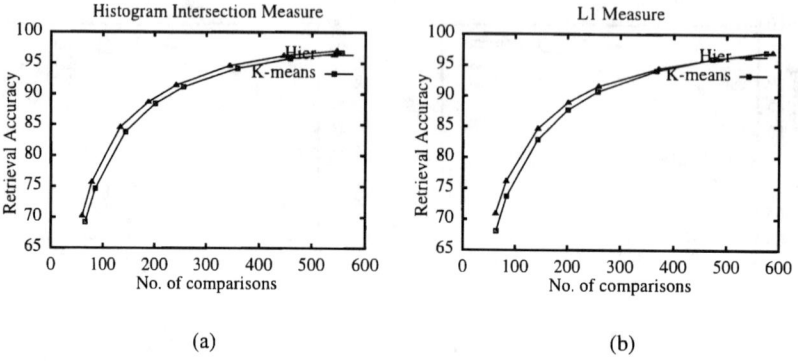

Figure 3: Comparison of average retrieval accuracies of the hierarchical and the K-means clustering with (a) the histogram intersection measure and (b) the L_1 measure, with N=10

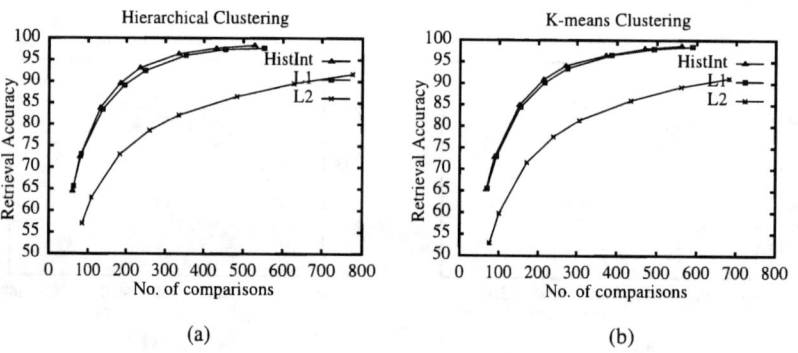

Figure 4: Average retrieval accuracy of (a) the hierarchical clustering and (b) the K-means clustering with N=10 (queries from outside the database).

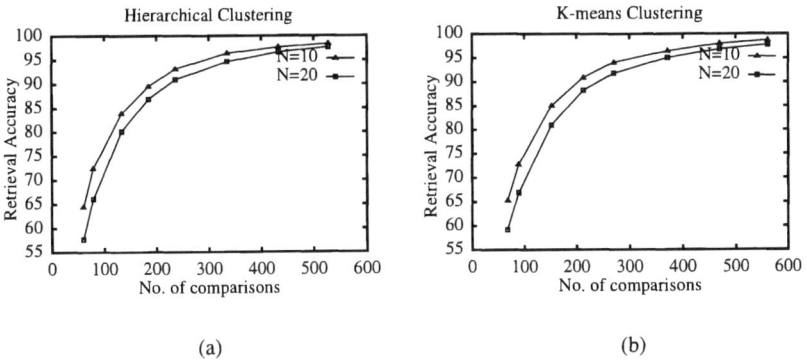

Figure 5: Average retrieval accuracy of (a) the hierarchical clustering and (b) the K-means clustering with the histogram intersection measure for different values of N (queries from outside the database).

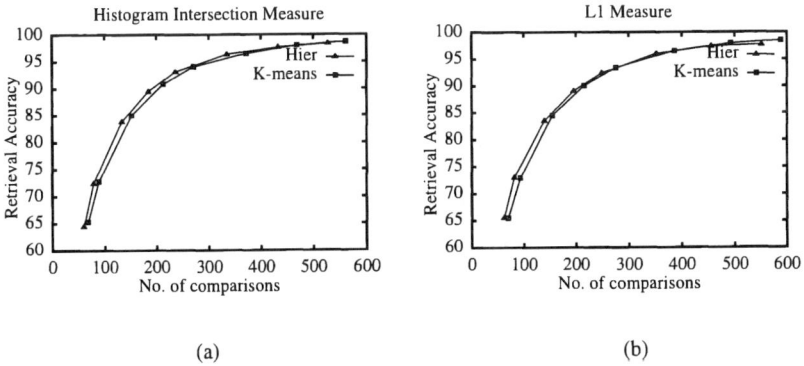

Figure 6: Comparison of average retrieval accuracies of the hierarchical and K-means clustering with (a) the histogram intersection measure and (b) the L_1 measure, with N=10 (queries from outside the database).

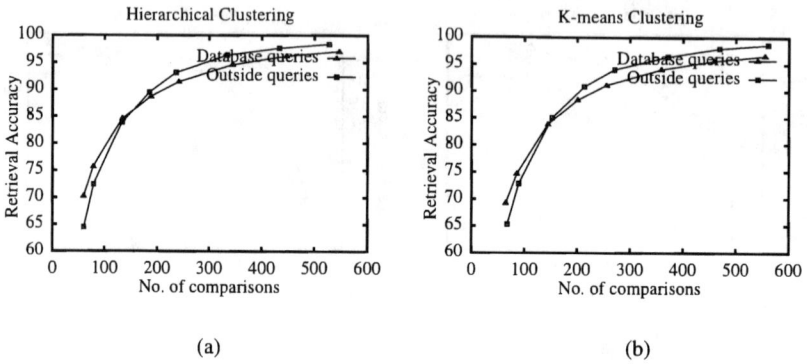

Figure 7: Comparison of average retrieval accuracies when the queries come from inside and outside the database for (a) the hierarchical clustering and (b) the K-means clustering with the histogram intersection measure, with N=10.

Analysis of PCA-based Face Recognition Algorithms

Hyeonjoon Moon[1] and P. Jonathon Phillips[2]

[1]Department of Electrical and Computer Engineering
State University of New York at Buffalo, Amherst, NY 14260
email; moon@acsu.buffalo.edu

[2]National Institute of Standards and Technology
Gaithersburg, MD 20899
email; jonathon@nist.gov

Abstract

Principal component analysis (PCA) based algorithms form the basis of numerous algorithms and studies in the face recognition literature. PCA is a statistical technique and its incorporation into a face recognition system requires numerous design decisions. Some of these decision are documented in the literature, but many are not. We explicitly state the design decisions by introducing a generic modular PCA-algorithm. We experiment with different implementations of each module, and evaluate the different implementations using the September 1996 FERET evaluation protocol. We experiment with (1) changing the illumination normalization procedure; (2) studying effects on algorithm performance of compressing images using JPEG and wavelet compression algorithms; (3) varying the number of eigenvectors in the representation; and (4) changing the distance function in the classifier. We perform two experiments. In the first experiment, we report performance results on the standard September 1996 FERET large gallery image sets. In the second experiment, we examine variation in algorithm performance, by computing algorithm performance on 100 randomly generated image sets (galleries) of the same size.

1 Introduction

Over the last several years, numerous face recognition algorithms have been developed. Among these are algorithms based on principal component analysis (PCA). PCA is popular because it is easy to implement, is a natural dimensionality reduction method, and achieves reasonable performance levels [16, 18, 19]. Because it is easy to implement, it serves as a strawman for comparison with new algorithms [3, 21, 25]. Because it reduces the dimensionality of the data, it is the bases for algorithms [2, 7, 11, 12, 15, 22], and serves as a computational model in psycho-physics [9, 23, 24]. The PCA algorithm has been applied in broad spectrum of studies including face detection [13, 20], face recognition [4, 5, 9, 22] and sex classification [1, 6, 14].

PCA is a statistical method for reducing the dimensionality of high dimensional data, where the data is represented as a vector. In designing an algorithm around PCA, a number of critical design issues have to be addressed. Each of these design decisions has an impact on the overall performance of the algorithm. Some of these design decisions have been explicitly stated in the literature; for example, the distance function in the nearest neighbor classifier. However, a large number of decisions are not mentioned and are passed from researcher to researcher by word of mouth. For example, illumination normalization and number of eigenfeatures included in the representation.

Because the design details are not explicitly stated, a reader cannot assess the merits of a particular implementation and the associated claims. This can unnecessarily caste a shadow on performance claims of a new algorithm when a PCA-based algorithm is used as a strawman. Knowledge of the basic strengths and weaknesses of different implementations can provide insight and guidance in developing algorithms that build on PCA.

In this paper, we present a generic modular PCA-based face recognition algorithm. The algorithm consists of preprocessing, PCA decomposition, and recognition modules. Each module consists of a series of basic steps, where the purpose of each step is fixed. However, we systematically vary the algorithm in each step. For example, the classifier step will always recognize a face, but we will experiment with different classifiers. The selection of which algorithm is in each step is a design decision.

Based on the generic model for PCA-based algorithms, we evaluate different implementations. Because we use the generic model, we can change the implementation in an orderly manner and assess the impact on performance of these modification.

The algorithms are evaluated with the September 1996 FERET testing procedure [16]. Experiment one does a detailed evaluation of variations in the implementation. Results are reported for standard galleries and probe sets described in Phillips et al. [16]. The *gallery* is the set of known individuals. An image of unknown face presented to the algorithm is called a *probe*, and the collection of probes is called the *probe set*. By testing on standard galleries and probe sets, the reader can compare the performance of our PCA implementations with the algorithms tested under the FERET program.

In experiment one, we vary the illumination normalization procedure, the number of eigenvectors in the representation, and the distance in the classifier; and we study the effects of compressing facial images on algorithm performance. The effects of image compression on recognition has not previously been studied. This is of interest in applications where image storage space or image transmission time are critical parameters.

In algorithm evaluation, the following two critical questions are often ignored. First, how does performance vary with different galleries and probe sets. Second, when is the difference in performance between two algorithms significant. In experiment two, we look at this question by randomly generating 100 galleries of the same size. We then calculate performance on each of the galleries against two different categories of probes (probes taken on the same day as the gallery images and probes taken on different days than the gallery images). Because we have 100 scores for each category of probe, we can examine the range of scores, and the overlap in scores among different implementations of the PCA algorithm.

2   PCA-based Face Recognition System

2.1   Principal component analysis (PCA)

Principal component analysis (PCA) is a statistical dimensionality reduction method, which produces the optimal linear least squared decomposition of a training set. Kirby and

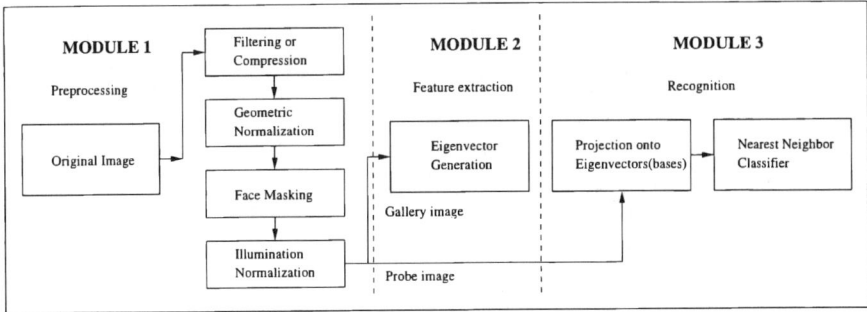

Figure 1. Block Diagram of PCA-based Face Recognition System

Sirovich [11] applied PCA to representing faces and Turk and Pentland [22] extended PCA to recognizing faces. (For further details on PCA, see Fukunaga [8] or Jolliffe [10]).

In PCA, an image is represented as a point in $\Re^{n \times m}$, where the image is n by m pixels. The input to the PCA algorithm is a training set of images where the ensemble mean is zero. From a training set of N images, PCA generates a set of $N - 1$ eigenvectors and eigenvalues. (In the literature, the eigenvectors are sometimes referred to as *eigenfaces*.) We normalize the eigenvectors so that they are orthonormal. The eigenvectors are ordered so that $\lambda_i > \lambda_{i+1}$, where λ_i are the eigenvalues. The λ_i's are equal to the variance of the projection of the training set onto the ith eigenvector. Thus, the low order eigenvectors encode the larger variations in the training set (low order refers to the index of the eigenvectors and eigenvalues).

In a PCA-based face recognition algorithm, the face is represented by its projection onto a subset of M eigenvectors, which we will call *face space*. This is represented as a point in $\Re^M$. A gallery of K individuals is represented as K points g_j in face space. Let p be the representation of a probe in face space. The probe is identified as person j^* in the gallery, if distance $d(p, g_{j^*}) = \min_j d(p, g_j)$; i.e., the probe is identified as the person that minimizes the distance between p and the g_j's. Selecting the distance function d is one of the algorithm design decisions.

2.2 System modules

Our face recognition system consists of three modules and each module is composed of a sequence of steps (see figure 1). The first module preprocesses the input image. The goal of the preprocessing is to transform the facial image into a standard format that removes variations that can effect recognition performance. This module consists of four steps. Figure 2 shows the input and output of some of the steps in the preprocessing module. The first step filters or compresses the original image. The image is filtered to remove high frequency noise in the image. An image is compressed to save storage space and reduce transmission time. The second step places the face in a standard geometric position by rotating, scaling, and translating the center of eyes to standard locations. The goal of this step is to remove variations in size, orientation, and location of the face. The third step masks out background pixels, hair, and clothes. The masked regions vary from day to day and can interfere with identification. The fourth step removes some of the variations in illumination between images. Changes in illumination are critical factors in algorithm

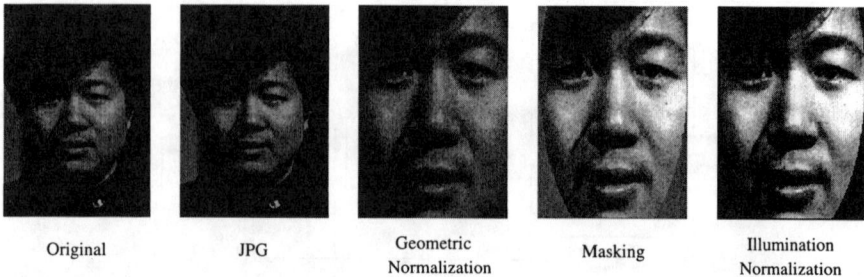

| Original | JPG | Geometric Normalization | Masking | Illumination Normalization |

Figure 2. Input and output of several steps of the preprocessing module.

performance [16].

The second module performs the PCA decomposition on the training set. This produces the eigenvectors (eigenfaces) and eigenvalues.

The third module identifies the face from a preprocessed image, and consists of two steps. The first step projects the image onto the eigenvectors that represent the face. The critical parameter in this step is the subset of eigenvectors that represent the face. The second step recognizes faces using a nearest neighbor classifier. The critical parameter in this step is the distance function in the classifier.

3 Test Design

3.1 FERET Database.

The FERET database provides a common database of facial images for both development and testing of face recognition algorithms and has become the de facto standard for face recognition of still images [16, 18].

In the FERET database, images of individual were acquired in sets of 5 to 11 images. Each set includes two frontal views (**fa** and **fb**); a different facial expression was requested for the second frontal image. For 200 sets of images, a third frontal image was taken with a different camera and different lighting (**fc**). (The remaining images were non-frontal images and were not used in this study.)

One emphasis of the database collection was on duplicate image sets. *Duplicate* is defined as an image of a person whose corresponding gallery image was from a different image set (usually taken on a different date). The database contains 365 duplicate sets of images. For 91 duplicate sets, the time between the first and last sittings was at least 18 months. In an effort to maintain a degree of consistency throughout the database, the photographer used the same physical setup in each session. However, because the equipment had to be reassembled for each session, there was variation from session to session. This results in variations in scale, pose, expression, and illumination of the face (figure 3). For details of the FERET database, refer to Phillips et al. [17, 18].

3.2 Design rule.

To obtain a robust comparison of algorithms, it is necessary to calculate performance on a large number of galleries and probe sets. To allow scoring on multiple galleries and

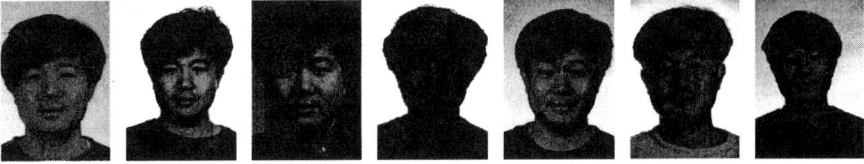

Figure 3. Duplicate images (examples of variations between collections).

probe sets, we have to adopt an appropriate protocol. In the new protocol, an algorithm is given two sets of images: the *target set* and the *query set*. We introduce this terminology to distinguish these sets from the galleries and probe sets that are used in computing performance statistics. The target set is given to the algorithm as the set of known facial images. The images in the query set are the unknown facial images to be identified.

- *Definition.* For each image q_i in the query set Q, an algorithm reports the similarity $s_i(k)$ between q_i and each image t_k in the target set T. The key property, which allows for greater flexibility in scoring, is that for any two images s_i and t_k, we know $s_i(k)$. From the output files, algorithm performance can be computed for virtual galleries and probe sets. Thus, we can create and score results for any gallery $G \subset P$ and any probe set $P \subset Q$. For a given gallery G and probe set P, the performance scores are computed by examination of the similarity measures $s_i(k)$ such that $q_i \in P$ and $t_k \in G$. We refer to such galleries and probe sets as virtual galleries and virtual probe sets (because they are subsets of the target set and query set).

In our experiments, we report performance for four different categories of probes. The **FB** probes are the second frontal images from the same session as the frontal images that are in the gallery. (In a gallery, either the **fa** or **fb** images are placed in the gallery, and the other image is placed in the probe set. We will denote those frontal images that are placed in the probe set by **FB**.) The **fc** probes are the **fc** image from the same session as the frontal image in the gallery. (Note: there are only 200 sets of images with **fc** images.) The *duplicate I* probes are the duplicate frontal images of the images in the gallery (frontal images from different image sets—usually different days). The *duplicate II* probes are images where there is at least a year and half between the acquisition of the gallery and probe images.

We report identification results using a closed universe model. In the closed universe, every probe is in the gallery. The complement to the closed universe is the open universe where some probes are not in the gallery. The open universe model is used in verification and authentification scenarios. The results for these scenarios can be found in Rizvi et al. [19].

For the closed universe, we report performance as cumulative match scores. The rank is plotted along the horizontal axis, and the vertical axis is the percentage of correct matches. The cumulative match scores can be calculated for any subset of the probe set. We calculated this score to evaluate an algorithm's performance on different categories of probes. The computation of the score is quite simple. Let P be the number of probes to be scored and R_k the number of these probes in the subset that are in the top k. The fraction is reported as R_k/P. The top rank match (or score) is R_1 (the fraction of probes correctly identified).

4   Experiment 1

The purpose of this experiments is to examine the effects of changing the algorithms in the generic PCA-based face recognition system. We do this by establishing a baseline algorithm and then varying the implementation of select steps one at a time. Ideally, we would test all possible combination of variations. However, because of the number of combination, this is not practical.

The baseline algorithm has the following configuration: The images are not filtered or compressed. Geometric normalization consists of rotating, translating, and scaling the images so the center of the eyes are on standard pixels. This is followed by masking the hair and background from the images. In the illumination normalization step, the non-masked facial pixels were processed by a histogram equalization algorithm. Then, the facial pixels were transformed so that the mean was equal to 0.0 and variance equal to 1.0. The geometric normalization and masking steps are not varied in either experiment. The training set for the PCA consists of 501 images (one image per person), which produces 500 eigenvectors. The images in the training set are not varied in this chapter. In the recognition module, a face is represented by its projection onto the first 200 eigenvectors and the classifier uses the L_1 norm.

4.1   Test sets, galleries, and probe sets

All images are from the FERET database, and the testing was done with the September 1996 FERET protocol. In this protocol, the target set contained 3323 images and the query set 3816 images. All the images in the target set were frontal images. The query set consisted of all the images in the target set plus rotated images and digitally modified images. (The rotated and digitally modified images were not included in our analysis.) Thus, the output contained every target image matched with itself. This allowed for a detailed analysis on multiple galleries and probe sets.

We report results for four different probe categories. First is the **FB** probes. For this category, the gallery consisted of 1196 individuals with one image per person, and there were 1195 probes. The second was the 722 duplicate I probes. The third was the 194 **fc** probes. For both the duplicate I and **fc** categories, the gallery was the **FB** gallery. The fourth category was the 234 duplicate II probes. For this category the gallery consisted of 864 individuals and was a subset of the **FB** gallery.

Table 1. Performance results for illumination normalization methods. Performance score are the top rank match.

	Probe category			
Illumination normalization method	duplicate I	duplicate II	**FB** probe	**fc** probe
Baseline	0.35	0.13	0.77	0.26
Original image	0.32	0.11	0.75	0.21
Histogram Eq. only	0.34	0.12	0.77	0.24
$\mu = 0.0,\ \sigma^2 = 1.0$	0.33	0.14	0.76	0.25

Table 2. Performance score for low pass filter, JPEG, and wavelet compressed images (0.5 bits/pixel compression). Performance scores are the top rank match.

	Probe category			
Preprocessing	duplicate I	duplicate II	**FB** probe	fc probe
Baseline	0.35	0.13	0.77	0.26
JPEG	0.35	0.13	0.78	0.25
Wavelet	0.36	0.15	0.79	0.25
LPF	0.36	0.15	0.79	0.24

4.2 Variations in the preprocessing module

- *Illumination normalization.* We experimented with three variations to the illumination normalization step. First, the non-masked pixels were not changed (original pixel values). Second, the non-masked facial pixels were processed with a histogram equalization algorithm. Third, the non-masked facial pixels were transformed so that the mean was equal to 0.0 and variance equal to 1.0. The results for these variations and the baseline algorithm are presented in table 1.

- *Compressing and filtering the images.* We examined the effects of JPEG and wavelet compression, and low pass filtering (LPF) on recognition. For this experiment, the original images were compressed and then uncompress prior to being feed into the geometric normalization step of the preprocessing module. For both compression methods, the images were compressed approximately 16:1 (0.5 bits per pixel). We experimented with other compression ratios and found that performance was comparable. The results are for eigenvectors generated from non-compressed images. We found that performance in this case was slightly better than on eigenvectors trained from compressed images. Because compression algorithms usually are low pass filters, we decided to see the effects on performance of low pass filtering the original image. The filter was a 3x3 spatial filter with a center value of 0.2 and the remaining values equal to 0.1. Table 2 reports performance for the baseline algorithm, JPEG and wavelet compression, and low pass filtering.

4.2.1 Variations in the recognition module

- *Number of low order eigenvectors.* The higher order eigenvectors encode small vari-

Table 3. Performance score with low order eigenvectors removed. Performance scores are the top rank match.

Number of low order	Probe category			
eigenvectors removed	duplicate I	duplicate II	**FB** probe	fc probe
0 (Baseline)	0.35	0.13	0.77	0.26
1	0.35	0.15	0.75	0.38
2	0.34	0.14	0.74	0.36
3	0.31	0.14	0.72	0.37
4	0.20	0.09	0.50	0.22

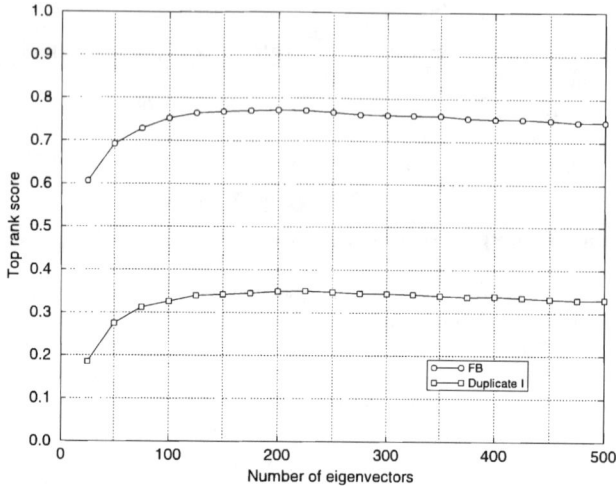

Figure 4. Performance on duplicate I and FB probes based on number of low order eigenvectors used. (Number of images in gallery = 1196; Number of FB images in probe set = 1195, Number of duplicate I in probe set = 722)

ations and noise among the images in the training set. One would expect that the higher order eigenvectors would not contribute to recognition. We examined this by computing performance as a function of the number of low order eigenvectors in the representation. Figure 4 shows top rank score for **FB** and duplicate I probes as the function of the number of low order eigenvectors included in the representation. The representation consists of $e_1, \ldots, e_n$, $n = 50, 100, \ldots, 500$, where e_i are the eigenvectors generated by the PCA decomposition.

- *Removing low order eigenvectors.* The low order eigenvectors encode gross differences among the training set. If they encode variable information such as lighting changes, performance may improve if they are not included in the representation. Table 3 reports change in performance of not including the first 1, 2, 3, and 4th eigenvectors in the representation; i.e., the representation consisted of $e_i, \ldots, e_{200}$, $i = 1, 2, 3, 4$. For **fc** probes, cumulative match scores are given in figure 5.

- *Nearest neighbor classifier.* We experimented with seven distances for the classifier. They are listed in table 4, along with the results. Details of the distances are given in appendix A. The results for **fc** probes are presented in figure 6.

4.3   Discussion

In the preprocessing module, we experimented with varying the illumination normalization and compression steps. The results show that performing an illumination normalization step improves performance, but which implementation that is selected is not critical.

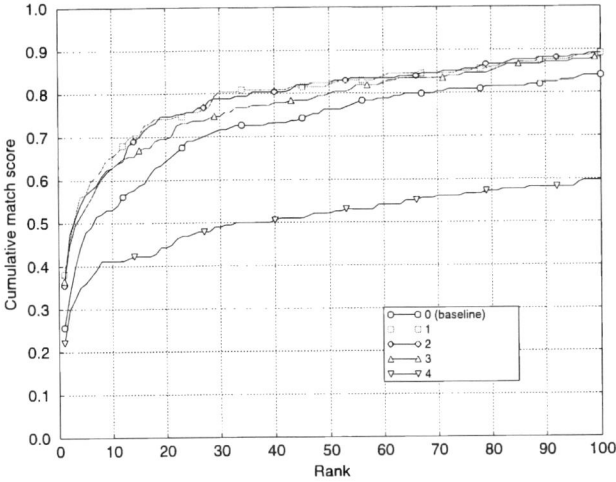

Figure 5. Performance on fc probes with first 1, 2, 3, and 4 eigenvectors removed.

The results also show that compressing or filtering the images does not significantly effect performance.

In the recognition module, we experimented with three classes of variations. First, we varied the number of low order eigenvectors in the representation from 50 to 500 by steps of 50. Figure 4 shows that performance increases until approximately 200 eigenvectors are in the representation and then performance decreases slightly. Representing faces by the first 40% of the eigenvectors is consistent with results on other facial image sets that the authors have seen.

Second, low order eigenvectors were removed. Table 3 shows that removing the first eigenvector resulted in an overall increase in performance. The largest increase was ob-

Table 4. Performance scores based on different nearest neighbor classifier. Performance scores are the top rank match.

Nearest neighbor classifier	Probe category			
	duplicate I	duplicate II	**FB** probe	**fc** probe
Baseline (L_1)	0.35	0.13	0.77	0.26
Euclidean (L_2)	0.33	0.14	0.70	0.04
Angle	0.34	0.12	0.70	0.07
Mahalanobis	0.42	0.17	0.74	0.23
L_1 + Mahalanobis	0.31	0.13	0.73	0.39
L_2 + Mahalanobis	0.35	0.13	0.77	0.31
Angle + Mahalanobis	0.45	0.21	0.77	0.24

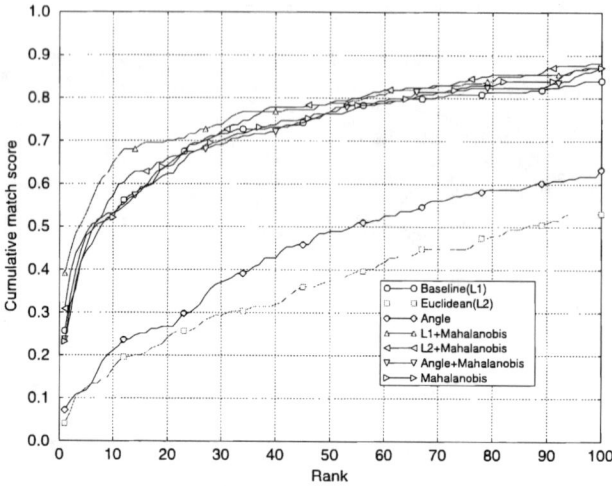

Figure 6. Effects of nearest neighbor classifier for face recognition. Performance scores for fc probes.

served with the **fc** probes. This increase is further highlighted in figure 5. The low order eigenvectors encode the greatest variations among the training set. The most significant difference between the **fc** probes and the gallery images was a change in lighting. If the low order eigenvectors encode lighting differences, then this would explain the substantial increase in performance by removing the first eigenvector.

Third, the distance in the classifier was changed. This variation showed the largest range of performance. For duplicate I probes, performance ranged from 0.31 to 0.45, and for **fc** probes the range was from 0.07 to 0.39. For duplicate I, duplicate II and **FB** probes, the angle+Mahalanobis distance performed the best. For the **fc** probes, the L_1+Mahalanobis distance performed the best. But, this distance was the worst for the duplicate I probe.

Because of the range of performance, it is clear that selecting the distance function for the classifier is the critical decision in designing a PCA-based face recognition system. However, decision of which distance to select is dependent on the type of images in the galleries and probe sets that the system will process.

5   Experiment 2

The comparison among algorithms in experiment one are based on performance on four probe sets, and the performance among the different probe sets cannot be directly compared. For most of the variations in the preprocessing module, performances are comparable. However, for the classifier distances, there is a considerable range in performance. The natural question is, when is the performance between two classifiers significant?

To address this question, we randomly generated 100 galleries of 200 individuals, with one

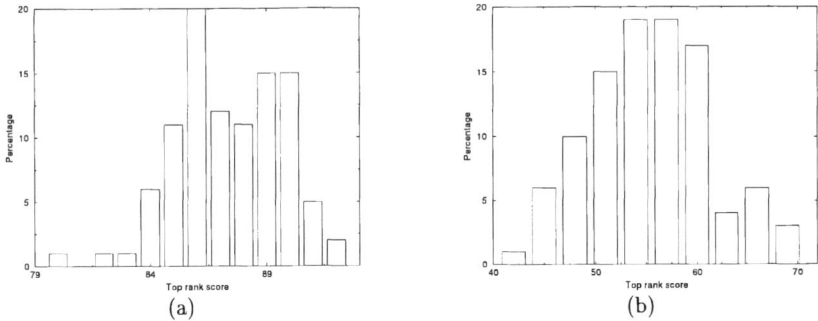

(a)   (b)

Figure 7. Histogram of top rank scores of the angle+Mahalanobis classifier. (a) FB probes and (b) duplicate I probes.

frontal image per person. The galleries were generated without replacement from the **FB** gallery of 1196 individuals in experiment one. Then we scored each of the galleries against the **FB** and duplicate I probes for each of the seven classifiers in experiment one. (There were not enough **fc** and duplicate II probes to compute performances for these categories.) For each gallery, the corresponding **FB** probe set consisted of the second frontal image for all images in the gallery; the duplicate I probe set consisted of all duplicate images in the database for each image in the gallery. We measured performance by the top rank score (the fraction of probes that were correctly identified).

For each classifier and probe category, we had 100 different scores. Figure 7 presents the histogram of top rank scores for angle+Mahalanobis distance for both **FB** and duplicate I probe sets. This shows a range in performance from 0.80 to 0.93 for **FB** probe, from 0.42 to 0.70 for duplicate I probe.

In figure 8, we present a truncated range of top rank scores for the classifiers. Figure 8(a) shows the range for **FB** probes and figure 8(b) for duplicate I probes. We mark the median by $\times$, the 10th percentile by $+$, and 90th percentile by $*$. We plotted these values because they are robust statistics, and selected the 10th and 90th percentile because they mark a robust range of scores and outliers are ignored.

5.1 Discussion

The main goal of experiment two is to get a rough estimate of when the difference in performance is significant. From figure 8, the range in scores is approximately ± 0.05 about the median for all 14 runs. This suggests a reasonable threshold for measuring significant difference in performance for the classifiers is ~ 0.10.

The results for duplicate I probes in experiment two are consistent with the results in experiment one. In table 4, the top classifiers were the Mahalanobis and angle+Mahalanobis and these two classifiers have the two highest scores in table 8. In both experiments, the L_1+Mahalanobis received the lowest scores. This suggest that for duplicate I scores that the angle+Mahalanobis or Mahalanobis distance should be used, and that the results on

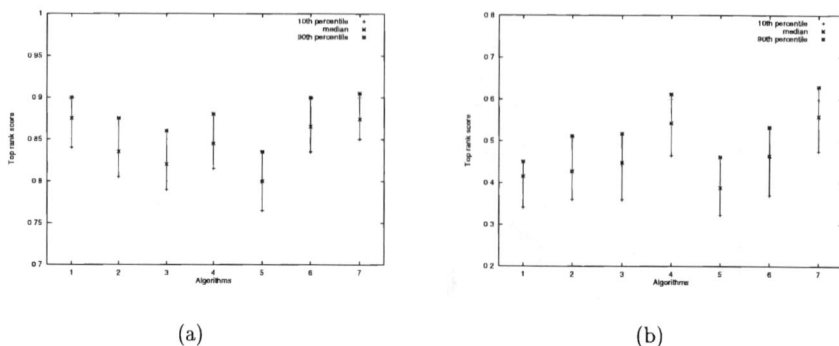

(a)   (b)

Figure 8. The nearest neighbor classifiers presented are: (1) L_1, (2) L_2, (3) Angle, (4) Mahalanobis, (5) L_1+Mahalanobis, (6) L_2+Mahalanobis, and (7) Angle+Mahalanobis. (a) FB probes and (b) duplicate I probes.

the smaller gallery are predictive of the scores on larger galleries.

For **FB** probes, there is not as sharp a division between classifiers. One possible explanation is that in experiment one, the top match scores for the **FB** probes did not vary as much as the duplicate I scores. There is consistency among the best scores (L_1, L_2+Mahalanobis, and angle+Mahalanobis). The remaining classifiers' performances can be grouped together. The performance scores of these classifiers are within each other's error margins. This suggests that either the L_1, L_2+Mahalanobis, or angle+Mahalanobis distance should be used.

6   Conclusion

From the experiments described in these chapter, we have come to four important conclusion. First, JPEG and wavelet compression algorithms do not degrade performance. This is important because it suggests that compressing images to save transmission time and storage costs will not reduce algorithm performance.

Second, selection of the nearest neighbor classifier is the critical decision in designing a PCA-based algorithm. The proper selection of distance metric is essential to improve performance scores. Furthermore, our experiments show that distance functions that achieve the best performance are not generally considered in the literature.

Third, the performance scores vary among the probe categories, and that the selection of components in the algorithm need to consider the type of images that the algorithm will process. The **FB** and duplicate I probes are least sensitive to system design decisions, while **fc** and duplicate II probes are the most sensitive.

Fourth, the performance within a category of probes can vary greatly. This recommends that when comparing algorithms, that performance scores from a set of galleries and probe sets need to be examined.

In our evaluation process, we introduced a modular design for PCA-based face recognition systems. This allowed us to systematically vary the components and measure the impact of these variations on performance. Like PCA, the majority of the face recognition algorithms in the literature are view-based and have the same basic architecture as our PCA-based system. By following the evaluation procedure presented in this chapter, an algorithm designers can determine the optimal configuration of their face recognition system.

A   Appendix

In this appendix, we describe mathematically the distances used in the nearest neighbor classifier. The variables x, y, z are k-dimensional vectors and x_i, y_i, and z_i are the ith components of the vectors.

A.1 L_1 distance.

$$d(x, y) = |x - y| = \sum_{i=1}^{k} |x_i - y_i|$$

A.2 L_2 distance.

$$d(x, y) = \|x - y\|^2 = \sum_{i=1}^{k} (x_i - y_i)^2$$

A.3   Angle between feature vectors.

$$d(x, y) = -\frac{x \cdot y}{\|x\|\|y\|} = -\frac{\sum_{i=1}^{k} x_i y_i}{\sqrt{\sum_{i=1}^{k} (x_i)^2 \sum_{i=1}^{k} (y_i)^2}}$$

A.4   Mahalanobis distance.

$$d(x, y, z) = -x \cdot y \cdot z = -\sum_{i=1}^{k} x_i y_i z_i$$

$$z_i = \sqrt{\frac{\lambda_i}{\lambda_i + \alpha^2}} \simeq \frac{1}{\sqrt{\lambda_i}}, \alpha = 0.25$$

where λ_i = eigenvalue of ith eigenvector.

A.5 L_1 + Mahalanobis distance.

$$d(x, y, z) = |x - y| \cdot z = \sum_{i=1}^{k} |x_i - y_i| z_i$$

A.6 L_2 + Mahalanobis distance.

$$d(x, y, z) = \|x - y\|^2 \cdot z = \sum_{i=1}^{k} (x_i - y_i)^2 z_i$$

A.7   Angle + Mahalanobis distance.

$$d(x, y, z) = -\frac{x \cdot y \cdot z}{\|x\|\|y\|} = -\frac{\sum_{i=1}^{k} x_i y_i z_i}{\sqrt{\sum_{i=1}^{k} (x_i)^2 \sum_{i=1}^{k} (y_i)^2}}$$

References

[1] H. Abdi, D. Valentin, B. Edelman, and A. J. O'Toole. More about the difference between men and women. *Perception*, 24:539–562, 1995.

[2] M. S. Barlett, H. M. Lades, and T. J. Sejnowski. Independent component representations for face recognition. In *Preceedings of the SPIE Symposium on Electronic Imaging; Science and Technology; Conference on Human Vision and Electronic Imaging III*, in press 1998.

[3] P. Belhumeur, J. Hespanha, and D. Kriegman. Eigenfaces vs fisherfaces: Recognition using specific linear projection. *IEEE Trans. PAMI*, 19(7):711–720, 1997.

[4] R. Brunelli and T. Poggio. Face recognition: Features versus templates. *IEEE Trans. PAMI*, 15(10):1041–1052, 1993.

[5] G. W. Cottrell and M. Fleming. Face recognition using unsupervised feature extraction. In *Proc. Int. Conf. Neural Network*, pages 322–325. Kluwer, Dordrecht, 1990.

[6] G. W. Cottrell and J. Metcalfe. Empath: Face, emotion, and gender recognition using holons. In D. Touretzky, editor, *Advances in Neural Information Processing*, pages Vol. 3, 564–571, San Mateo, CA, 1991. Morgan Kaufman.

[7] K. Etemad and R. Chellappa. Discriminant analysis for recognition of human face images. *J. Opt. Soc. Am. A*, 14:1724–1733, August 1997.

[8] K. Fukunaga. *Introduction to statistical pattern recongition*. Academic Press, Orlando, FL, 1972.

[9] P. Hancock, A. M. Burton, and V. Bruce. Face processing: Human perception and principal component analysis. *Memory and Cognition*, pages 26–40, 1996.

[10] I. T. Jolliffe. *Principal Component Analysis*. Springer-Verlag, 1986.

[11] M. Kirby and L. Sirovich. Application of the karhunen-loeve procedure for the characterization of human faces. *IEEE Trans. PAMI*, 12(1):103–108, 1990.

[12] B. Moghaddam and A. Pentland. Face recognition using view-based and modular eigenspaces. In *Proc. SPIE Conference on Automatic Systems for the Identification and Inspection of Humans*, volume SPIE Vol. 2277, pages 12–21, 1994.

[13] B. Moghaddam and A. Pentland. Maximum likelihood detection of faces and hands. In M. Bichsel, editor, *International Workshop on Automatic Face and Gesture Recognition*, pages 122–128, 1995.

[14] A. O'Toole, T. Vetter, N. Troje, and H Bulthoff. Sex classification is better with three-dimensional head structure than with image intensity information. *Perception*, 26:75–84, 1997.

[15] P. Penev and J. Atick. Local feature analysis: a general statistical theory for object representation. *Network: Computation in Neural Systems*, 7(3):477–500, 1996.

[16] P. J. Phillips, H. Moon, P. Rauss, and S. Rizvi. The FERET evaluation methodology for face-recognition algorithms. In *Proceedings Computer Vision and Pattern Recognition 97*, pages 137–143, 1997.

[17] P. J. Phillips, P. Rauss, and S. Der. FERET (face recognition technology) recognition algorithm development and test report. Technical Report TR # 995, U.S. Army Research Laboratory, 1996.

[18] P. J. Phillips, H. Wechsler, J. Huang, and P. Rauss. The FERET database and evaluation procedure for face-recognition algorithms. *Image and Visual Computing*, in press 1998.

[19] S. Rizvi, P. J. Phillips, and H. Moon. The FERET verification testing protocol for face recognition algorithms. In *Proceedings Computer Vision and Pattern Recognition 98*, in press 1998.

[20] K. K. Sung and T. Poggio. Example-based learning for view-based human face detection. *IEEE Trans. PAMI*, 20(1):39–51, 1998.

[21] D. Swets and J. Weng. Using discriminant eigenfeatures for image retrieval. *IEEE Trans. PAMI*, 18(8):831–836, 1996.

[22] M. Turk and A. Pentland. Eigenfaces for recognition. *J. Cognitive Neuroscience*, 3(1):71–86, 1991.

[23] D. Valentin, H. Abdi, and B. Edelman. What represents a face: A computational approach for the integration of physiological and psychological data. *Perception*, 26:1271–1288, 1997.

[24] Tim Valentine, editor. *Cognitive and Computational Aspects of Face Recognition*. Routledge, 1995.

[25] J. Wilder, P. J. Phillips, C. Jiang, and S. Wiener. Comparison of visible and infrared imagery for face recognition. In *2nd International Conference on Automatic Face and Gesture Recognition*, pages 182–187, 1996.

Performance Assessment by Resampling: Rigid Motion Estimators

Bogdan Matei[1]   Peter Meer[1]   David Tyler[2]

[1]Electrical and Computer Engineering Department
[2]Statistics Department
Rutgers University, Piscataway, NJ, 08855-0909, USA
matei, meer, dtyler@caip.rutgers.edu

Abstract

Quantitative assessment of performance in image understanding tasks with real data is difficult since the data is complex and the different computational modules most often interact. Employing modern statistical techniques we have developed a set of numerical tools which provide rigorous performance measures derived solely from the given input. Covariance matrices and confidence intervals are computed for the estimated parameters and individually for the corrected data points. As an example, the proposed methodology is applied to compare rigid motion estimators.

1: Performance assessment in image understanding

The lack of universally accepted, rigorous performance assessment methodology is considered by many as one of the major bottlenecks of progress in image understanding. In a recent paper Christensen and Förstner [7] discuss several objections against the widespread use of evaluation techniques. Most of these objections are well justified.

- Performance for real data often cannot be reliably predicted from the controlled experiments with simple, synthetic inputs.

- A practical system contains several interacting computational modules, and traditional techniques (like error propagation) may be unfeasible.

- Empirical testing often yields incompatible performance measures and thus ignores the very goal of the procedure, enabling comparison with other techniques.

- The number of experiments required to obtain highly accurate statistical measures is prohibitively large [14], and for real data it is not well understood how the input space can be parameterized.

In this paper we present a new, very general performance assessment methodology which hopefully provides at least partial response to these objections.

The new method is an extension of the *resampling paradigm* widely used in statistics, and has a solid theoretical basis. The method is entirely numerical and derives statistically valid measures *solely from the given input*. To obtain similar measures using more traditional methods a large set of distinct inputs has to be available. Being data driven the new method belongs to the class of empirical techniques. It retains the advantage of the empirical techniques (they are useful in real situations), but eliminates their often ad-hoc nature. When applied in analytically tractable cases they will reproduce (within the computational accuracy) the theoretically predicted results. However, the often considerable mathematical sophistication needed to obtain these results is no longer necessary. Being a numerical method, it can be applied to any complex system as long as the underlying mild assumptions about the measurement errors are satisfied.

The first use of the resampling paradigm in a low-level image understanding problem (edge detection) is given in [6]. Here we introduce a more general approach which can be applied to performance assessment in most vision tasks. In Section 2 a short review of the employed resampling methods is provided. The proposed performance assessment tools are developed in Section 3. In Section 4 the problem of rigid motion estimation is analyzed and three different algorithms are discussed. The performance of these three algorithms is compared in Section 5.

2: Review of resampling techniques

Let the data set contain n independent, identically distributed (i.i.d.) measurements $z_i \in R^m$, $Z = \{z_1, z_2, \cdots, z_n\}$, with an unknown probability distribution F. The distribution is partially characterized by the value of some parameter of interest $\theta = t(F) \in R^p$ which is estimated from the data by the estimator $\hat{\theta} = S[Z]$. Resampling techniques are numerical methods to determine from the available *single* data set statistical measures (like bias or variance) of $\hat{\theta}$, usually without any assumption about F.

It is important to notice that the data in image understanding may not satisfy the i.i.d. condition. For example, in conic fitting and tasks involving the epipolar geometry, the estimate of a bilinear form is computed from non-i.i.d. data, and the quasi-optimal estimators must take this into account [21]. The application of resampling methods to non-i.i.d. data can lead to inconsistent results unless additional safeguards are included. In the following we review only material relevant to the topic of the paper.

2.1: Bootstrap

The bootstrap methodology was proposed by Bradley Efron in 1979. An excellent review of bootstrap can be found in [10] and additional material in [2][23][32]. Let $\hat{F}_n$ be the empirical distribution function of Z, i.e. obtained by associating equal probability $\frac{1}{n}$ with each measurement z_i. New data sets $Z^{*b} = \{z_1^{*b}, z_2^{*b}, \cdots, z_n^{*b}\}$ called *bootstrap samples* are formed by sampling with replacement from Z. For each bootstrap sample Z^{*b}, $b = 1, 2, \ldots, B$, the statistics $\hat{\theta}^{*b} = S\left[Z^{*b}\right]$ is computed. The ensemble of $\hat{\theta}^{*b}$ is used instead of the sampling distribution of $\hat{\theta}$ and the accuracy of $\hat{\theta}$ as an estimator of θ is inferred from this ensemble.

In a bootstrap sample Z^{*b} each z_i appears k_i times, such that $\sum_{i=1}^{n} k_i = n$. Thus the bootstrap samples can also be generated by keeping Z fixed and varying the probability of each component z_i. Therefore, Z^{*b} is characterized by the vector $p^{*b} = \left[\frac{k_1}{n} \ \frac{k_2}{n} \ \cdots \ \frac{k_n}{n}\right]^T$, with np^{*b} obtained by sampling from a multinomial distribution with n draws and equal class probabilities [10, pp. 285–287]. The bootstrap statistics $\hat{\theta}^{*b}$ can be expressed either as a function of the probability vector $\hat{\theta}^{*b} = t\left(\hat{F}_n^{*b}\right) = T\left[p^{*b}\right]$, or as a function of the input data $\hat{\boldsymbol{\theta}}^{*b} = \boldsymbol{S}\left[\boldsymbol{Z}^{*b}\right]$. Similarly, $\hat{\boldsymbol{\theta}}$ can be written as $\hat{\boldsymbol{\theta}} = \boldsymbol{t}\left(\hat{F}_n\right) = \boldsymbol{T}\left[\boldsymbol{p}_0\right]$, where $\boldsymbol{p}_0 = \left[\frac{1}{n}\ \frac{1}{n}\ \cdots\ \frac{1}{n}\right]^{\top}$. Both formulations will be used in the following sections.

The bootstrap estimate of the covariance matrix of $\hat{\boldsymbol{\theta}}$ is

$$\hat{\boldsymbol{C}}_{BO} = \frac{1}{B-1}\sum_{b=1}^{B}\left[\hat{\boldsymbol{\theta}}^{*b} - \bar{\boldsymbol{\theta}}^*\right]\left[\hat{\boldsymbol{\theta}}^{*b} - \bar{\boldsymbol{\theta}}^*\right]^{\top}, \quad \bar{\boldsymbol{\theta}}^* = \frac{1}{B}\sum_{b=1}^{B}\hat{\boldsymbol{\theta}}^{*b}. \tag{1}$$

For large values of B and n the bootstrapped covariance estimator is essentially unbiased. In practice B should be at least 100 with higher values required if n is small (say less than 30). In confidence interval computations, where the tails of the sampling distribution have to be accurately estimated, larger values for B (in the range $1000 - 2000$) have to be used. It should be emphasized that the lower bound on B for satisfactory performance depends on the nature of $\hat{\boldsymbol{\theta}}$ and the number of available measurements n.

The methodology presented so far is for i.i.d. observations from a single population, such as in estimating a single population mean. However, it can be extended to regression problems with i.i.d. errors by *bootstrapping the residuals*. Under the general *errors in variables* (EIV) linear regression model, the true values must satisfy the constraint

$$y_{io} = \boldsymbol{\beta}^{\top}\boldsymbol{x}_{io} + \alpha , \tag{2}$$

where $\boldsymbol{x}_{io} \in R^{p-1}$, $y_{io} \in R$ and $\boldsymbol{\beta}, \alpha$ are the regression parameters. Let $\boldsymbol{z}_{io} = \left[\boldsymbol{x}_{io}^{\top}\ y_{io}\right]^{\top}$ and $\boldsymbol{\theta} = \left[\boldsymbol{\beta}^{\top}\ \alpha\right]^{\top}$. The available measurements $\boldsymbol{z}_i$ are the additively corrupted true values $\boldsymbol{z}_{io}$

$$\boldsymbol{z}_i = \boldsymbol{z}_{io} + \Delta\boldsymbol{z}_i ,$$

where the i.i.d. errors $\Delta\boldsymbol{z}_i$ have zero-mean and covariance $E[\Delta\boldsymbol{z}_i\Delta\boldsymbol{z}_i^{\top}] = \boldsymbol{C}_z$, known up to a constant σ^2. The parameter estimate $\hat{\boldsymbol{\theta}}$ and the corrected measurements $\hat{\boldsymbol{z}}_i$ are obtained by the Generalized Total Least Squares (GTLS) procedure, in which the sum of squared Mahalanobis distances between the measurements and the true (unknown) points

$$[\hat{\boldsymbol{\theta}}, \hat{\boldsymbol{z}}_i] = \arg\min_{\theta, z_{io}}\sum_{i=1}^{n}\left(\boldsymbol{z}_i - \boldsymbol{z}_{io}\right)^{\top}\boldsymbol{C}_z^{-1}\left(\boldsymbol{z}_i - \boldsymbol{z}_{io}\right) = \arg\min_{\theta, z_{io}}\sum_{i=1}^{n}\|\Delta\boldsymbol{z}_i\|_{C_z}^2$$

is minimized subject to the constraints

$$y_{io} = \alpha + \boldsymbol{\beta}^{\top}\boldsymbol{x}_{io}, \quad i = 1\cdots n .$$

When the errors are normal $\|\Delta\boldsymbol{z}_i\|_{C_z}^2$ is distributed as χ_p^2 and the GTLS is also the maximum likelihood estimator. The corrected data points $\hat{\boldsymbol{z}}_i$ are the projections of $\boldsymbol{z}_i$ on the hyperplane defined by $\hat{\boldsymbol{\theta}}$ and the estimate of the error variance is computed from the residuals $\Delta\hat{\boldsymbol{z}}_i = \boldsymbol{z}_i - \hat{\boldsymbol{z}}_i$ as

$$\hat{\sigma}^2 = \frac{1}{n-p}\sum_{i=1}^{n}\|\Delta\hat{\boldsymbol{z}}_i\|_{C_z}^2 . \tag{3}$$

Under normality assumption $\|\Delta\hat{\boldsymbol{z}}_i\|_{C_z}^2$ has an approximate distribution χ_1^2. For homogeneous and i.i.d. errors we have $\boldsymbol{C}_z = \sigma^2\boldsymbol{I}$ and the *bootstrap of residuals* can be used to obtain a consistent estimate of $\boldsymbol{C}_\theta$. From the components of the residual vectors

$\{\Delta \hat{z}_{ij} \mid i = 1 \cdots n, \; j = 1 \cdots p\}$, "new" residuals $\{\Delta \hat{z}_{ij}^{*b} \mid i = 1 \cdots n, \; j = 1 \cdots p\}$ are bootstrapped and used to generate the data sets $\hat{z}_{ij}^{*b} = \hat{z}_{ij} + \sqrt{p}\Delta \hat{z}_{ij}^{*b}$. The constant $\sqrt{p}$ compensates the difference between the distributions of $\|\Delta \hat{z}_i\|^2_{C_z}$ and $\|\Delta z_i\|^2_{C_z}$ for a consistent bootstrap covariance estimation. For every $b = 1, \cdots, B$ bootstrap sample the GTLS solution $\hat{\theta}^{*b}$ is computed and $\hat{C}_\theta$ is estimated using (1). A better approach that takes into account the difference in variance between the populations of residual components on each direction is currently under investigation.

When the error distribution $F(./\sigma)$ is known *parametric bootstrap* can be used [5, pp. 275–276]. By drawing new residuals $\{\Delta \hat{z}_{ij}^{*b}\}$ from distribution $F(./\hat{\sigma})$ the bootstrap samples are generated as $\hat{z}_{ij}^{*b} = \hat{z}_{ij} + \Delta \hat{z}_{ij}^{*b}$.

2.2: The jackknife

The jackknife was originally introduced by Quenouille in 1956 as a bias reduction method. Tukey proposed in 1958 the use of jackknife for variance and confidence intervals estimation [24]. The resampling technique used is simpler than that of the bootstrap and consists of deleting one measurement at a time from the data and then computing the statistic of interest. The number of jackknife samples available thus is fixed and is equal to the number of available measurements n.

Let $\hat{\theta}_{(i)}$ be the jackknife estimate of $\hat{\theta}$ with the ith observation z_i deleted from Z. Using the probability vector formulation introduced before, the ith jackknife estimate is then $\hat{\theta}_{(i)} = S[z_1, \cdots, z_{i-1}, z_{i+1}, \cdots, z_n] = T[p_{(i)}]$, where

$$p_{(i)} = \left[\frac{1}{n-1}, \cdots, 0, \frac{1}{n-1}, \cdots, \frac{1}{n-1}\right]^{\top} = \frac{n}{n-1}p_0 - \frac{1}{n-1}e_i,$$

and $e_i = [0 \; \cdots \; 1 \; \cdots \; 0]^{\top}$ is a vector having all but the ith component equal to 0.
The jackknife estimate of the covariance matrix is

$$\hat{C}_{JK} = \frac{n-1}{n}\sum_{i=1}^{n}\left[\hat{\theta}_{(i)} - \bar{\theta}_{JK}\right]\left[\hat{\theta}_{(i)} - \bar{\theta}_{JK}\right]^{\top}, \quad \bar{\theta}_{JK} = \frac{1}{n}\sum_{i=1}^{n}\hat{\theta}_{(i)}. \tag{4}$$

The jackknife methodology works very well when $\hat{\theta}$ has a linear dependence in the probability vector p, i.e.,

$$\hat{\theta} = T[p] = T[p_0] + A[p - p_0], \tag{5}$$

where A is $p \times n$ matrix with $\sum_{j=1}^{n}A_{ij} = 0$, $i = 1, \cdots, p$ [10, pp. 287–291]. This is the case when moments of the data are to be estimated. It can be shown that for linear statistics the jackknife covariance estimate $\hat{C}_{JK}$ agrees up to a multiplicative constant $\frac{n-1}{n}$ with the bootstrap estimate $\hat{C}_{BO}$ [10, pp. 287–291]. Thus the jackknife can be regarded as a first order approximation of bootstrap. The accuracy of the jackknife estimates can often be improved by deleting more than one observations at a time giving the so-called *delete-d jackknife* [27][31]. The weighted delete-d jackknife proposed by Wu [31] gives consistent estimates for linear regression even when the errors are non-i.i.d..

Stromberg [27] analyzed the robustness of bootstrap and jackknife covariance estimates for the case when $\hat{\theta}$ has breakdown point 0.5. He proved that bootstrap has breakdown point $1/n$, i.e. it is nonrobust, while delete-d jackknife has breakdown point $\frac{n-d}{2n}$, i.e. it is robust. By deleting more than one data point at a time, the quality of the jackknife covariance estimation improves, however, at the cost of lowering the breakdown point of the estimation process. The bootstrap methodology does not allow the presence of outliers (the measurements not obeying the model) in the data. As a general rule the outliers should be eliminated before the application of resampling techniques.

2.3: Empirical influence function

The influence function is a tool proposed by Hampel in 1968 to study the infinitesimal behavior of an estimator and is a central concept in the theory of robust estimators [13]. The amount of change in the estimate $\hat{\theta} = t(\hat{F}_n)$, due to an infinitesimal contamination in probability at point z_i can be measured by the *empirical influence function* (EIF) computed at $z = z_i$

$$EIF(z_i, t, \hat{F}_n) = \lim_{\varepsilon \to 0} \frac{t\left[(1-\varepsilon)\hat{F}_n + \varepsilon\delta_{z_i}\right] - t(\hat{F}_n)}{\varepsilon}. \tag{6}$$

Using the probability vector formulation introduced in Section 2.1 the empirical influence function can be expressed also as

$$EIF(z_i, T, p_0) = \lim_{\varepsilon \to 0} \frac{T\left[(1-\varepsilon)p_0 + \varepsilon\delta_{z_i}\right] - T(p_0)}{\varepsilon}. \tag{7}$$

In practice (6) and (7) are difficult to compute. Depending on various choices of ε, different approximations for $EIF(z_i, T, p_0)$ can be obtained [13, pp. 93–95]. If $\varepsilon = -\frac{1}{n-1}$ we get

$$\widehat{EIF}(z_i, T, p_0) = \frac{T\left[(1+\frac{1}{n-1})p_0 - \frac{1}{n-1}e_i\right] - T(p_0)}{-\frac{1}{n-1}} = (n-1)\left[\hat{\theta} - \hat{\theta}_{(i)}\right], \tag{8}$$

where $\hat{\theta}_{(i)}$, the jackknife estimate, and e_i have been defined in Section 2.2. The covariance matrix estimate computed from the empirical influence function is

$$\hat{C}_{EIF} = \frac{1}{n^2}\sum_{i=1}^{n}\widehat{EIF}(z_i, T, p_0) \cdot \widehat{EIF}(z_i, T, p_0)^{\top} = \left[\frac{n-1}{n}\right]^2\sum_{i=1}^{n}\left[\hat{\theta} - \hat{\theta}_{(i)}\right]\left[\hat{\theta} - \hat{\theta}_{(i)}\right]^{\top}. \tag{9}$$

Since $\bar{\theta}_{JK} \approx \hat{\theta}$ we obtain that $\hat{C}_{EIF} = \frac{n-1}{n}\hat{C}_{JK}$. For large n the empirical influence function approximated with (8) and the jackknife give very similar covariance estimates.

3: Tools for performance assessment

Two related questions have to be answered when the quality of an estimate obtained at the output of an image understanding algorithm is to be assessed.

- Did the computation include measurements not obeying the assumed model?
- What is the confidence in the obtained result?

A positive answer to the former question implies the presence of outliers and/or leverage points (observations far away from the bulk of the data). These gross errors severely corrupt the estimate and should be discarded *before* the estimate is computed. Removal of most outliers and leverage points is automatically achieved by using robust estimation techniques, an approach which is already extensively employed in image understanding. Nevertheless, for performance assessment it is important to obtain quantitative measures about the role of each data point in the computation.

3.1: Data diagnostics tools

Two different approaches can be used to evaluate the data. The first approach makes use of robust estimators, while the second is based on the empirical influence function.

3.1.1: The RZ diagram

A versatile technique for classifying the measurements was proposed by Rousseeuw and von Zomeren in 1990 for linear regression [26]. The Rousseeuw and von Zomeren (RZ) diagram is built in two steps. First the regression problem is solved using a high breakdown point estimator, usually the *least median of squares* (LMedS) [25]. The residuals of the robust fit are then computed and standardized by dividing with a robust scale estimate of the noise corrupting the inliers. The second step analyzes the structure of the independent variables x_i with a high breakdown point multivariate location estimator, such as the *minimum volume estimator* (MVE) [25]. The robust distances are computed via robust Mahalanobis distances from the center. Points with large distances are possible leverage points.

In our EIV example we followed the linear regression convention, though there is not a clear distinction between the dependent, y_i, and the independent, x_i variables. Note that the MVE algorithm performed on one dimensional data, $x_i \in R$ is equivalent to LMedS.

Under the assumption of normally distributed inlier noise and independent variables x_i, the outliers are defined as yielding standardized residuals larger than 2.5. The leverage points are classified in relation to the χ^2_{p-1} distribution.

In Figure 1 an EIV regression example with $n = 30$ observations is presented. Normal noise with zero-mean and standard error $\sigma = 1$ affects both x_i and y_i. The data does not contain gross errors, however, due to the noise some of the data points can become more influential. The RZ diagram shows that point 30 is a leverage point, while 3, 24, 28, 29 are outliers, having standardized residuals greater than 2.5. The threshold for robust distances was taken equal to $\chi^2_{1,0.95}$.

3.1.2: EIF diagnostics

Another approach used extensively in statistics is the Cook diagnostics [25, p. 227]. Though Cook diagnostics was initially proposed for linear regression, by using the empirical influence values defined in Section 2.3 the procedure can be extended to more general models. Similar to [13, pp. 228–229] we define the *self standardized empirical influence values* s_i^2, $i = 1, \ldots, n$,

$$s_i^2 = \frac{\widehat{EIF}(z_i, T, p_0)^\top \hat{C}_{EIF}^{-1} \widehat{EIF}(z_i, T, p_0)}{n} . \quad (10)$$

The s_i^2 are self-standardized since their sum is constant. Using equation (9)

$$n \sum_{i=1}^{n} s_i^2 = \sum_{i=1}^{n} \text{trace}(\widehat{EIF}_i^\top \hat{C}_{EIF}^{-1} \widehat{EIF}_i) = \text{trace}(\hat{C}_{EIF}^{-1} \sum_{i=1}^{n} \widehat{EIF}_i \cdot \widehat{EIF}_i^\top) = n^2 p , \quad (11)$$

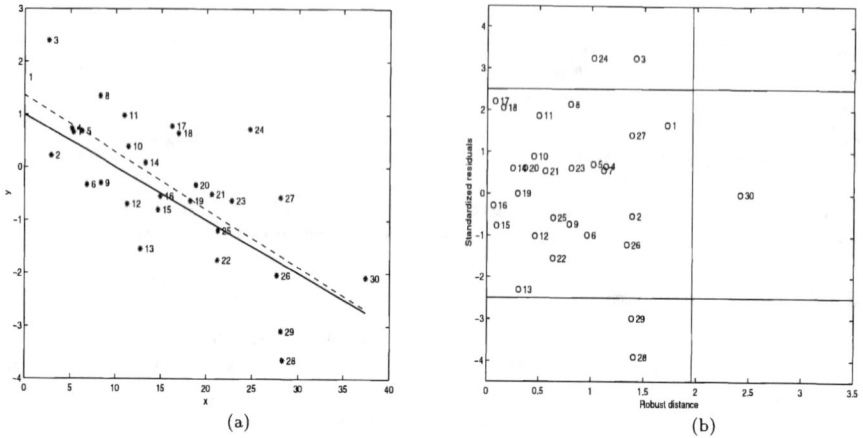

Figure 1. An EIV regression diagnostics example. (a) The data: the true fit $[-0.1\ 1]^{\top}$ (continuous line) and the GTLS estimate (dashed line). (b) The RZ diagram of the data.

where $\widehat{EIF}_i$ is a shortcut for $\widehat{EIF}(z_i, T, p_0)$. A large value of s_i^2 means that the observation i^{th} is influential, i.e. it can be either an outlier, or a leverage point. When the estimate $\hat{\theta}$ is approximately normal, the $\widehat{EIF}_i$ is also approximately normal from (8), so s_i^2 will be χ_p^2 distributed.

The main problem in the analysis using the empirical influence function is the *masking effect*. Because the sum of s_i^2 is constant, a bad outlier or leverage point will lower all the values associated with the other points, in spite of some of them being still influential. Therefore in practice it is recommended to eliminate all the gross errors with a robust estimator before using the EIF based diagnostics.

Recently, Lu et al. [22] introduced an influence function based diagnostics technique using the spectral decomposition of a self standardized influence matrix. The values s_i^2 are an approximation to the diagonal elements of this matrix.

Using the same data as in Section 3.1.1 we computed the corresponding self standardized empirical influence values s_i, $i = 1 \cdots n$. The ten most influential measurements are shown in Figure 2. The measurements 3, 24, 28 and 29 declared outliers in the RZ diagram are ranked as 2, 3, 1 and 4 respectively, according to their standardized scores. The measurement 30 (a leverage point in the RZ diagram) is ranked 8-th. Note that the influential points detected by the EIF diagnostics are placed on the convex hull of the data, i.e. are the farthest from the obtained fit.

Another aspect of interest is to determine how much the estimate will change its value when a measurement is altered with a known displacement. The potentially influential points can be also unraveled using the theory of influence function. If a point z_i is moved to a nearby location z_i^j, then the effect on the estimate can be measured by the following quantity, related to the *standardized influence displacement* [22]

$$d_{ij} = \frac{\left[\widehat{EIF}(z_i^j) - \widehat{EIF}(z_i)\right]^{\top} \hat{C}_{EIF}^{-1} \left[\widehat{EIF}(z_i^j) - \widehat{EIF}(z_i)\right]}{\hat{\sigma}^2}, \tag{12}$$

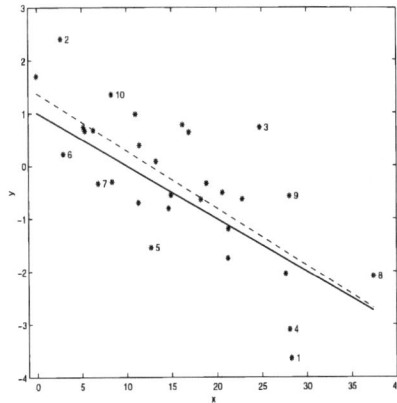

Figure 2. The first ten influential points of the data in Figure 1a. The label 1 corresponds to the most influential measurements.

where $\widehat{EIF}(z_i^j)$ is the empirical influence function computed at point z_i^j, with all the other measurements $z_1, \cdots, z_{i-1}, z_{i+1}, \cdots, z_n$ unmodified. It is convenient to choose the points z_i^j on the hyperellipsoid

$$(z_i^j - z_i)^\top C_z^{-1}(z_i^j - z_i) = 1 , \tag{13}$$

where C_z is the known covariance of the i.i.d. measurements.

The procedure of assessing the sensitivity of the estimate $\hat{\theta}$ is summarized in the following algorithm:

1. Compute the estimate of the error variance $\hat{\sigma}^2$. For each measurement z_i generate J points on the hyperellipsoid defined by (13).

2. Compute $d_{ij}, j = 1, \cdots, J$, using (12). Determine $d_i = \frac{1}{J} \sum_{j=1}^J d_{ij}$.

3. Normalize d_i to obtain the normalized sensitivities $\bar{d}_i = \frac{d_i}{\sum_{k=1}^n d_k}$.

In analogy to the RZ diagram, a plot of s_i versus $\bar{d}_i$, i.e. the square root of the self-standardized empirical influence value of a data point against the sensitivity of the estimate when the point is perturbed, can be obtained. The new diagram for the previously used example is shown in Figure 3a. Following the practice of Cook diagnostics the threshold used in declaring a point sensitive ($\bar{d}_i$) was chosen $2/n$. The threshold to declare a point influential (s_i) was $\sqrt{\chi^2_{2,0.95}}$. The points 1 and 30 declared as very sensitive by EIF diagnostics are the farthest from the center of the x_i. The points 3 and 28 declared very influential are also close to the bounds of the data range and have large residuals. In fact there is a high degree of correlation between the sensitivity of a measurement and the robust distances from RZ diagram, as it can be seen from Figure 3b.

Both diagrams can give useful insights about the data and the estimator used. The robustness of RZ diagram offers a major advantage over the EIF diagnostics. However a shortcoming of RZ diagram may appear when the number of available measurements is not large enough in order to render a stable MVE computation (typically $n < 5p$). The EIF diagnostics, though not robust in the presence of gross outliers analyzes the data by

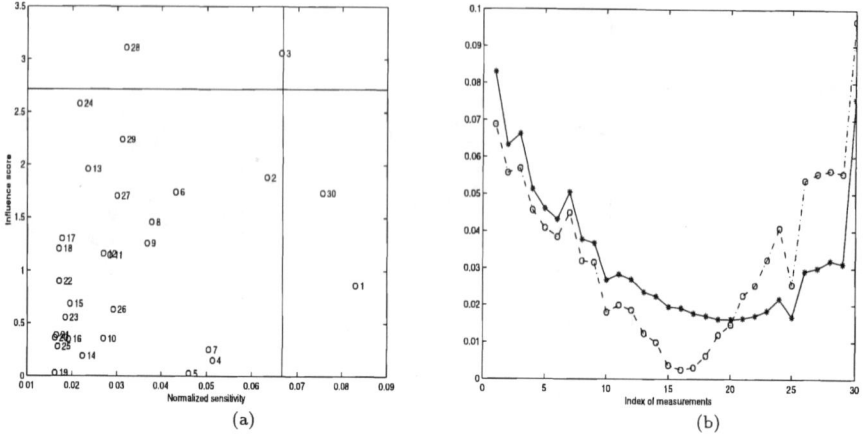

(a)   (b)

Figure 3. EIF based diagnostics. (a) Self standardized empirical influence values s_i vs. normalized sensitivities $\tilde{d}_i$. (b) Normalized sensitivities $\tilde{d}_i$ (continuous line) compared with normalized robust distances (dash-dot line).

assessing the effect of the data points on the estimate $\hat{\theta}$. It also allows to investigate the influence only on a particular component $\hat{\theta}_j$ unlike the RZ diagram.

3.2: Constructing confidence regions

Two issues must be distinguished when building confidence regions. One deals with the confidence in the parameter estimate $\hat{\theta}$, and the other with the uncertainty associated with the corrected data $\hat{z}_i$. Both issues are addressed below.

3.2.1: Confidence regions in the parameter space

The bootstrap methodology can be used to construct the confidence region for the estimate $\hat{\theta}$. For confidence intervals of univariate $\hat{\theta}$, several methods based on bootstrap (percentile, ABC, bootstrap-t) were proposed [10]. For multivariate estimates $\hat{\theta}$ with components $\hat{\theta}_i$, confidence intervals can be determined for each component. It is known that if $\left[\theta_i^{lo}, \theta_i^{up}\right]$ are the componentwise confidence intervals, each with coverage probability $(1 - \alpha)$, then $\bigcup_{i=1}^{p}\left[\theta_i^{lo}, \theta_i^{up}\right]$ has a simultaneous confidence of $(1 - p\alpha)$ [4, p. 107]. Componentwise confidence intervals however do not exploit the structure of $\hat{\theta}$ and elliptical confidence region $\mathcal{D}_{1-\alpha}$ with coverage probability $(1 - \alpha)$ is defined as

$$\mathcal{D}_{1-\alpha} = \left\{\boldsymbol{\theta} \,\middle|\, \left[\boldsymbol{\theta} - \hat{\boldsymbol{\theta}}\right]^\top C_\theta^{-1} \left[\boldsymbol{\theta} - \hat{\boldsymbol{\theta}}\right] \le \gamma_{1-\alpha}\right\}, \quad \text{Prob}\left\{\boldsymbol{\theta} \in \mathcal{D}_{1-\alpha} \,\middle|\, \hat{\boldsymbol{\theta}}\right\} = 1 - \alpha, \quad (14)$$

is more recomended. When $\hat{\boldsymbol{\theta}}$ can be assumed normal, γ has a χ_2^2 distribution, so $\gamma_{1-\alpha} = \chi_{p,1-\alpha}^2$. In the general case the bootstrap percentile method determines $\gamma_{1-\alpha}$ as the $100(1 - \alpha)$ percentile of the probability distribution of $\left[\hat{\boldsymbol{\theta}}^* - \hat{\boldsymbol{\theta}}\right]^\top \hat{C}_\theta^{-1} \left[\hat{\boldsymbol{\theta}}^* - \hat{\boldsymbol{\theta}}\right]$, where $\hat{\boldsymbol{\theta}}^*$ is a bootstrap estimate of $\hat{\boldsymbol{\theta}}$ and $\hat{C}_\theta$ is the bootstrap estimate of C_θ.

The number of bootstraps for the percentile method must be taken at least 1000 in order to give sufficiently good estimates. Though other more accurate bootstrap based techniques exist, the percentile method suffices for computer vision applications where the main source of errors is usually caused by a poor model.

3.2.2: Uncertainty visualization in the input space

Since $\mathcal{D}_{1-\alpha}$ is a confidence region in the parameter space defined according to equation (14) with probability $(1 - \alpha)$ we have $\boldsymbol{\theta} \in \mathcal{D}_{1-\alpha}$. The following algorithm generates individual confidence regions in the input space for the corrected data points $\hat{z}_i$, i.e. visualizes the uncertainty associated with each $\hat{z}_i$.

1. Generate B_1 parameter vectors $\hat{\boldsymbol{\theta}}^{*j}$, $j = 1, \cdots B_1$, inside the confidence region $\mathcal{D}_{1-\alpha}$.

2. Using the bootstrap of residuals generate B_2 bootstrap samples $\boldsymbol{Z}^{*b}, b = 1, \cdots B_2$.

3. Project each bootstrap sample $\boldsymbol{Z}^{*b}$ onto the hyperplane $\hat{\boldsymbol{\theta}}^{*j}$ obtaining $\hat{\boldsymbol{Z}}^{*jb}$.

4. Compute

$$\hat{\boldsymbol{C}}_{z_i} = \frac{1}{B_1 B_2} \sum_{b=1}^{B_2} \sum_{j=1}^{B_1} \left[\hat{z}_i^{*bj} - \hat{z}_i\right] \left[\hat{z}_i^{*bj} - \hat{z}_i\right]^\top, \quad i = 1, \cdots, n . \tag{15}$$

5. With probability $(1 - \beta)(1 - \alpha)$ we have $\boldsymbol{z}_{io} \in \Gamma_i, \ \forall i = 1, \cdots, n$, where

$$\Gamma_i = \left\{ \boldsymbol{z} \in R^p \,\middle|\, (\boldsymbol{z} - \hat{z}_i)^\top \hat{\boldsymbol{C}}_{z_i}^{-1} (\boldsymbol{z} - \hat{z}_i) \leq \chi_{p,1-\beta}^2 \right\} .$$

For convenience at Step 5 spherically distributed errors were assumed. However, if this assumption is strongly violated nonparametric confidence regions can be also determined from the sets $\hat{z}_i^{*bj}$.

In Figure 4a the individual confidence regions are plotted together with the true fit and the noisy measurements for the regression example used throughout Section 3. As it is well known for linear regression, the confidence regions increase towards the boundaries of the range of the data. In spite of using an EIV model the result is similar to Scheffe's method applied to ordinary linear regression [4, pp. 330–335]. Note that the true data points $\boldsymbol{z}_{io}$ lie inside the confidence region centered on the corrected measurements $\hat{z}_i$. The size of the confidence regions in Figure 4a, however, is misleading since the scale of y is much larger than that of x.

The number of data points n has a strong influence on the confidence region. In the asymptotic case (n very large) the parameter estimate $\hat{\boldsymbol{\theta}}$ tends toward the true value $\boldsymbol{\theta}$ since the GTLS is an unbiased estimator. The confidence region of the corrected data point $\hat{z}_i$ will collapse onto the $p - 1$ manifold of the fit since only the uncertainty along the orthogonal direction to the fit can be eliminated. In Figure 4b (drawn at the same scale as Figure 4a) we can see that as the number of measurements increases ($n = 120$), the confidence regions become more elongated in the direction of the fit and decrease in size as a secondary effect. At the limit ($n = \infty$) the regions become intervals defined along the fit.

4: A case study: rigid motion

Rigid motion estimation, i.e., estimating the rotation and translation between two sets of matched points, is one of the fundamental problems of computer vision. Several iterative

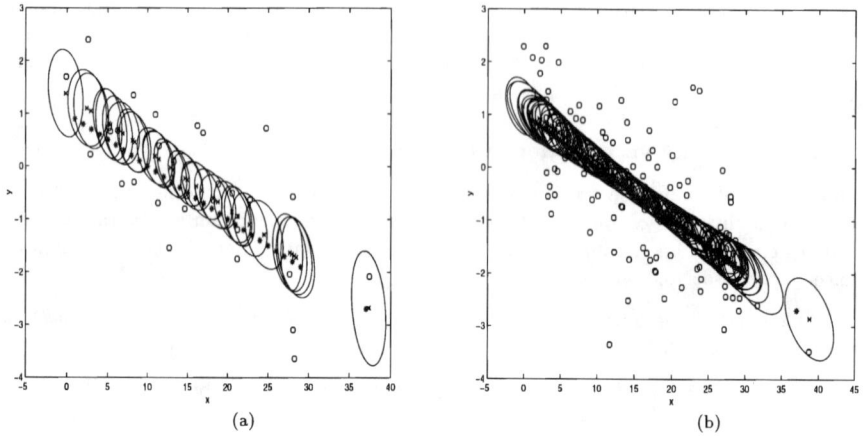

Figure 4. Individual confidence regions in the input space. (a) The data from Figure 1a: the measurements z_i "o", the true values z_{io} "*" and the corrected data $\hat{z}_i$ "x". (b) Another example with the same true fit and noise level, but with $n = 120$.

and noniterative methods were proposed in the literature. An extensive comparison using Monte Carlo simulations of four widely used methods appeared recently [11]. We will now apply the performance assessment techniques introduced in the previous section to compare three different algorithms. We have found, similar to Eggert et al. [11] that the quaternion and singular value decomposition (SVD) have identical performance. Therefore in the sequel we will present only the results for the SVD method. The third estimator analyzed is based on the *linear subspace method* proposed by Wang and Jepson [29].

We compute the covariance matrices for the rotation and the translation estimates and define confidence regions both in the parameter space (for the rotation and the translation) and in the input space (for the corrected data points). It should be emphasized that the performance assessment is accomplished using *only the available single input*, the set of matched data points. Monte Carlo simulations were also performed to obtain the ground truth.

4.1: Error model for rigid motion

Let the two sets of measurements be $X = \{x_1, x_2, \cdots, x_n\}$ and $Y = \{y_1, y_2, \cdots, y_n\}$, with $x_i, y_i \in R^3$, corrupted by noise

$$x_i = x_{io} + \Delta x_i \quad \text{and} \quad y_i = y_{io} + \Delta y_i , \tag{16}$$

having zero-mean and covariance $E[\Delta x_i \Delta x_i^T] = C_x$, and $E[\Delta y_i \Delta y_i^T] = C_y$, respectively. The one-to-one correspondence between the true data points in X_o and Y_o satisfy

$$y_{io} = Rx_{io} + t \tag{17}$$

where R is the 3×3 rotation matrix and t is the translation. Using (16) and (17) yields

$$\begin{aligned} y_i &= Rx_i + t - R\Delta x_i + \Delta y_i \\ &= Rx_i + t + \varepsilon_i . \end{aligned} \tag{18}$$

Then $E[\boldsymbol{y}_i] = \boldsymbol{y}_{io}$ and

$$C = \text{cov}(\boldsymbol{\varepsilon}_i) = \boldsymbol{R}E[\Delta\boldsymbol{x}_i\Delta\boldsymbol{x}_i^\top]\boldsymbol{R}^\top + E[\Delta\boldsymbol{y}_i\Delta\boldsymbol{y}_i^\top] = \boldsymbol{R}\boldsymbol{C}_x\boldsymbol{R}^\top + \boldsymbol{C}_y . \tag{19}$$

All three methods (to be discussed in Section 4.2) assume that the errors in both data sets are i.i.d. with covariance matrices $\boldsymbol{C}_x = \sigma_x^2\boldsymbol{I}$ and $\boldsymbol{C}_y = \sigma_y^2\boldsymbol{I}$. Then (19) becomes

$$C = \sigma_x^2\boldsymbol{R}\boldsymbol{R}^\top + \sigma_y^2 = (\sigma_x^2 + \sigma_y^2)\boldsymbol{I} = \sigma^2\boldsymbol{I} . \tag{20}$$

The estimation of the rotation and translation parameters requires the solution of a generalized least squares problem. The rotation estimate is obtained as

$$\hat{\boldsymbol{R}} = \arg\min_{R} \sum_{i=1}^{n} [\bar{\boldsymbol{y}}_i - \boldsymbol{R}\bar{\boldsymbol{x}}_i]^\top \boldsymbol{C}^{-1} [\bar{\boldsymbol{y}}_i - \boldsymbol{R}\bar{\boldsymbol{x}}_i] , \tag{21}$$

where $\bar{\boldsymbol{x}}_i$ and $\bar{\boldsymbol{y}}_i$ are the centered data points. Thus, as long as the sum of the variances $\sigma^2 = \sigma_x^2 + \sigma_y^2$ is constant the same estimate $\hat{\boldsymbol{R}}$ is obtained, i.e., one can consider that the noise with variance σ^2 affects only one set. The result is similar to that of Goryn and Hein [12]. If the noise is anisotropic, or inhomogeneous the estimators analyzed here will be statistically biased. In order to remove the bias, advanced techniques like Kanatani's renormalization [20] or the method proposed by Leedan and Meer [21] must be employed.

The rotation and translation components of the rigid motion obey different group structures. The former belong to the multiplicative group, while the latter to the additive group [18]. Following the usual approach in the literature we will define separate covariance matrices for the rotation and translation estimates.

A 3-D rotation matrix $\boldsymbol{R}$ is characterized by three independent parameters. In the *angle-axis representation*, using the notations from [18, pp. 100–103], $\boldsymbol{R}$ is represented by the vector $\boldsymbol{r} = \Omega\boldsymbol{l}$ where

$$\Omega = \|\boldsymbol{r}\| = \arccos\left[\frac{\text{trace}(\boldsymbol{R}) - 1}{2}\right] \quad \text{and} \quad \boldsymbol{l} = \mathcal{N}\begin{bmatrix} R_{32} - R_{23} \\ R_{13} - R_{31} \\ R_{21} - R_{12} \end{bmatrix} , \tag{22}$$

with $\mathcal{N}[.]$ being the *normalization operator*. Will denote this operation by $\boldsymbol{r} = \text{rot}(\boldsymbol{R})$. The rotation matrix $\boldsymbol{R}$ can be uniquely recovered from $\boldsymbol{r}$ [18, pp. 102-103].

Since the rotations form a multiplicative group, the error associated with the rotation component cannot be defined in the standard way, i.e., $\hat{\boldsymbol{R}} - \boldsymbol{R}$. The rotation error is defined as the relative rotation $\Delta\hat{\boldsymbol{r}} = \text{rot}(\hat{\boldsymbol{R}}\boldsymbol{R}^\top)$ between the true (unknown) rotation matrix $\boldsymbol{R}$, and the rotation estimate $\hat{\boldsymbol{R}}$. The covariance matrix of the rotation component is then [18, pp. 305-307]

$$\boldsymbol{C}_R = E[\Delta\hat{\boldsymbol{r}}\Delta\hat{\boldsymbol{r}}^\top] . \tag{23}$$

The covariance of the translation estimate is computed in the usual way.

4.2: Rigid motion estimators

We briefly present now the three rigid motion estimators analyzed in our study. Two widely used estimators are based on the singular value decomposition (SVD) of the cross-correlation matrix of the two data sets, and on the quaternion representation. Both methods first eliminate the translation component by centering the data and then estimate the rotation $\hat{\boldsymbol{R}}$ by (21). With $\hat{\boldsymbol{R}}$ known the translation $\hat{\boldsymbol{t}}$ is determined. In the third, linear subspace method, the translation is estimated first and the result is used to obtain $\hat{\boldsymbol{R}}$.

4.2.1: Singular value decomposition based estimator

Arun et al. [1] were the first to propose a rigid motion estimator using SVD. Later Umeyama [28] refined the method to always return a proper rotation matrix. The SVD method first eliminates the translation by centering the data

$$\tilde{x}_i = x_i - \bar{x}, \qquad \bar{x} = \frac{1}{n} \sum_{i=1}^{n} x_i$$

$$\tilde{y}_i = y_i - \bar{y}, \qquad \bar{y} = \frac{1}{n} \sum_{i=1}^{n} y_i \; . \tag{24}$$

The SVD of the cross-correlation matrix K of the centered data is then defined

$$K = \sum_{i=1}^{n} \tilde{y}_i \tilde{x}_i = U \Sigma V^\top \tag{25}$$

and the estimated rotation matrix $\hat{R}$, the solution of (21), is given by

$$\hat{R} = U \begin{bmatrix} 1 & & \\ & 1 & \\ & & \det[U V^\top] \end{bmatrix} V^\top \; . \tag{26}$$

The translation estimate is obtained from $\hat{R}$ as

$$\hat{t} = \bar{y} - \hat{R}\bar{x} \; . \tag{27}$$

4.2.2: The quaternion method

The quaternion method was proposed by Horn [16][17]. Using the formulation of Kanatani, we define for the ith measurement a 3×4 matrix M_i as

$$M_i = \begin{bmatrix} \tilde{y}_i - \tilde{x}_i & (\tilde{y}_i + \tilde{x}_i) \times I \end{bmatrix} \; , \tag{28}$$

where the vector product between a vector a and a matrix B is the matrix with columns given by the vector products between a and the columns of B. Let $q = [q_0 \; q_1 \; q_2 \; q_3]^\top$ be the quaternion representing the unknown rotation. It can be shown that $\hat{q}$ corresponds to the minimum eigenvalue of the 4×4 matrix M

$$M = \sum_{i=1}^{n} M_i^\top M_i. \tag{29}$$

The estimated rotation angle $\hat{\Omega}$ and the rotation axis $\hat{l}$ are determined from the quaternion $\hat{q}$ as [18, pp. 114–115]

$$\hat{\Omega} = \begin{cases} 2 \arccos(\hat{q}_0) & \text{if } \hat{q}_0 \geq 0 \\ 2(\pi - \arccos(q_0)) & \text{if } \hat{q}_0 < 0 \end{cases} \tag{30}$$

$$\hat{l} = \begin{cases} \hat{q}_r / \|\hat{q}_r\| & \text{if } \hat{q}_0 \geq 0 \\ -\hat{q}_r / \|q_r\| & \text{if } \hat{q}_0 < 0 \end{cases} , \qquad \hat{q}_r = [\hat{q}_1 \; \hat{q}_2 \; \hat{q}_3]^\top \; . \tag{31}$$

The rotation $\hat{R}$ is computed from $\hat{r} = \hat{\Omega}\hat{l}$ and the translation $\hat{t}$ using equation (27). Since the quaternion method is only a different formalization to solve (21), the obtained solution is identical with the SVD solution up to machine precision. The study of Eggert et. al [11] reported the same result.

4.2.3: Linear subspace methods

Wang and Jepson [29] proposed to compute first the translation by algebraically eliminating the rotation component. Multiplying (18) with an arbitrary constant c_i and summing over all the data points we obtain

$$\sum_{i=1}^{n} c_i y_i = R \sum_{i=1}^{n} c_i x_i + t \sum_{i=1}^{n} c_i + \sum_{i=1}^{n} c_i \varepsilon_i . \tag{32}$$

Let c be the vector defined with c_i, $i = 1, \cdots, n$. For c_i satisfying

$$\sum_{i=1}^{n} c_i = 1 \quad \text{and} \quad \sum_{i=1}^{n} c_i x_i = 0 , \tag{33}$$

the translation estimate is obtained by finding the minimum norm vector c, with the components satisfying the constraints (33). Indeed, then (32) can be written as

$$t = \sum_{i=1}^{n} c_i y_i - \sum_{i=1}^{n} c_i \varepsilon_i \tag{34}$$

and from (18) and (34) the covariance matrix of the error which is also the covariance of the translation estimate is

$$C_t = C_\varepsilon = \sum_{i=1}^{n} c_i^2 E\left[\varepsilon_i \varepsilon_i^\top\right] = \|c\|^2 \sigma^2 I . \tag{35}$$

Thus the least squares translation estimate is

$$\hat{t} = \sum_{i=1}^{n} c_i y_i . \tag{36}$$

In [29] the constants c_i were found using the Gram-Schmidt orthogonalization process. Here we present an equivalent solution using the SVD decomposition. The c_i must span the null space of the $3 \times n$ matrix $X = [x_1 \ x_2 \cdots x_n]$ and an orthonormal basis of this null space is given by the vectors $\{v_4, v_5, \cdots, v_n\}$, where $V = [v_1, \cdots, v_n]$ is obtained by the SVD decomposition of $X = U\Sigma V^\top$. Thus we have to find the $\{a_i, i = 4, \ldots, n\}$, such that

$$c = \sum_{i=4}^{n} a_i v_i. \tag{37}$$

But $\|c\|^2 = \sum_{i=4}^{n} a_i^2$ and $\sum_{i=1}^{n} c_i = \sum_{i=4}^{n} a_i m_i$, with $m_i = \sum_{j=1}^{n} v_{ij}$, and using the Langrange multipliers

$$\mathcal{J} = \sum_{i=4}^{n} a_i^2 - \lambda \left[\sum_{i=4}^{n} a_i m_i - 1 \right] \tag{38}$$

we obtain

$$a_i = \frac{m_i}{\sum_{j=4}^{n} m_j^2}, \quad i = 4, \cdots, n . \tag{39}$$

With $\hat{t}$ determined one may estimate the rotation matrix $\hat{R}$ either by shifting the data set Y by $\hat{t}$ and solving (21), or by centering the data and computing $\hat{R}$ directly as before. We used the former method, given that better results were reported by Wang and Jepson [29].

4.3: The estimation of the covariance matrices

To assess the performance of the estimation processes the covariance matrices of the parameters (rotation, translation) and of the corrected data points should be computed. Taking $z_{io} = \left[y_{io}^{\top} \; x_{io}^{\top} \right]^{\top}$, the rigid motion model (17) can be written as

$$\left[\; I \quad -R \; \right] z_{io} = N^{\top} z_{io} = t \;. \tag{40}$$

The rigid motion solution thus defines a 3-D manifold in a 6-D space. If $\hat{R}, \; \hat{t}$ are the parameter estimates, then the projection of the noisy measurements $z_i = [y_i \; x_i]^{\top}$ onto the manifold $\hat{N}^{\top} z_i = \hat{t}$ can be computed by means of the projection matrix [20, pp. 34-35]

$$\hat{P}_N = I_6 - \sum_{i=1}^{3} \frac{\hat{n}_i \hat{n}_i^{\top}}{\|\hat{n}_i\|^2}, \quad \hat{N} = \begin{bmatrix} \hat{n}_1^{\top} \\ \hat{n}_2^{\top} \\ \hat{n}_3^{\top} \end{bmatrix} . \tag{41}$$

From (40) $\hat{n}_i = \left[e_i^{\top} \; -\hat{r}_i^{\top} \right]^{\top}$ and using the orthogonality of $\hat{R}$ we obtain

$$\hat{P}_N = \frac{1}{2} \begin{bmatrix} I & \hat{R} \\ \hat{R}^{\top} & I \end{bmatrix} .$$

The corrected data points $\hat{x}_i$ and $\hat{y}_i$ can now be computed as

$$\hat{z}_i = \hat{P}_N \check{z}_i + \left[\hat{t}^{\top} \; 0^{\top} \right]^{\top} \quad \text{with} \quad \check{z}_i = \left[(y_i - \hat{t})^{\top} \; x_i^{\top} \right] \;,$$

or

$$\hat{x}_i = \frac{1}{2} \left[x_i + \hat{R}^{\top}(y_i - \hat{t}) \right] \quad \text{and} \quad \hat{y}_i = \frac{1}{2} \left[y_i + \hat{R} x_i + \hat{t} \right] . \tag{42}$$

Thus, under the assumed error model ($C_x = \sigma_x^2 I$, $C_y = \sigma_y^2 I$) the corrected data points are the average between the measurement in one data set and its pair from the other set transformed back onto the first set. Since only $\sigma^2 = \sigma_x^2 + \sigma_y^2$ can be observed, from (42) we obtain the covariance matrices for the corrected data $\hat{x}$ and $\hat{y}$ as

$$C_{\hat{x}} = C_{\hat{y}} = \frac{\sigma^2}{4} I. \tag{43}$$

The residuals $\Delta \hat{x}_i = x_i - \hat{x}_i$ and $\Delta \hat{y}_i = y_i - \hat{y}_i$ are used to estimate σ_x^2 and σ_y^2

$$\hat{\sigma}_x^2 = \hat{\sigma}_y^2 = \frac{1}{3(n-3)} \sum_{i=1}^{n} \|\Delta \hat{x}_i\|^2 = \frac{1}{3(n-3)} \sum_{i=1}^{n} \|\Delta \hat{y}_i\|^2 \;.$$

Then $\hat{\sigma}^2 = \hat{\sigma}_x^2 + \hat{\sigma}_y^2 = 2\hat{\sigma}_x^2$, or for n very large $\hat{\sigma}^2 = \frac{\sigma^2}{2}$. In practice, the values of σ_x^2 and σ_y^2 must be obtained through an unrelated procedure since the rigid motion estimator will always yield $\hat{\sigma}_x^2 = \hat{\sigma}_y^2$.

Sampling with replacement from the set of residuals we get $\Delta \hat{X}^{*b} = \left\{ \Delta \hat{x}_{ij}^{*b} \right\}$, $\Delta \hat{Y}^{*b} = \{\Delta \hat{y}_{ij}\}$ and the bootstrap data samples are formed using

$$\hat{x}_{ij}^{*b} = \hat{x}_{ij} + \sqrt{2} \Delta \hat{x}_{ij}^{*b} \quad \text{and} \quad \hat{y}_{ij}^{*b} = \hat{y}_{ij} + \sqrt{2} \Delta \hat{y}_{ij}^{*b} \;. \tag{44}$$

The multiplicative factor $\sqrt{2}$ assures consistent covariance estimates. For each bootstrap sample pair $\left(\hat{X}^{*b}, \hat{Y}^{*b}\right)$, the rotation $\hat{R}^{*b}$ and translation $\hat{t}^{*b}$ are determined. As was discussed in Section 4.1 we will assume that the translation and rotation components are independent. The covariance for the rotation component is given by

$$\hat{C}_R = \frac{1}{B-1} \sum_{b=1}^{B} \left[\Delta \hat{r}^{*b}\right] \left[\Delta \hat{r}^{*b}\right]^\top, \quad \Delta \hat{r}^{*b} = \mathrm{rot}\left(\hat{R}^{*b} \hat{R}^\top\right) . \tag{45}$$

The covariance for translation is defined similar to (1)

$$\hat{C}_t = \frac{1}{B-1} \sum_{b=1}^{B} \left[\hat{t}^{*b} - \hat{t}\right] \left[\hat{t}^{*b} - \hat{t}\right]^\top . \tag{46}$$

Because under the assumed error model only the sum of variances is observable we can assume that $\sigma_x^2 = 0$ and $\sigma_y^2 = \sigma^2$. In this case the bootstrap of residuals can be applied only to the Y dataset. The residuals $\Delta \tilde{y}_i = y_i - \tilde{y}_i$, with $\tilde{y}_i = \hat{R} x_i - \hat{t}$, are used to compute the estimate for the total noise variance $\hat{\sigma}^2$

$$\hat{\sigma}^2 = \frac{1}{3(n-3)} \sum_{i=1}^{n} \|\Delta \tilde{y}_i\|^2 .$$

For very large n we have $\hat{\sigma}^2 = \sigma^2$. By bootstrapping from the population $\Delta \tilde{Y} = \{\Delta \tilde{y}_{ij}\}$ new data sets Y^{*b} are generated as $\tilde{y}_{ij}^{*b} = \tilde{y}_{ij} + \Delta \tilde{y}_{ij}^{*b}$. The rotation $\hat{R}^{*b}$ and the translation $\hat{t}^{*b}$ are determined for each bootstrap sample $\left(X, Y^{*b}\right)$ and used in (45) and (46).

We will also use the EIF approach from Section 2.3 for rigid motion. The data is resampled from the population $Z = \{z_1, z_2, \cdots, z_n\}$, where $z_i = \left[y_i^\top \; x_i^\top\right]^\top$. The ith jackknife estimate $(\hat{R}_{(i)}, \hat{t}_{(i)})$ is computed with z_i deleted from Z. The covariance matrix for the rotation is estimated by EIF method as

$$\hat{C}_R = \left[\frac{n-1}{n}\right]^2 \sum_{i=1}^{n} \left[\Delta \hat{r}_{(i)}\right] \left[\Delta \hat{r}_{(i)}\right]^\top, \quad \Delta r_{(i)} = \mathrm{rot}\left(\hat{R}_{(i)} \hat{R}^\top\right)$$

and for the translation by (9) with $\hat{\theta}_{(i)}$, being substituted by $\hat{t}_{(i)}$.

4.4: Performance assessment for rigid motion estimators

The performance assessment techniques introduced in Section 3 can be directly applied to the rigid motion problem. The data points can be classified using the RZ diagram and the EIF based diagnostics. In both cases the analysis assumed that all the errors were associated with the Y data set. As was discussed in the previous section the assumption is without loss of generality for the employed i.i.d. error model.

For a given data set the confidence regions with coverage probability $(1 - \alpha)$ for the translation and rotation, $\mathcal{D}_R$ and $\mathcal{D}_t$, are determined using equation (14). Thus the simultaneous coverage probability is $(1 - 2\alpha)$. Since only $\sigma_x^2 + \sigma_y^2$ is observable, simultaneous confidence regions for both X and Y can be constructed only if the ratio σ_x/σ_y is known.

To apply the algorithm described in Section 3.2.2, rotation and translation pairs $(\hat{R}^j, \hat{t}^j)$, $j = 1, \cdots, B_1$, are chosen from $\mathcal{D}_R$ and $\mathcal{D}_t$. In the input space B_2 bootstrap samples $(\hat{\boldsymbol{Y}}^{*b}, \hat{\boldsymbol{X}}^{*b})$ are generated using the bootstrap of residuals. The bootstrapped datasets $(\hat{\boldsymbol{Y}}^{*b}, \hat{\boldsymbol{X}}^{*b})$ are projected onto the manifold defined by $(\hat{\boldsymbol{R}}^j, \hat{\boldsymbol{t}}^j)$ using (42) and yielding $(\hat{\boldsymbol{Y}}^{*bj}, \hat{\boldsymbol{X}}^{*bj})$. The covariance matrices $\hat{\boldsymbol{C}}_{x_i}$ and $\hat{\boldsymbol{C}}_{y_i}$ can now be estimated using (15). With probability $(1 - 2\alpha)(1 - \beta)$ we have $\boldsymbol{x}_{io} \in \Gamma_{x_i}$ and $\boldsymbol{y}_{io} \in \Gamma_{y_i}$, $\forall\ i = 1, \cdots, n$, where Γ_{x_i} and Γ_{y_i} are defined as

$$\Gamma_{u_i} = \left\{ \boldsymbol{u} \in R^3 \,\middle|\, (\boldsymbol{u} - \hat{\boldsymbol{u}}_i)^\top \hat{\boldsymbol{C}}_{u_i}^{-1} (\boldsymbol{u} - \hat{\boldsymbol{u}}_i) \leq \chi^2_{3,1-\beta} \right\}, \text{ with } u \text{ standing for } x, \text{ or } y . \quad (47)$$

The volume of each confidence region $V(\Gamma_{u_i}) \sim \left| \hat{\boldsymbol{C}}_{u_i} \right|^{\frac{1}{2}}$ can be used in performance assessment. It is important to emphasize that the confidence regions in the input space capture not only the a priori uncertainty about the data (as defined by $\hat{\sigma}^2$), but also the uncertainty due to the estimation process (as definded by $\hat{\boldsymbol{C}}_\theta$). Because the rigid motion is a 3-D manifold in a 6-D space, as the number of measurements n increases, the ellipsoids in the input space (47) will never degenerate and continue to reflect the uncertainty along the fit which cannot be decreased.

5: Experimental results

We compare now the three rigid motion estimators presented in Section 4. Synthetic data is used to have ground truth available in both the parameter and the input space. The methodology remains however identical when the data points are derived from images. The measurements $\{\boldsymbol{x}_{io}, i = 1, \cdots, n\}$ with n taken between 20 and 100 were generated randomly inside a cube with side 100 centered on $(50, 50, 50)$. The corresponding dataset $\boldsymbol{Y}_o$ was determined by choosing random translations with components selected from the interval $[-10, 10]$ and unconstrained random rotations. In all the experiments normal noise with zero-mean and equal variance $\sigma_x^2 = \sigma_y^2 = 4$ was added to $\boldsymbol{X}_o$ and $\boldsymbol{Y}_o$ according to the error model (16). It should be emphasized that the nature of the noise is never used in the experiments, all the computations being data driven. The simulations were done using MATLAB 5.1.

5.1: Data diagnostics

The SVD estimator was used to illustrate the diagnostics tools proposed in Section 3.1. For any other estimator the procedure can be applied in the same way. The two data sets contained $n = 50$ points. The true parameter vectors were $\boldsymbol{t} = [4.1443\ 7.1991\ 5.2919]^\top$ and $\boldsymbol{r} = [-0.1571\ -0.0735\ -1.6406]^\top$. The RZ diagram and EIF diagnostics are compared for this data set in Figure 5.

The RZ diagram is based on the residuals. No outliers are present in the data, but some points have larger residuals (like points 5, 47 and 50). The points 20 and 34 are leverage point having a robust distance greater than the chosen threshold $\chi^2_{3,0.95}$.

In the EIF diagnostics the influence of each point on the translation and rotation components of the rigid motion can be assessed separately. The points 27 and 50 are mostly influential for the rotation estimate (Figure 5b), while for the translation estimate the points 3 and 27 have the highest influence (Figure 5c). The threshold used in declaring a measurement influential was set equal to $\sqrt{\chi^2_{6,0.95}}$. The sensitivity of each measurement can also be assessed from Figure 5b and c. The translation estimate is most sensitive to

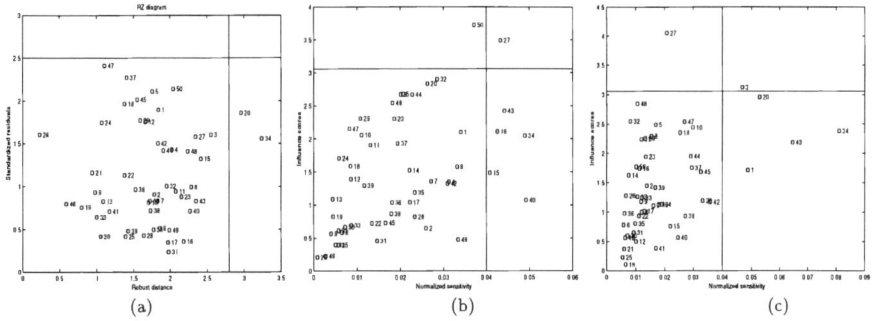

Figure 5. Diagnostics for the rigid motion problem. (a) RZ diagram. (b) EIF diagnostics for rotation. (c) EIF diagnostics for translation

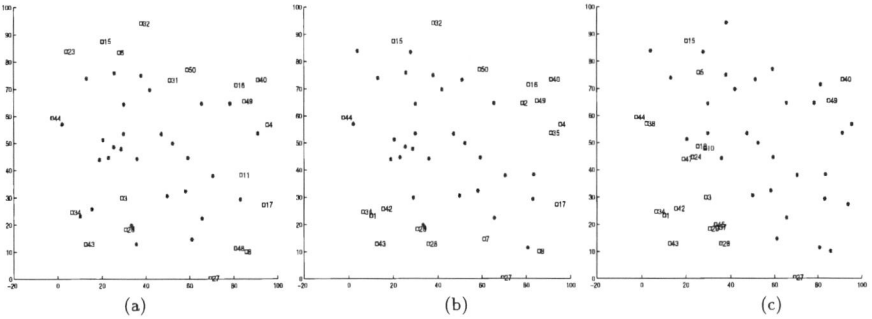

Figure 6. Projection of the 3-D set X onto the horizontal plane. (a) The largest 20 robust distances. (b) The most sensitive 20 points for the rotation estimate. (c) The most sensitive 20 points for the translation estimate.

perturbations at points 1, 20, 34, 43 while the rotation estimate is sensitive to perturbations at the points 15, 16, 27, 34, 40, 43. The point 34, declared a leverage point in the RZ diagram, also appears as the most sensitive in the rotation/translation estimates, in spite of having a small influence. To see the location of the points of interest in the input space in Figure 6 the 3-D data set X is projected onto the horizontal plane. Points far from the bulk of data (large robust distances in the RZ diagram) are the most sensitive in the rotation and translation estimates.

5.2: Estimation of the covariance matrices

The traditional approach in performance assessment with synthetic data uses Monte Carlo trials. The true data sets are corrupted with noise having a known p.d.f. obtaining (X^k, Y^k) for which the rotation and translation estimates are computed. The rotation and translation errors, are defined as $\|r - \hat{r}_{MC}\|$ and $\|t - \hat{t}_{MC}\|$ respectively, with $\hat{r}_{MC} = \frac{1}{K}\sum_{k=1}^{K}\hat{r}^k$ and $\hat{t}_{MC} = \frac{1}{K}\sum_{k=1}^{K}\hat{t}^k$. This approach is used in the comparative study of Eggert

Table 1. Covariance for translation $\hat{C}_t$ in SVD method.

n	Monte Carlo			Bootstrap			EIF		
20	1.0845	0.3868	−0.3137	1.0794	0.1917	−0.4643	0.4899	0.4437	0.1232
	0.3868	1.6117	0.8046	0.1917	1.1222	0.6890	0.4437	3.5028	1.3982
	−0.3137	0.8046	1.6638	−0.4643	0.6890	1.4611	0.1232	1.3982	1.1367
60	0.4661	0.1680	−0.1640	0.4338	0.1700	−0.1454	0.4560	0.3075	−0.0416
	0.1680	0.5598	0.1953	0.1700	0.4762	0.1410	0.3075	0.7776	0.2589
	−0.1640	0.1953	0.3993	−0.1454	0.1410	0.3421	−0.0416	0.2589	0.3402
100	0.2852	0.0989	−0.1144	0.2855	0.0905	−0.1711	0.3198	0.0538	−0.2175
	0.0989	0.3558	0.1513	0.0905	0.2946	0.1010	0.0538	0.3037	0.1777
	−0.1144	0.1513	0.3248	−0.1711	0.1010	0.3474	−0.2175	0.1777	0.4949

et al. [11]. In our experiments the Monte Carlo trials ($K = 500$) were used only to generate the ground truth.

The main disadvantage of Monte Carlo methods is that the noise p.d.f. and the true fit must be known prior to the estimation. In practice however, these two conditions are hard to fulfill. To obtain relevant information the methodology presented here uses only the available, *single* matched pair of data sets (X, Y). The only requirement is to have i.i.d. errors.

The three rigid motion estimators considered are unbiased under the assumed error model. Our comparative study focuses on covariance and confidence regions construction for the parameter estimates and the corrected data points. Under asymptotic conditions (large n, K and B) the Monte Carlo and the bootstrap estimates agree

$$\text{trace}(\hat{C}_t) \approx \left\| t - \hat{t}_{MC} \right\|^2 \quad \text{and} \quad \text{trace}(\hat{C}_R) \approx \left\| r - \hat{r}_{MC} \right\|^2 .$$

As expected the quaternion and SVD gave the same estimates up to the machine precision for all the tests performed. Therefore we will present next only the results for the SVD and linear subspace based estimators.

The bootstrap was implemented with two populations of residuals (44), using $B = 100$ bootstrap samples unless otherwise stated. In Tables 1 and 2 the translation, and in Tables 3 and 4 the rotation covariance determined using Monte Carlo, bootstrap and EIF methods are shown. The parameters used were $t = [4.1443 \ 7.1991 \ 5.2919]^{\top}$ and $r = [-0.1571 \ -0.0735 \ -1.6406]^{\top}$. Experiments were performed with three data sets containing $n = 20, 60$ and 100 measurements.

The bootstrap estimates are fairly accurate for all n, even if the number of bootstraps is relatively low. The accuracy of EIF covariance estimates is poor for small number of measurements, but improves as the number of measurements increases. Because the EIF is only an approximation to the bootstrap, the latter should always be preferred in covariance estimation and confidence region construction. Our simulations confirm the theoretical result of Kanatani [20, pp. 67–68], that the rotation covariance is proportional to the identity matrix.

For the linear subspace method the theoretical expression for the translation covariance is available (35). As expected the bootstrap estimates improve as the number of bootstraps is increased, as shown in Table 5. The increased computational burden is the price to pay for higher accuracy.

In a second group of experiments for twenty different rigid motion parameters the translation and rotation errors were computed from the bootstrapped covariance matrices. For the linear subspace method the theoretical translation error $\sqrt{3} \left\| c \right\| \sigma$, derived from (35)

Table 2. Covariance for translation $\hat{C}_t$ in the linear subspace method.

n	Theoretical			Bootstrap			EIF		
	3.7650	0	0	4.0265	0.0205	−0.2635	4.3203	−1.4445	2.5010
20	0	3.7650	0	0.0205	2.3238	−0.1981	−1.4445	3.7736	0.3233
	0	0	3.7650	−0.2635	−0.1981	2.4725	2.5010	0.3233	3.2917
	1.2396	0	0	1.4296	0.1917	0.0323	1.2067	0.2577	0.0331
60	0	1.2396	0	0.1917	1.1580	0.0047	0.2577	1.6261	0.0115
	0	0	1.2396	0.0323	0.0047	1.1420	0.0331	0.0115	1.0545
	0.8523	0	0	0.9260	−0.0205	0.0371	0.9944	−0.1228	0.1153
100	0	0.8523	0	−0.0205	0.9322	−0.2022	−0.1228	0.4980	0.1681
	0	0	0.8523	0.0371	−0.2022	0.8245	0.1153	0.1681	0.9246

Table 3. Covariance for rotation $\hat{C}_r$ in SVD method.

n	Monte Carlo ($\times 10^{-3}$)			Bootstrap($\times 10^{-3}$)			EIF($\times 10^{-3}$)		
	0.2511	0.0333	−0.0335	0.2124	0.0173	−0.0649	0.4132	0.0810	−0.3399
20	0.0333	0.3679	−0.0794	0.0173	0.2159	−0.0494	0.0810	0.2816	−0.1119
	−0.0335	−0.0794	0.2662	−0.0649	−0.0494	0.2429	−0.3399	−0.1119	0.3472
	0.0883	−0.0073	−0.0113	0.1096	−0.0075	−0.0119	0.0893	−0.0210	−0.0268
60	−0.0073	0.0853	−0.0062	−0.0075	0.0992	−0.0343	−0.0210	0.0645	−0.0111
	−0.0113	−0.0062	0.0868	−0.0119	−0.0343	0.0854	−0.0268	−0.0111	0.0995
	0.0496	0.0036	−0.0014	0.0523	0.0079	0.0057	0.0515	0.0103	−0.0064
100	0.0036	0.0466	−0.0035	0.0079	0.0557	0.0058	0.0103	0.0717	−0.0044
	−0.0014	−0.0035	0.0480	0.0057	0.0058	0.0455	−0.0064	−0.0044	0.0425

Table 4. Covariance for rotation $\hat{C}_r$ in the linear subspace method.

n	Monte Carlo ($\times 10^{-3}$)			Bootstrap($\times 10^{-3}$)			EIF($\times 10^{-3}$)		
	0.5112	0.0261	−0.1658	0.3561	−0.0090	−0.1057	0.3681	0.1509	−0.1462
20	0.0261	0.6082	0.0005	−0.0090	0.6494	0.0479	0.1509	0.4484	−0.0140
	−0.1658	0.0005	0.3894	−0.1057	0.0479	0.3054	−0.1462	−0.0140	0.1378
	0.1361	−0.0015	−0.0309	0.1192	−0.0060	−0.0414	0.1486	−0.0088	−0.0588
60	−0.0015	0.1501	−0.0091	−0.0060	0.1808	0.0117	−0.0088	0.1184	0.0089
	−0.0309	−0.0091	0.1082	−0.0414	0.0117	0.1167	−0.0588	0.0089	0.1493
	0.0953	0.0142	−0.0196	0.0869	0.0020	−0.0175	0.0756	0.0160	−0.0087
100	0.0142	0.0970	0.0107	0.0020	0.0939	0.0146	0.0160	0.0997	0.0085
	−0.0196	0.0107	0.0619	−0.0175	0.0146	0.0603	−0.0087	0.0085	0.0536

Matei, Meer, Tyler

Table 5. Dependence of $\hat{C}_t$ on the number of bootstraps in the linear subspace method.

n	Theoretical			$B = 500$			$B = 1000$		
	3.7650	0	0	2.9832	0.1234	0.0329	3.3666	0.0373	−0.0208
20	0	3.7650	0	0.1234	3.0051	−0.0475	0.0373	3.1578	0.0400
	0	0	3.7650	0.0329	−0.0475	3.3349	−0.0208	0.0400	3.3730
	1.2396	0	0	1.1019	−0.0303	0.0518	1.1475	0.0214	−0.0007
60	0	1.2396	0	−0.0303	1.0846	−0.0072	0.0214	1.0867	−0.0312
	0	0	1.2396	0.0518	−0.0072	0.9928	−0.0007	−0.0312	1.1646
	0.8523	0	0	0.9478	−0.0155	0.0487	0.9058	0.0199	0.0097
100	0	0.8523	0	−0.0155	0.8587	0.0292	0.0199	0.8307	−0.0017
	0	0	0.8523	0.0487	0.0292	0.7893	0.0097	−0.0017	0.8520

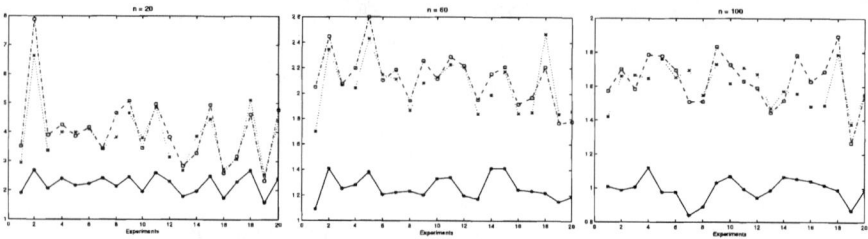

Figure 7. Translation error $\left[\text{trace}(\hat{C}_t)\right]^{\frac{1}{2}}$ for SVD (continuous line), linear subspace method (dash-dot) and theoretical value for linear subpace (dotted line). Twenty experiments were performed for $n = 20, 60, 100$.

was also computed. The experiments were performed for $n = 20$, 60 and 100. In Figures 7 and 8 the results are shown. The linear subspace method always gives larger errors, while closely agreeing with the theoretical value. Beside being less accurate the linear subspace method is also more computation intensive requiring a Gram-Schmidt orthogonalization (or SVD computation) in the translation estimate. We conclude that the SVD (or quaternion) method should be prefered in rigid motion estimation.

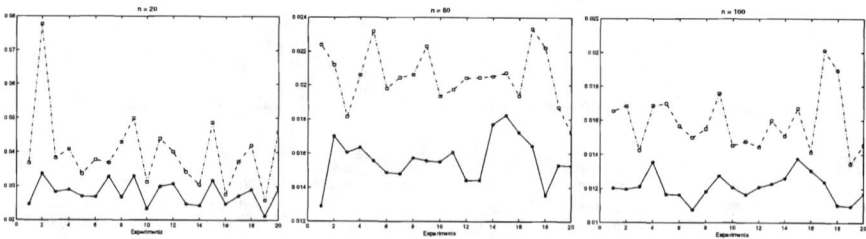

Figure 8. Rotation error $\left[\text{trace}(\hat{C}_R)\right]^{\frac{1}{2}}$ for SVD (continuous line) and linear subspace (dash-dot). Twenty experiments were performed for $n = 20, 60, 100$.

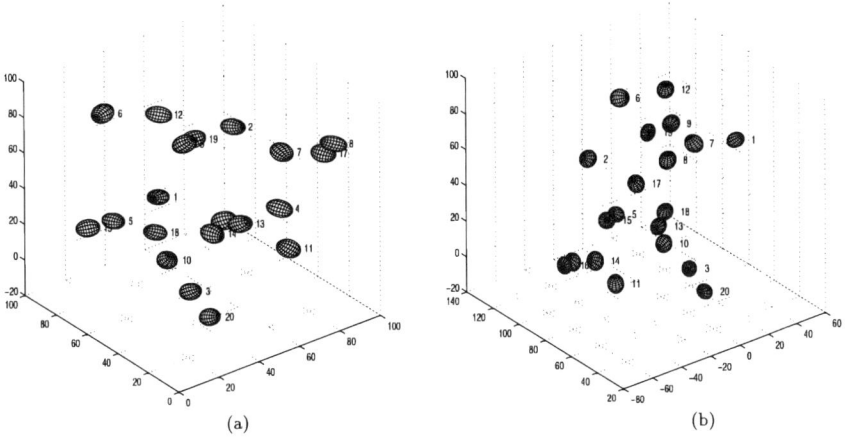

(a)   (b)

Figure 9. Input-space confidence regions for SVD method. (a) For data set X. (b) For data set Y. Note different scales and viewpoints.

5.3: Confidence regions and uncertainty visualization

The confidence regions for the parameter estimates were determined using the percentile bootstrap method with $B = 2000$. As mentioned before, confidence regions constructed for rotation and translation with coverage probability 95% will yield a 90% simultaneous coverage probability. To visualize the uncertainty in the input space the procedure from Section 4.4 was applied with $B_1 = 100$ and $B_2 = 50$, assuming that noise with equal variance corrupted both sets. Otherwise, to construct separate confidence regions for X and Y the ratio σ_x/σ_y must be known a priori.

Input space confidence regions for *each* corrected data point are shown in Figures 9 and 10 for the SVD and the linear subspace methods. The larger parameter covariances of the latter technique yielded larger confidence regions (ellipsoids of confidence with higher volumes) at the same coverage probability. The visualization procedure has the advantage that combines the uncertainty from the input space with the confidence regions computed in the parameter space and provides in one measure all the relevant information about the estimator as applied to the given data.

6: Conclusion

We have introduced and validated a new performance assessment paradigm for image understanding tasks. We have shown using synthetic data that the provided measures capture all the relevant information about the output. Since these measures are obtained entirely from the input, the technique is the same for image data.

The proposed paradigm is very general and we intend to apply it to a wide variety of tasks with real data. Providing reliable confidence regions in the input space can be helpful in developing "self-tuning" vision algorithms. In these algorithms the output would be iteratively refined through integrating a priori knowledge together with performance measures.

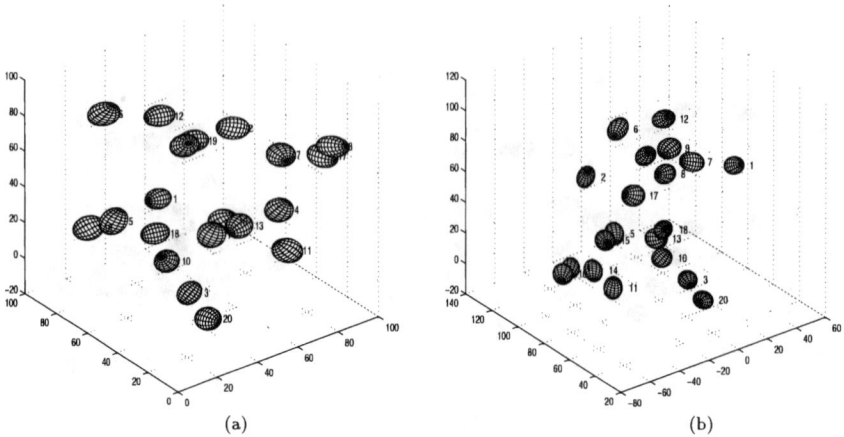

(a)   (b)

Figure 10. Input-space confidence regions for the linear subspace method. (a) For data set X. (b) For data set Y. Note different scales and viewpoints.

7: Acknowledgments

The authors gratefully acknowledge the support by the National Science Foundation under grant IRI-9530546.

References

[1] K. S. Arun, T.S. Huang and S. D. Blostein, "Least-squares fitting of two 3-D point sets," *IEEE Trans. Pattern Analysis and Machine Intelligence.* Vol. 9, pp. 698–700, 1987.

[2] G. J. Babu and C. R. Rao, "Bootstrap methodology," *C. R. Rao, ed., Handbook of Statistics*, Vol. 9, pp. 627–657, Elsevier Science Publishers, 1993.

[3] P. J. Bickel and D. A. Freedman, "Some asymptotic theory for the bootstrap," *Annals of Statistics* Vol. 9, pp. 1196–1217, 1981.

[4] L. Breiman, *Statistics: With a View Toward Applications*, Houghton Mifflin Company–Boston, 1973.

[5] R. J. Carroll, D. Ruppert and L.A. Stefanski, *Measurement error in nonlinear models*, Chapman & Hall, 1996.

[6] K. Cho, P. Meer and J. Cabrera, "Performance assessment through bootstrap," *IEEE Trans. Pattern Analysis and Machine Intelligence*, Vol. 19, pp. 1185–1198, 1997.

[7] H.I. Christensen and W. Förstner, "Performance characteristics of vision algorithms," *Machine Vision and Applications*, Vol. 9, pp.215–218, 1997.

[8] B. Efron and G. Gong, "A leisurely look at the bootstrap, the jackknife and cross-validation," *The American Statistician.*Vol. 37, pp. 36–48, 1983.

[9] B. Efron and R. Tibshirani, "Bootstrap methods for standard errors, confidence intervals, and other measures of statistical accuracy," *Statistical Science* Vol. 1, pp. 54–77, 1986.

[10] B. Efron and R. J. Tibshirani, *An Introduction to the Bootstrap*, Chapman & Hall— New York, 1993.

[11] D.W. Eggert, A. Lorusso and R.B. Fisher, "Estimating 3-D rigid body transformations: A comparison of four major algorithms," *Machine Vision and Applications*, Vol. 9, pp. 272–290, 1997.

[12] D. Goryn and S. Hein, "On the estimation of rigid body rotation from noisy data," *IEEE Trans. Pattern Analysis and Machine Intelligence*, vol.17, pp. 1219–1220, 1995.

[13] R. Hampel, E.M. Ronchetti, P.J. Rousseeuw and W.A. Stahel, *Robust Statistics. The approach Based on Influence Function*, Wiley & Sons — New York, 1986.

[14] R.H. Haralick, "Performance assessment of near-perfect machines", *Machine Vision and Applications*, Vol. 2, pp.1–16, 1989.

[15] D. J. Heeger and A. D. Jepson, "Subspace methods for recovering rigid motion I: Algorithm and implementation," *International Journal of Computer Vision*, Vol. 7, pp. 95–117, 1992.

[16] B.K.P. Horn, H.M. Hilden and S. Negahdaripour, "Closed-form solution of absolute orientation using orthonormal matrices," *J. Opt. Soc. Am.* Vol 5, pp. 1127–1135, July 1988.

[17] B.K.P. Horn, "Relative orientation revisited," *J. Opt. Soc. Am.* Vol 8, pp. 1630–1638, 1991.

[18] K. Kanatani, *Geometric Computation for Machine Vision*, Oxford Science Publications, 1993.

[19] K. Kanatani, "Analysis of 3-D rotation fitting," *IEEE Trans. Pattern Analysis and Machine Intelligence*, Vol. 16, No. 5, pp. 543–549, 1994.

[20] K. Kanatani, *Statistical Optimization for Geometric Computation: Theory and Practice*, Elsevier Science, 1996.

[21] Y. Leedan and P. Meer, "Estimation with bilinear constraints in computer vision," *6th International Conference on Computer Vision*, Bombay, India, pp. 733–738, January, 1998.

[22] J. Lu, D. Ko and T. Chang, "The standardized influence matrix and its applications," *J. American Stat. Assoc.*, Vol. 92, pp. 1570–1580, 1997.

[23] W.C. Manteiga, J. M. P. Sanchez and J. Romo, "The bootstrap — A review," *Computational Statistics*, Vol. 9, pp. 165–205, 1994.

[24] S.D. Peddada, "Jackknife variance estimation and bias reduction," *C. R. Rao, ed., Handbook of Statistics*, Vol. 9, pp. 723–744, Elsevier Science Publishers, 1993.

[25] P.J. Rousseeuw and A.M. Leroy, *Robust regression and outlier detection*, John Wiley & Sons, 1987.

[26] P.J. Rousseeuw and B.C. van Zomeren, "Unmasking multivariate outliers and leverage Points," *J. American Stat. Assoc.*, Vol. 85, pp. 633 –639, 1990.

[27] A.J. Stromberg, "Robust covariance estimates based on resampling". *Journal of Statistical Planning and Inference*, Vol. 57, pp. 321–334, 1997.

[28] S. Umeyama, "Least-squares estimation of transformation parameters between two point patterns," *IEEE Trans. Pattern Analysis and Machine Intelligence*, Vol. 13, pp. 376–380, 1991.

[29] Z. Wang and A. Jepson, "A new closed-form solution for absolute orientation," *Proceedings of CVPR 94*, Seattle, pp. 129–134, June, 1994.

[30] J. Weng, T.S. Huang and H. Ahuja, *Motion and Structure from Image Sequences*, Springer-Verlag, 1993.

[31] C.F.J. Wu, "Jackknife, bootstrap and other resampling methods in regression analysis," *Annals of Statistics*, Vol. 14, pp. 1261–1295, 1986.

[32] G.A. Young, "Bootstrap: More than a stab in the dark?," *Statistical Science*, Vol. 9, No. 3, pp. 382–415, 1994.

[33] Z. Zhang and O. Faugeras, *3D Dynamic Scene Analysis*, Springer–Verlag, 1992.

Sensor Errors and the Uncertainties in Stereo Reconstruction

Gerda Kamberova and Ruzena Bajcsy
E-mail: kamberov@cis.upenn.edu, bajcsy@cis.upenn.edu
Department of Computer and Information Science
University of Pennsylvania
Philadelphia, PA 19104

Abstract

An important objective in the evaluation of algorithms with sensory inputs is the development of measures characterizing the intrinsic errors in the results. Intrinsic are those errors which are caused by noise in the input data. The particular application which we consider is 3-D reconstruction from stereo. We demonstrate that a radiometric correction of the images could improve significantly the accuracy. We propose a confidence interval approach for quantifying the precision. We also illustrate the use of the confidence intervals for the rejection of unreliable 3D points.

1 Introduction

1.1 Motivation

Our work is inspired by the call in the computer vision community for performance evaluation methods [11, 12, 32]. In this paper we discuss errors in 3-D reconstructions from stereo induced by sensor noise. Such errors we call *intrinsic*. Any undesired feature causing discrepancies in the digital image is considered noise. Systematic noise (errors) effect the **accuracy** of the vision algorithms, i.e., the closeness of the results to the true values. Random errors primary effect the **precision**, i.e., the variability in the results due to random noise in the digital images. Both the accuracy and the precision of vision systems should be investigated.

For establishing the **accuracy**, the results must be compared with ground truth models. General ground truth models are difficult to obtain. Usually simple test examples are used. The best strategy is to compare results with widely accepted ground truth models, if such models exist.

A complete characterization of the **precision** consists of a probability distribution for the output noise. Such distributions are even more difficult to get than ground truth models since there are usually too many factors influencing the results. A popular approach is to report the variance in the output. In that approach, and others related to it (factor analysis, performance analysis, variance components), it is implicitly assumed that the input noise is normally distributed, and that the input-output transformation is well approximated by a linear transformation. Unfortunately, one often uses these methods without verifying the validity of the model assumptions.

A methodology for performance evaluation should include characterization not only of the errors depending on the environmental conditions (i.e. extrinsic errors), but also on the sensor characteristics. *For specifying the performance metrics of a vision system with various hardware configurations, it is necessary to quantify the intrinsic errors.* Our goal in this paper is to stress the importance of understanding the sensor and its effects on the performance of the algorithms.

In the rest of this section we, first, review the video sensor and the stereo algorithm (Section 1.2), and second, overview our contribution and the results presented in this paper (Section 1.3). Section 1.4 outlines the organization of the paper.

1.2   The Sensor and the Stereo Algorithm: An Overview

1.2.1   The Video Sensor

The video sensor consists of a lens, a CCD camera, and a framegrabber. The image is formed on the CCD array of identical sensing elements (*sels* [23]), and then transfered by the framegrabber, in a linear order (line-by-line, sel-by-sel), to the computer memory. The geometric and radiometric uncertainties and discrepancies in the digital image are due to the optics, the CCD camera, and the joint operation of the camera, the framegrabber, and other electronic components.

Camera related noise
The *total random noise* of the CCD has three major components: photon (shot) noise, read noise, and fixed pattern noise. The total noise is dominated by the read noise at low illumination levels, by the fixed pattern noise at high levels, and by the photon shot noise in between [18].

The source of the *photon (shot) noise* is external. This noise is due to fluctuations of the photon flux, and is always present in the data.

The sources of the *read noise* are internal to the CCD camera. The read noise is related to the physics of the camera and the measurement process. Major factors contributing to the read noise are the background noise and output amplifier noise. Noticeable components of the background noise are the dark current (thermally generated charges), and the internal luminescence. In general, dark current doubles with every 8° increase in temperature [5]. To reduce dark current cameras are cooled. Dark current can originate at different locations in the CCD but has, in all cases, to do with the irregularities in the crystal structure of the silicon. This contributes to a fixed pattern noise (FPN) in dark current. Because the background noise is temperature dependent, cameras must be warmed up before use.

The individual sels have their own physical characteristics. The difference in these characteristics is manifested in the FPN which can be observed in dark images[1] and flat fields[2]. The photoresponse nonuniformity (PRNU) of the sels is observed in flat fields. In this paper, we correct for systematic, FPN, in dark current and PRNU. Flat fields and dark images are used for a radiometric calibration of the sensors, Section 4.

Framegrabber related noise
Geometric discrepancies due to the digitization in the framegrabber are aliasing, linejitter,

---

[1] A dark images is the response (readout) of the camera with no access of light to the CCD array.
[2] A flat field is the response of the sensor to uniform light illumination.

and systematic fall-off of the intensity in a line. An effect in radiometric distortion due to interlacing[3] is the shift in gray level between odd and even fields in a frame. Severe distortions occur when the sampling theorem[4] is violated. The sel spacing puts limitation on the highest frequency in the input beyond which severe distortions occur, and the area of the CCD chip limits the lowest frequencies which can be detected. Another radiometric noise component is modeled by the quantization error [39].

Modeling the sensor

The detailed modeling of the sensor noise is a very hard problem. For that reason, an approximation using a linear model with normally distributed noise is usually used. As we demonstrate, in Section 5, even with this simple model, the precision of the stereo algorithm can be satisfactory quantified, and the accuracy can be significantly improved.

Other radiometric models are presented in [38, 14]. The model in [38] is very attractive with its simplicity: the noise is divided into three components: (a) signal-dependent; (b)signal-independent and position-dependent; and (c) signal- and position-independent. It is used in the context of Bayesian framework for image restoration. The model in [14] is detailed. It capturing the various camera and framegrabber related noise sources. But, the actual procedure for the practical estimation of the various noise parameters, and for the radiometric correction of images is very much simplified. After the simplifications, the model is practically reduced the popular class of linear models with additive noise (similar to the one we use).

References

CCD array, cameras, and displays are analyzed in [5, 40, 16, 18]. Distortions related to the optics are addressed in [37]. Discretization effects are in the center of the signal processing literature [17]. A brief and clear overview of the video signal formats is given in [23, 2]. Various sources of geometric distortions in the video sensor are analyzed in [23]. A detailed study of different framegrabber architectures and synchronization mechanisms, and related geometric and radiometric distortions is given in [4], and a comprehensive discussion of linejitter problem, its sources and detection, in [23, 2, 4].

1.2.2 The Stereo System

By the late 17th century the use of a pair of images obtained from different view points (from both eyes) in perceiving 3-D scenes was recognized [15]. But it was not until the 1970s that the power of stereo in estimating 3-D, even without recognizable features in the images, was established [19], and a computational theory of stereo proposed [25, 26]. Since, stereo reconstruction has been an active area of research in computer vision. For a review on stereo and matching see [9], or more recently [33, 30, 3].

The main principle underlying 3-D reconstruction from stereo is a simple one. Given the pinhole camera model[5], the camera parameters, and the configuration geometry of the pair of cameras: if a correspondence is established between two points of interest, F_L and F_R, in the left and the right images, the position of the 3-D feature corresponding to F_L and

---

[3]Interlacing is a standard mode of operation of cameras.

[4]The input signal can be fully reconstructed from the samples, if the input signal frequency, f, is at most half of he sampling frequency.

[5]Images are formed on the image plane by perspective projection with center of projection the optical center of the camera.

F_R is the intersection of the optical rays $F_L C_L$ and $F_R C_R$, where C_L and C_R denote the optical centers of the two cameras. Depending on the types of features used for establishing correspondences, matching algorithms, sensor suites, and environmental assumptions vary.

The stereo algorithm we used was developed and implemented by Radim Šára, [34]. It uses: area-based matching; multiple pairs of verging, strongly calibrated cameras; assumes lambertian, opaque surfaces, and enforces the usual ordering and disparity gradient limit constraints [9, 41]. The features in area-based matching are the pixel intensities.

An intrinsic geometric property of stereo images is that the projections of the same 3-D point in left and right images lie on corresponding *epipolar lines* (resulting from the intersection of the two image planes with the plane defined by the 3-D point and the optical centers of the cameras),[27]. If the cameras are in a normal (standard) configuration[6], the epipolar lines coincide with the horizontal (u) coordinate lines[7]. Thus for a given pixel at position (u, v) in the left image, the search for its corresponding pixel in the the right image is in 1-D, among the pixels at positions (u, i), where only i varies. Single pixel values are unstable as matching primitives, so small windows (neighborhoods) around the pixels are selected. For a fixed pixel (u, v) in the left image, the matching procedure establishes the pixel, (u, v^*), in the right image whose neighborhood intensity pattern is most "similar" to that of (u, v). This correspondence is established provided that the 3-D feature projected at (u, v) is in the field of view of both cameras. The similarity is measured by a modified normalized cross correlation [29],

$$\mathrm{Cor}(\boldsymbol{I}_L^w, \boldsymbol{I}_R^w) \stackrel{def}{=} \frac{2\mathrm{Cov}(\boldsymbol{I}_L^w, \boldsymbol{I}_R^w)}{\mathrm{Var}(\boldsymbol{I}_L^w) + \mathrm{Var}(\boldsymbol{I}_R^w)}, \tag{1}$$

where $\boldsymbol{I}_L^w$ and $\boldsymbol{I}_R^w$ denote the pixel intensities over the correlation windows in the left and right images respectively, and Var and Cov denote the spatial sample variances and covariance over the windows. The *integer disparity*, $\boldsymbol{D}(u, v)$, at (u, v) is defined as the shift $(v - v^*)$.

The major steps in the stereo reconstruction are: (1) integer disparity map computation; (2) subpixel disparity map recovery; and (3) reconstruction.

(1) The result of the matching procedure is an *integer disparity map*, $\boldsymbol{D}$: for each (u, v) in the left image if the correspondence is resolved, $\boldsymbol{D}(u, v)$ is assigned the disparity value, otherwise it is marked as *undefined*.

(2) For many applications, $\boldsymbol{D}$ is too coarse (a surface recovered from integer disparities would be step-like), so a subpixel disparity, $\bar{\boldsymbol{D}}$, is approximated. In [34] $\bar{\boldsymbol{D}}$ is estimated based on an approach given in [8]. The mapping between the intensity values of corresponding positions in the image planes is modeled by:

$$I_R(\boldsymbol{x}) = B I_L(T_x(\boldsymbol{x} - d(\boldsymbol{x}))) + M \tag{2}$$

$\boldsymbol{x} = (u, v)$ denotes image position, $d(\boldsymbol{x})$ the subpixel disparity, T_x models the local perspective distortion between corresponding neighborhoods, and M and B are scalar parameters accounting for radiometric differences in the cameras over the windows of interest. Under the assumption of local planarity of the surface being imaged, and linearization of the

---

[6]i.e., the cameras are oriented fronto-parallel in the same direction, the image planes are coplanar, and cameras are aligned so that the horizontal coordinate lines in both image planes are collinear, and the optical centers are at the same distance from the corresponding image planes.

[7]Since we use verging cameras, the images were first rectified,[1], thus brought to a standard configuration.

functions involved, in local coordinates, for (u, v) in the window, (2) is reduced to

$$I_R(u, v) = (d_\varepsilon \ B, \gamma, \ M)\left(\tfrac{\partial I_R}{\partial v}(u, v), \ I_L(u, v), \ \tfrac{\partial I_L}{\partial v}(u, v)v, \ 1\right)^t, \tag{3}$$

where d_ε is the error in the subpixel disparity correction for the center of the window in the left image, γ relates to the distortion T_x (as in (2)), and t denotes transpose of a vector. The derivatives $\tfrac{\partial I_R}{\partial v}(u, v)$ and $\tfrac{\partial I_L}{\partial v}(u, v)$ are approximated by finite differences. Equation (3) is applied over the reference window, and then solved by least squares for $\boldsymbol{\theta} = (d_\varepsilon, B, \gamma, M)$. Thus, $\tilde{\boldsymbol{D}}(u, v) = \boldsymbol{D}(u, v) + d_\varepsilon$.

(3) The 3-D reconstruction is a linear transformation in a projective space. Let Q be the projective 4x4 matrix recovering the 3-D point in projective space, $(\tilde{X}, \tilde{Y}, \tilde{Z}, \tilde{K})$, i.e., $(\tilde{X}, \tilde{Y}, \tilde{Z}, \tilde{K})^t = Q * (u, v, v - \boldsymbol{D}(u, v), 1)^t$ We rewrite this in the form

$$(\tilde{X}, \tilde{Y}, \tilde{Z}, \tilde{K})^t = S * (u, v, 1, \tilde{\boldsymbol{D}}(u, v))^t \tag{4}$$

where the ith row of S is $S(i, :) = (Q(i, 1), \ Q(i, 2) + Q(i, 3), \ Q(i, 4), \ -Q(i, 3))$, $i = 1, 2, 3, 4$. Then the Euclidean coordinates of the reconstructed point which was projected at (u, v) are calculated from

$$(X, Y, Z)^t = \frac{1}{\tilde{K}}(\tilde{X}, \tilde{Y}, \tilde{Z})^t. \tag{5}$$

1.3   Characterizing the Intrinsic Errors in Stereo: Our Contributions

The literature on the errors in 3-D reconstruction from stereo has been primarily focused on the effects of the systematic geometric discrepancies in the digital images (not radiometric), the camera configuration geometry, and the camera calibration parameters, [6, 7, 10, 36]. We address two new issues:

1. The effect of the *random sensor noise on the precision* which we quantify using confidence intervals. (Some necessary terminology and a summary of the approach are given in Section 1.3.1, and details in Section 3.)

2. The effect of the *systematic radiometric sensor errors on the accuracy.* (We show that the accuracy can be improved by the procedure for radiometric correction given in Section 4.)

1.3.1   The Confidence Intervals Approach in Stereo

Definition   A confidence interval for a parameter of interest, δ, is an interval $C = [C_{min}, C_{max}]$ with bounds C_{min} and C_{max} which are random variables. The values of the bounds are obtained as a result of an experiment. The theoretical probability of capture (TPC), $\Pr[\delta \in C]$, gives the reliability of the interval, while the size, $C_{max} - C_{min}$, represents the uncertainty in δ. The TPC characterizes the likelihood of computing δ within tolerance bounds even prior to observing the outcome of the experiment.

For example, in the context of stereo, the experiment is the sampling of left and right images and the computation of the 3-D points. If δ denotes a true depth of a 3-D point, the bounds C_{min} and C_{max} could be functions of the disparity. They will be random because of the random noise in the disparity induced from the image noise.

Previous uses of confidence intervals in stereo for quantifying the accuracy
In stereo vision confidence intervals have been used for characterizing the *accuracy* of the estimate of an *average* distance to a plane [24]. The restricted geometric domain, and the particular application in this case, made possible: the development of a ground truth model for the true depth measurement and the design of a probabilistic model for the errors in the average depth estimate from stereo. Under these conditions, it is possible to derive optimal (in minimax sense) confidence intervals for the true depth [21]. These intervals are of fixed size, for a given depth range, and have guaranteed TPC.

The use of confidence intervals for quantifying the precision
In the current paper, we are interested in *dense point set* recovery, and a *wide class of scenes* (like general office scenes, human faces, hands, machine parts, etc), thus the method used in [24] is infeasible. We take a different approach here.

In order to isolate the effect of sensor errors from other noise sources, we induce a *strong random dot texture* on the scenes by using an overhead projector. Thus we restrict the domain of inputs, but do not impose farther conditions on the geometry and the material of the surfaces. In the restricted domain, the integer disparity map based on (1) is stable. For strong random texture scenes the main component of the intrinsic errors is the random one. Our experiments clearly show that the random noise in the digital images has little effect the integer disparity (for $70 - 95\%$ of the pixels, the integer disparities were constant over multiple runs of the matching algorithm on fixed scenes each time taking new pairs of images; the integer disparity is unstable in areas where the fronto-parallel planarity assumption of matching is violated).

We characterize the precision in the reconstruction with confidence intervals (one in each coordinate direction).

Algorithmic steps in deriving the intervals:

- We start with a simple image model with normal noise (described in Section 2).
- Next: (a) we propagate the variance from the images to the disparity using the method from [13]; (b) for a selected TPC, we compute confidence intervals for the disparity variance; and (c) we map these intervals to intervals in the coordinates of the reconstructed points. In Section 3 we review the steps (a)-(c).

1.4 Organization

In Section 2 we present the models, and in Section 3, the confidence intervals quantifying the uncertainties in the reconstruction. Section 4 outlines the radiometric correction procedure used for reducing the intrinsic systematic errors. In Section 5 we show the experimental results, and in Section 6 we summarize the result and draw conclusions regarding the use of the intervals for data reduction and fusion.

2 The Models

2.1 The Sensor Model

In this section we present a sensor model which is consistent with the assumptions made in the stereo algorithm, and which can be used to account for the intrinsic errors in stereo.

We do not attempt to give a general sensor model capturing all noise components[8].

We assume that the random noise per pixel is spatially varying over the whole image, but locally constant (over the correlation window size). For a pixel $(u, v), 1 \leq u \leq M, 1 \leq v \leq N$, where MN is the image size, the intensity is modeled by

$$I(u, v; \theta(u, v)) = \beta(u, v)\theta(u, v) + m(u, v) + W(u, v), \tag{6}$$

where $\theta(u, v)$ denotes the true, unobservable (ideal) intensity of the pixel, the gain factor $\beta(u, v)$ model the photoresponse nonuniformity (PRNU), the offset $m(u, v)$ models the fixed pattern noise (FPN), and $W(u, v)$ is a zero-mean normally distributed random variable with variance $\sigma^2(u, v)$. Note that this is the model under which in practice most of the radiometric correction procedures are done [14, 4].

We use this model in two ways: (1) to obtain estimates of the input variance, which later is used in the confidence intervals derivations (Section 3); and (2) to justify theoretically, in [22], the radiometric correction procedure given in Section 4.

Let $F_0^i, i = 1..n$ denote a sequence of dark images, and $F_\theta^i, i = 1, ..(n+1)$ — a sequence of flat fields taken at the same level of illumination. Then the estimates of $m(u, v)$ and $\sigma^2(u, v)$ are obtained using sample means and sample variances as follows,

$$\hat{m}(u, v) = \frac{1}{n} \sum_{i=1}^{n} F_0^i(u, v), \quad \hat{\sigma}^2(u, v) = \frac{1}{2(n-1)} \sum_{i=1}^{n} (F_\theta^i(u, v) - F_\theta^{n+1}(u, v))^2, \tag{7}$$

since the mean of the differences of two flat fields is $0_{M \times N}$, and the variance of the difference is two times the variance of the original flat fields.

In experiments, we have used sequences of 100-1000 images for the estimates. Under this noise model, the flat fields used for radiometric correction in Section 4 should be taken at highest possible illumination level. The reason being that the actual signal-dependent component of the noise has higher variance at higher illumination levels which we want to bound.

2.2   The Disparity Model and the Disparity Variance

As discussed in Section 1.2, given the left and right images, I_L and I_R, and an integer disparity map, $\boldsymbol{D}$, the subpixel disparity at $\bar{\boldsymbol{D}}(u, v)$ is obtained by

$$\bar{\boldsymbol{D}}(u, v) = \boldsymbol{D}(u, v) + \boldsymbol{D}_\varepsilon(u, v) \tag{8}$$

where $\boldsymbol{D}_\varepsilon(u, v)$ is the disparity correction computed as described in Section 1.2.2. As stated previously, we aided the matching by projecting strong random texture onto the scene, and have established that under this conditions, the first set of intermediate results where the random sensor noise has a significant effect is the subpixel disparity map. In (8), the integer disparity is deterministic, only the disparity correction is random, thus $\mathrm{Var}\bar{\boldsymbol{D}}(u, v) = \mathrm{Var}\boldsymbol{D}_\varepsilon(u, v)$. In [20], using the image variance estimates, and a propagation of variance method described in [13], we derived the estimates $\sigma_D^2(u, v)$ of $Var\boldsymbol{D}_\varepsilon(u, v)$. Thus, in our augmented algorithm, for every pixel in the left image we compute the disparity and its variance.

Under normal assumptions for the image noise, it follows that the subpixel disparity is normally distributed, i.e., $\bar{\boldsymbol{D}}(u, v)$ has a normal distribution with variance $\sigma_D^2(u, v)$.

---

[8]We would like to pursue such a model in the future.

Remark 2.1 Experimentally, we have confirmed that for the areas corresponding to pixels with a stable integer disparity, this is a reasonable model (the empirical subpixel disparity distributions are unimodal, symmetric). For about $5 - 30\%$ of the pixels in various experiments, we have observed bimodal empirical disparity distributions. We have concluded that the bimodality had resulted from ambiguities in matching since the shift between the modes was more than a pixel wide, and the locations of these pixels were in the areas where the integer disparity was unstable. This suggests that the matching procedure should be reexamined and improved.

3   Confidence Intervals and the Variability in 3-D Coordinates

Under the resulting normal model in disparity, for a given α, a confidence interval $C(\bar{\boldsymbol{D}}(u,v))$ for the subpixel disparity at the pixel (u,v) with theoretical probability of capture $TPC = 1 - \alpha$ is given by:

$$C(\bar{\boldsymbol{D}}(u,v)) = [\,\bar{\boldsymbol{D}}(u,v) - e(u,v)\,,\ \bar{\boldsymbol{D}}(u,v) + e(u,v)\,], \tag{9}$$

where $e(u,v) = \zeta(\alpha/2)\sigma_D(u,v)$, and $\zeta(\alpha/2)$ denotes the $\alpha/2$ quantile of the standard normal distribution, i.e.,

$$\alpha/2 = \frac{1}{\sqrt{2\pi}} \int_{-\infty}^{\zeta(\alpha/2)} \exp(-x^2/2)\,dx. \tag{10}$$

We remind the reader that $TPC = Pr[\bar{\boldsymbol{D}}(u,v) \in C(\bar{\boldsymbol{D}}(u,v))]$. In our experiments, we selected the TPC, and then computed the intervals. In general, either TPC (equivalently α) or the size of the intervals should be specified based on the requirements of the application. When one of these parameters is specified, the other is calculated from (9-10). Note that these intervals characterize the variability (precision) in the results, not accuracy (i.e. deviations from an absolute ground truth).

Next, we propagate the bounds of the intervals, carefully accounting for the singularities and for the sign of the Jacobian of the reconstruction transformation (4). Details are given in [20]. As a result of the interval propagation, we obtain confidence intervals along with the coordinates of the reconstructed points. For a pixel (u,v), for which the matching was resolved successfully with a disparity value $\bar{\boldsymbol{D}}(u,v) = \bar{d}$, the confidence interval for the Z coordinate of the reconstructed 3-D point is $C_Z(u,v) = [L_Z(u,v), U_Z(u,v)]$. The lower and upper confidence bounds are defined by

$$L_Z(u,v) = \frac{S(3,:) * (\,u,v,1,C_l(\bar{d})\,)^t}{S(4,:) * (\,u,v,1,C_l(\bar{d})\,)^t} \tag{11}$$

$$U_Z(u,v) = \frac{S(3,:) * (\,u,v,1,C_u(\bar{d})\,)^t}{S(4,:) * (\,u,v,1,C_u(\bar{d})\,)^t}, \tag{12}$$

where, $C_l(\bar{d})$ and $C_u(\bar{d})$ are the bounds of the confidence interval in disparity, $C(\bar{\boldsymbol{D}}(u,v))$. The confidence bounds in X and Y are obtained by substituting $S(3,:)$ in (11-12) with $S(1,:)$ and $S(2,:)$, respectively. Note that the intervals for the disparity and for each of the coordinates X, Y and Z have the same probabilities of capture (when well defined). Although the intervals in disparity are symmetric around the computed subpixel disparity values, the intervals for the 3-D coordinates are not symmetric around the reconstructed coordinates. Since the subpixel disparity is normally distributed, the the coordinates of the

3-D point in projective space are normally distributed (that is each of $\tilde{X}$, $\tilde{Y}$, $\tilde{Z}$, and $\tilde{K}$ is normal, with variances $S(i,4)^2\sigma_D^2(u,v), i = 1,2,3,4$, respectively), but the coordinates in Euclidean space are not normally distributed. Thus reporting a variance for the Euclidean coordinates is not a reasonable description of the variability (in a most dramatic example, the ratio of two independent random variable with standard normal distribution is a Cauchy distribution which has very heavy tails and nonexistent, i.e. infinite, variance).

The confidence intervals which we construct characterize the variability in the results, not the accuracy of the reconstruction algorithm. Thus experiments with ground truth models for 3-D positions are not a suitable test for these intervals. Such experiments will address directly the accuracy of the reconstruction algorithm, but not the precision. We present some quantitative data on the intervals in Section 5. The rationale which motivates us to use these intervals, despite the fact that they do not characterize the accuracy is based on the flowing observation: reconstruction results may be precise and inaccurate (the intervals do not characterize those), but we definitely want to reject as unreliable points which are not precise (have large variability), i.e. large interval sizes.

4   Radiometric Correction

The variation of intensity in dark images and in flat fields for different sensors clearly shows that sensors are not "equal", contrary to the assumptions made in many multisensor computer vision applications. Given, even, absolutely the same scene and illumination, physically different sensors "see" differently. Depending on the application and the algorithm these differences may or may not have a significant effect on the results. As we demonstrate (Section 5), for area-based stereo matching, if the scene imaged has a weak texture it is important to correct radiometricly. Even for a single sensor system, if the main cue is intensity (for example in shape from shading methods) radiometric correction is necessary.

There are different radiometric correction procedures in use, but they all assume a linear sensor model (even if not explicitly stated). In some of these procedures, the model parameters are estimated off-line, and images preprocessed on line. In others, parameters are not recovered directly rather the images are preprocessed so that the expected offset in the processed images is zero and the photoresponse is uniform. The process of such radiometric correction is called *flat fielding* or *shading correction* [31]. Intuitively, in order to achieve a corrected image with zero offset and constant gain, we subtract from the original image the offset observed in averaged dark images, and scale the resulting image inversely proportional to the photoresponse observed in averaged flat fields, i.e., [31]:

$$I_{correctd}(u,v) = \frac{I(u,v) - \hat{m}(u,v)}{F(u,v) - \hat{m}(u,v)} \left[\frac{1}{MN} \sum_{u=1}^{M} \sum_{v=1}^{N} (F(u,v) - \hat{m}(u,v)) \right]$$

where F is a pixelwise average of flat fields taken under fixed illumination (an estimate of PRNU at a fixed illumination close to saturation level), and $\hat{m}(u,v)$ is the estimate of the FPN in dark current, (7). The expression in the brackets on the right-hand side is the average response (over all pixels) after F has been adjusted to zero offset.

5 Experiments

In this section, first, we illustrate the sensor noise in dark images and flat fields and the variation in the noise characteristics between different sensor configurations (Section 5.2); second, we demonstrate with ground test scenes the effect of the radiometric correction procedure in improving the accuracy of the stereo reconstruction results (Section 5.3); and third, we show series of experiments pertaining to the use of the confidence intervals in assessing the precision and rejecting unreliable points (Section 5.4).

5.1 Experimental setup

We used the following types of black and white analog CCD cameras and 8-bit framegrabbers in our experiments:

Camera	CCD array size	Pixel clock frequency
HITACHI KP 230(231)	384(H)x485(V)	8MHz
SONY XC-77	768(H)X493(V)	14.318MHz
Framegrabber	Effective digital image size	Sampling frequency
DT1451	512(H)x480(V)	about 10MHz
S1V	640(H)x480(V)	not available
TIM40	756(H)x480(V)	set to 14.318MHz

Each camera was equipped with 25mm lens. For the stereo experiments: a pair of cameras were usually 6-10cm apart, non-verging, or verging at approximately 40°; the cameras were viewing a volume of approximately $30 \times 30 \times 30 \text{cm}^3$ at 80cm from the base line of the pair; each time a camera pair was reconfigured, it was strongly calibrated, [34]. In the experiments where strong texture was needed we induced it by using a slide projector and a slide representing random dots pattern.

5.2 Sensor Noise Observed in Dark Images and Flat Fields

In Section 5.2.1 we illustrate the noise in dark images, and in Section 5.2.2, the noise in flat fields.

5.2.1 Dark Images and the Background Noise

Figure 1 shows a typical dark image and a pixelwise average of 100 dark images for the camera/framegrabber configuration SONY XC-77/DT1451. In the left image (a representative dark image) the background random noise and internal luminance are noticeable. Systematic components in the background noise (offset and stripes) are prominent in the right image (the average of 100 dark images). The vertical stripes are primary due to aliasing resulting from the mismatch between the camera pixel clock and the framegrabber sampling frequency. When the camera pixel clock and the framegrabber sampling frequencies are well matched such aliasing does not occur (for example, in the configuration SONY XC-77/TIM 40, or for digital cameras where the digitization is done on the camera).

Figure 2 shows for 4 different sensor configurations, data based on the amplitude of the power spectra of the mean column intensities in dark images for each sensor. To enhance the display, we have applied the standard log-transform and have centered the results. For the configuration SONY XC-77/DT1451 the high frequencies are prominent, the camera lines

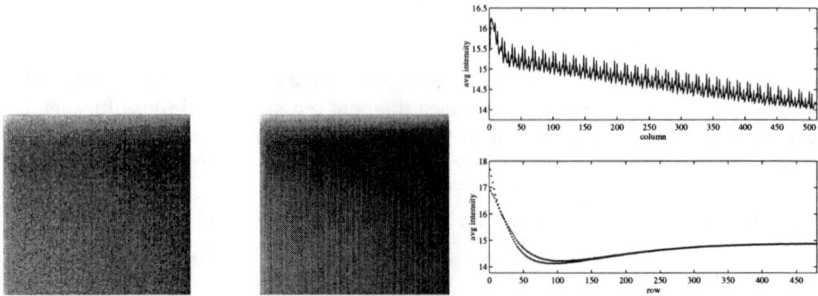

Figure 1. SONY XC-77/DT1451: **A single dark image (left) and a pixelwise average of 100 dark images (right). The graphs represent the average intensity in the right image: per column (top), and per row (bottom). The offset in intensity of odd and even fields, a result of interlacing, is clearly visible in the bottom graph.**

are under-sampled. Contrast this to SONY XC-77/TIM40, where the sampling is almost one-to-one. For each configuration, we tested several combinations of physically different cameras of the same type and different framegrabbers. Each physical configuration had its own dark image signature, but within a particular configuration type (same make), the background noise could be parametrized.

5.2.2 Flat Fields and Photoresponse Nonuniformities

The scene imaged was a white plane completely covering the field of view of the camera. To obtain flat fields we used a diffuser (a white glass filter) instead of the lens, achieving about 65% uniformity. Ideally, an integrating sphere should be used [28] which achieves 95% uniformity of the illumination.

The right pair of images in Figure 3 shows the pixelwise averages of 100 flat fields for each of two different cameras, configuration SONY XC-77/DT1451. The right pair of graphs underneath, Figure 4, represents the corresponding histograms. The acquisition of the flat fields without the use of a lens removes any nonuniformity which could have risen from lens vignetting (fall off in intensity from the center of the image to the boundary) or other lens distortions. The bright square frames noticeable in the two intensity images, are due to reflections from the square aperture in front of the sensor array (a part of the camera architecture) and the adapter holding the diffuser. Apart from that artifacts, the variation in intensity in the images is mostly due to fixed pattern noise of the photoresponse nonuniformities of the individual sel (the read noise is negligible at that high level, a gray value of about 220)[9].

5.3 The Effect of the Radiometric Correction on the Accuracy

When the texture is weak, correcting radiometricly the images reduce the number of false and ambiguous matches, and improves the accuracy of the stereo reconstruction. The purpose of the experiments in this section was to demonstrate that.

---

[9]The printing of the images has introduced minor defects as well.

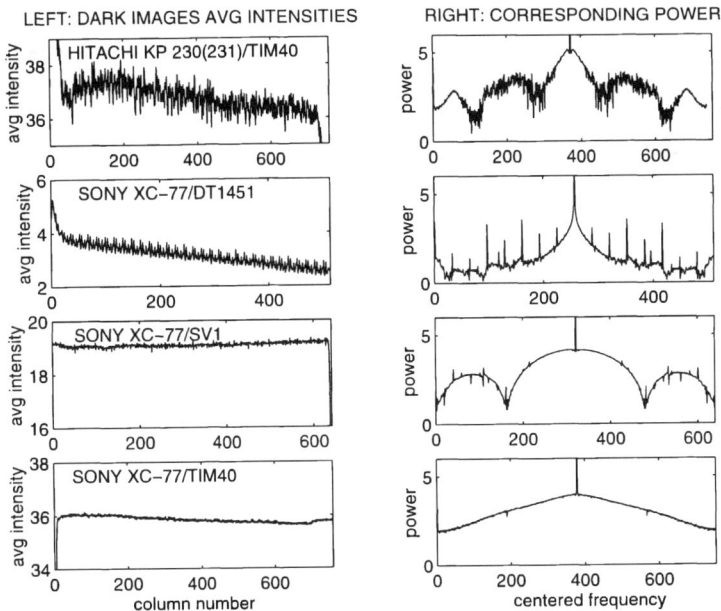

Figure 2. Data for single dark images obtained with different `camera/framegrabber` **configurations. The graphs represent average intensities per columns (left), and the corresponding amplitudes of the power spectra (right).**

For the radiometric correction procedure, we used average dark images and flat fields which were computed based on sequences of 100 images each. During the image acquisition process, caution should be paid to eliminate a possibility of image residues. The average dark image and the average flat field computation is one-time, off-line process (Figures 3-4).

On the day of the experiment, the cameras were warmed up; the lens parameters and illumination were controlled to prevent saturation; and the cameras were calibrated geometricly (intrinsic and extrinsic camera parameters, using least squares, were recovered).

Experiment 5.1 Flat Field Correction — Planar Example The target was a planar white card, lambertian reflectance surface[10]. A stereo pair of raw images was acquired (Figure 5, left pair of images), radiometrically corrected, and disparity maps based on the rectified uncorrected and on rectified corrected images were calculated. The rectification is a necessary step since the cameras were verging. Figure 6, shows the disparity maps as 3D meshes. The full size of the disparity maps is 256x256 since the rectification was done in half resolution. For the purpose of displaying the result, the 3D plots show the subsampled maps. "Holes" in the disparity maps can be observed, where the algorithm failed (matches were rejected). "Spikes" are outliers erroneously accepted as valid. The disparity map based on the radiometricly corrected images (right) has less holes and spikes. The two subpixel disparity maps were evaluated quantitatively by fitting a plane to each

---

[10]The material of the card was a standard poster board, there was some roughness texture, although very week, so the matching could be done.

Figure 3. A pair of SONY XC-77RR/DT1451 **sensors: the pair of average dark images (left) and the pair of average flat fields (right).**

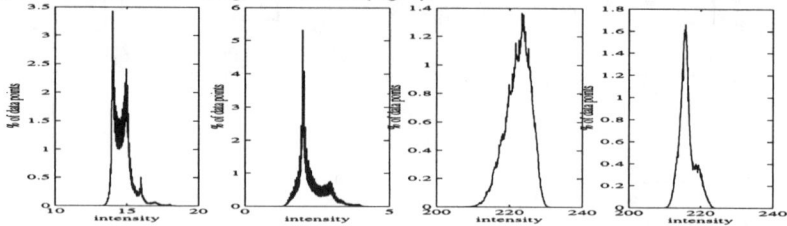

Figure 4. SONY XC-77RR/DT1451**: the histograms of the corresponding images given in Figure 3 (same left to right order).**

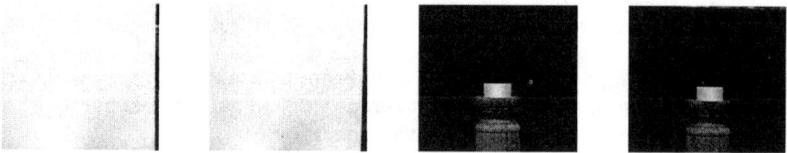

Figure 5. Original stereo pairs of images: a planar scene (left), a scene with a cylinder (right).

one of them, and comparing statistics of the residuals from the planar fit. Figure 7, shows the histogram for the residuals in the two cases. The standard deviation of the residuals, for the "uncorrected" map was 0.1668, and for the "corrected" was 0.0679, so a relative improvement of about 59% was obtained.

The experiment was repeated, testing the radiometric correction procedure at different flat field levels, approximately 60, 110, 160, 220. Subpixel disparity maps without and with flat-field correction were computed based on these, planes were fitted to each one, and statistics of residuals were compared. The dominance of the performance of the matching algorithm (in relative rms) with flat-field correction over the matching without flat-field correction is observed: from 26%, for the low intensity level flat field (level 60), to 59%, for the matching with flat field correction at level 220.

Experiment 5.2 Flat Field Correction – Non-planar Example. In this experiment, we used a scene which contained a cylindrical object, Figure 5, right pair of images. The target was a wooden cylinder, natural light color, it is the brightest object approximately in the center of the images. As in the previous experiment, the images were corrected radiometricly, and the 3D reconstructions based on the rectified uncorrected and on rectified corrected images were calculated. Figure 8, shows the reconstructed (x,y,z) points for the

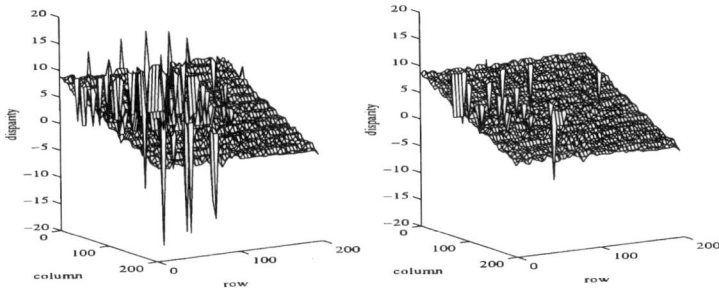

Figure 6. The planar scene: subpixel disparity maps shown as 3D mesh, based on original image pair (left), and on the radiometricly corrected pair (right).

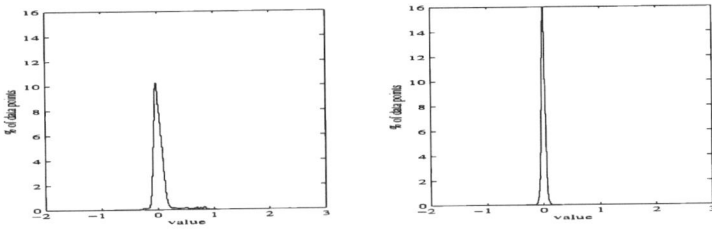

Figure 7. The planar scene: the histograms of the residuals from the planar fitting for the corresponding disparity maps shown in Figure 3 (same left to right order, uncorrected – left, and corrected – right).

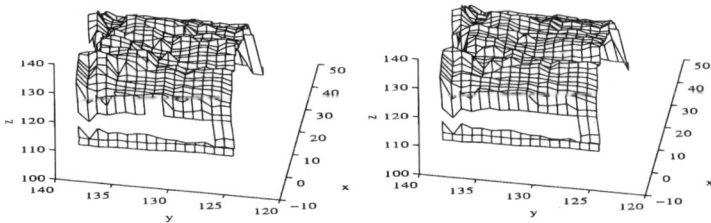

Figure 8. Cylinder: reconstruction results, shown as 3D mesh, based on the original image pair (left), and on the radiomericly corrected images (right).

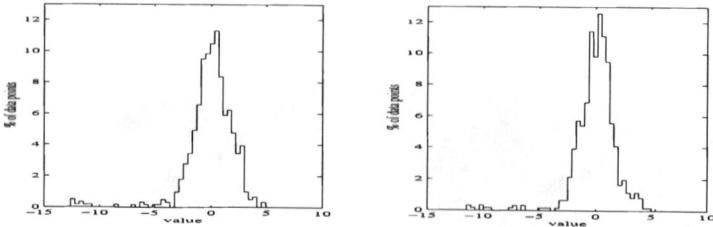

Figure 9. Cylinder: histograms of the residuals from fitting a cylinder to the reconstruction results given in Figure 8 (uncorrected — left, and corrected — right).

cylinder (only) as 3D meshes. The reconstruction based on the radiometricly corrected images (right) has smoother appearance. The two reconstructions were evaluated quantitatively by fitting a cylinder to each one of them, and comparing statistics of the residuals from the planar fit. Figure 9, shows the histogram for the residuals in the two cases. The standard deviation of the residuals, for the "uncorrected" map was 2.1477, and for the "corrected" was 1.9074, so a relative improvement of about 11% was obtained in this case.

5.4 Quantifying the Precision with Confidence Intervals

In this section we present a series of experiments regarding the confidence intervals for the precision. First we evaluated the intervals by comparing theoretical probabilities of capture (TPC) with empirical capture frequencies calculated during multiple runs of the stereo algorithm on tests scenes which were planar surfaces. We have also done experiments with more complex scenes, i.e. human faces and hands. Examples of the scenes are given on Figure 10. For the planar surfaces we present the quantitative data. For the human face and hands we did not evaluate the intervals quantitatively (the main reason being that the subjects could not keep still while sequences of 100 images needed for the empirical comparisons could be taken). We give the statistics of the intervals in the Z component of the reconstructed points, since this is of main interest in stereo. We have used standard normal distribution model for the image noise which amounts to an uniform scale of the intervals in subpixel disparity. The goal of these series of experiments was to assess the precision of the reconstruction algorithm in terms of the relative confidence interval sizes in depth, Z, and to compare these for several different stereo pairs of cameras. The test scene was a vertical planar surface coinciding with the coordinate plane $(X - Y)$. The plane was positioned at precise, and known position for Z.

A random texture was projected on the planar surface (Figure 10, left pair of images). A sequence of 100 images was acquired with each of four cameras SONY XC-77RR. During the evaluation process of the confidence intervals, camera pairs were formed (using the corresponding images which were already taken). This way the performance of the intervals for varieties of stereo configurations was examined. The reconstruction algorithm, [34], augmented with the confidence intervals estimation procedure described in this paper was run for one stereo pair images. Confidence intervals, C_X, C_Y, C_Z with TPC = 0.68 were computed. The reconstruction algorithm was run 100 times, on each of the 100 stereo pairs. For each stereo pair and for each pixel position in the left image of the pair, we counted how many times the newly computed depth Z for that pixel position fell within C_Z. The

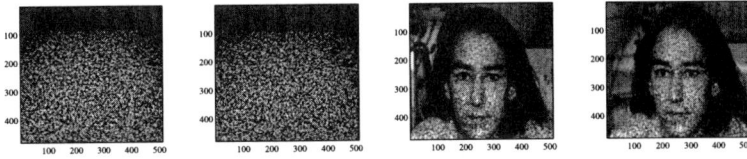

Figure 10. Test scene stereo pairs: plane (left), a face (right).

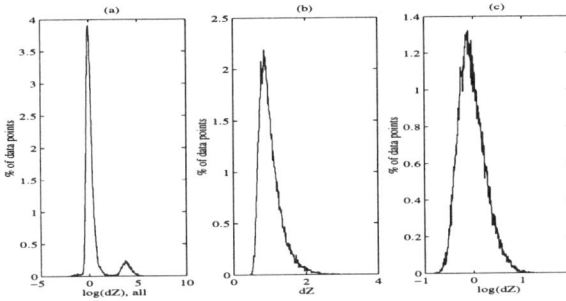

Figure 11. Pair(1,4): Histograms of interval sizes, dZ, in mm at 0.68 TPC. (a) a histogram over all intervals, log scale; (b) a histogram, only over intervals of recon-structed points, corresponding to image projection of the plane; (c) same as (b), but in log scale. We differentiate between (a) and (b)-(c) since the image of the plane does not coincide with the whole digital image.

final counts were divided by the total number of sample runs, thus the *empirical capture frequencies* (ECF) were computed.

Experiment 5.3 Quantitative data for the interval size in depth , Z, for a single camera pair. Figure 11 shows the histograms of the sizes, dZ, in mm for the confidence intervals with *theoretical probability of capture* (TPC) of 0.68, for one fixed stereo pair (labeled $Pair(1, 4)$) and the planar scene, Figure 10 left images. Figure 12 shows various statistics for the empirical capture frequency for $Pair(1, 4)$. The pixels which corresponds to the projections of the plane (with random texture) have better ECF. In the bottom graph the pixel positions for which the ECF is below TPC have non-satisfactory performance.

Experiment 5.4 Comparisons over 4 different camera pairs. Next we compare results for four different stereo pairs which we label Pair(1,2)', Pair(1,4)', Pair(2,3)' and Pair(3,4)'. The random texture plane was covering the complete view of all cameras, so consistent comparison could be drawn across pairs. Confidence intervals, with probability of capture 0.68, were computed for the reconstructed points for each pair, and the ECF were calculated during 100 runs of the polynocular stereo reconstruction algorithm [34]. Statistics for ECF, and for the confidence interval sizes, dZ, in mm, are summarized in Table 1. Note that the TPC is less than the actual ECF. TPC underestimates the performance of the intervals. Under the same model, deriving intervals with higher TPC also leads

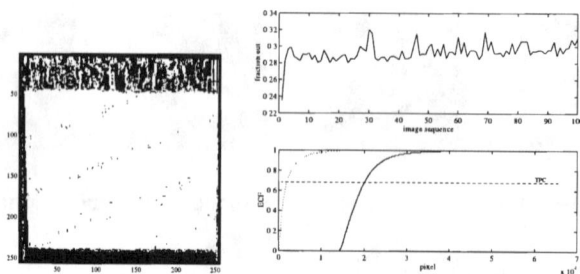

Figure 12. Pair(1,4): Data for ECF, planar scene. The image on the left shows in white the pixels for which the intervals for the corresponding 3-D depth have ECF$\geq$TPC= 0.68. The top graph on the right shows for each of the 100 reconstructions, the fraction of pixels for which the corresponding reconstructed Z is not captured by C_Z. The bottom graph shows sorted ECF of the intervals: the solid line is for the intervals corresponding to all reconstructed points; the doted line is for those corresponding to pixels projections of the plane; and the dashed line represents the TPC=0.68.

		ECF				dZmm		
Pair	*mean*	*std*	*median*	*min*	*max*	*mean*	*std*	*median*
(1,2)'	0.6838	0.3214	0.8000	0.1598	4.2830	0.9234	0.3043	0.8722
(1,4)'	0.7549	0.2842	0.8700	0.1180	226.5228	0.7082	0.9669	0.6666
(2,3)'	0.7327	0.2912	0.8500	0.1122	80.9347	0.6886	0.9303	0.6180
(3,4)'	0.6931	0.3154	0.8100	0.1725	4.9167	0.8375	0.2730	0.8035

Table 1. Statistics of ECF, and of confidence interval sizes, dZ in mm, across all points, for confidence intervals with TPC of 0.68, for 4 different stereo pairs of cameras

to intervals with larger sizes, meaning worse resolution and more uncertainty in the true position.

6   Conclusions

6.1   Summary

We demonstrated that by modeling the video sensor one can gain better understanding of the results of vision algorithms. We used as an illustration an algorithm for 3-D reconstruction from stereo. We showed that by modeling the sensor, and accounting for systematic and random sensor noise: (1) the accuracy could be improved, in particular, in cases where stereo usually fails, i.e. in weak texture areas (for example, one color, relatively smooth man made objects, like office walls, door, cabinets); and (2) that precision bounds could be provided in terms of confidence intervals. We have verified experimentally that the intervals meet the specified theoretical bounds. The interval sizes represent the degree of uncertainty, and the TPC degree of reliability.

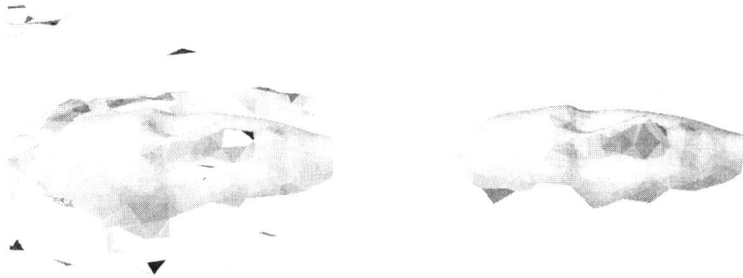

Figure 13. Left: A surface reconstruction based on all 3-D points. Right: A surface reconstruction based only on 50% of the 3-D points, only those with confidence interval in Z which have sizes smaller than the median interval size were used in the surface recovery.

6.2 Relevance of the Results to Other Vision Research Areas

The radiometric correction procedure is applicable not only to stereo, its effects are even more significant in other vision methods like shape from shading or photometric stereo where the intensity itself is of most importance. In fact, in any application, sensors should be carefully examined (using dark images and flat fields) to reveal possible defects and systematic and random noise characteristics. To obtain the intervals, we used a variance propagation method which is acceptable as far as algorithms or their modules are reasonably well approximated by linear transformations (i.e. in our case the subpixel disparity correction procedure was such). When linear approximations are not satisfactory, the possibilities are: (i) direct distribution model propagation (which could be very hard), or (ii) distribution models built directly for the results using ground test scenes.

6.3 Use of the Intervals in Surface Recovery

A feature of the intervals which supports their validity as quantifiers of the precision, is the fact that we have used them successfully for rejection of reconstructed points which have higher degree of uncertainty. We conclude here by presenting results from one such an experiment.

The test scene was a face. We have projected the random dot texture on it (Figure 10, right pair of images), and run the stereo algorithm augmented with the computation of the confidence intervals. Then, we have computed the median statistic of the interval sizes in the depth, Z, over all reconstructed points. Finally, we reconstructed the surface geometry and topology using the algorithm from [35] based on: (1) all reconstructed points, and (2) only those points for which the corresponding confidence interval sizes were less than the median. The results of the two reconstructed surfaces are shown on Figure 13. By the use of the confidence intervals for the rejection of points with high degree of uncertainty, we have reduced the set of the reconstructed points by 50%, and have still preserved the important features. We have removed outlier artifacts (particles hanging in front and around the face), and also have resolve ambiguities in the surface reconstruction stage (for example, the hole in the center of the face on the left is covered). The size of the confidence intervals is related to the scene attributes (texture and geometry) and the sensor noise parameters. The intervals are large when: (a) the signal-to-noise ratio of the image is locally low (as in weak texture areas, in areas overlaid by highlights where the texture contrast is low, and in dark or shadowed areas); and (b) when the match is wrong (as in case of highlights, in areas which does not image as a regular surface, like hair for example, and near occlusions where the matches are notoriously bad in all area-based stereo matching algorithms [33]). The intuition which may explain (a) is the dependence of the intervals size not only on the pixel noise levels, but also on the image gradients.

Currently we are exploring the use of the intervals for data fusion of redundant 3-D points reconstructed from different views or from multiple stereo pairs.

Acknowledgements

This work is supported by, or in part by, The National Science Foundation under grant numbers: MIP94-20397 A03 SUB, IRI93-3980, IRI93-07126, GER93-55018, and the U.S. Army Research Office under grant numbers: P-34150-MA-AAS DAAH04-96-1-0007 ARO/DURIP DAAG55-97-1-0064.

We thank Radim Šára and Reyes Enciso for helping in the preparation of this document.

References

[1] N. Ayache and C. Hansen, "Rectification of Images for Binocular and Trinocular Stereovision", *Proc. of 9th International Conference on Pattern Recognition*, 1, pp. 11-16, Italy, 1988.

[2] H. Beyer, "Linejitter and geometric calibration of CCD cameras", *ISPRS Journal of Photogrammetry and Remote Sensing*, Vol. 45, pp. 17-32, 1990.

[3] P. Belhumeur, "A Bayesian Approach to Binocular Stereopsis", *Intl. J. of Computer Vision*, 19(3), pp 237-260, 1996.

[4] H. Beyer, "Geometric and radiometric analysis of a CCD-camera based photogrammetric close-range system", *Dissertation, Institut für Geodäsie und Photogrammetrie, Zürich*, 1992.

[5] E. Beynon and D. Lamb,edt., *Charge-coupled devices and their applications*, McGraw-Hill, 1980.

[6] S. Blostein and T. Huang, "Error Analysis in Stereo Determination of 3-D Point Positions", *IEEE Trans. on Pattern Analysis and Machine Intelligence*, 9(6), pp. 752-765, 1987.

[7] Das and Ahuja, "Performance Analysis of Stereo, Vergence, and Focus as Depth Cues for Active Vision", *IEEE Trans. on Pattern Analysis and Machine Intelligence*, 17(12), pp 1213-1218, 1995.

[8] F. Devernay, "Computing Differential Properties of 3-D Shapes from Stereoscopic Images without 3-D Models", INRIA, RR-2304, Sophia Antipolis, 1994.

[9] U. Dohond and J. Aggrawal, "Structure from Stereo: a Review", *IEEE Transactions on Systems, Man, and Cybernetics*, 19(6), pp 1489-1510, 1989.

[10] G.Florou and R. Mohr, "What accuracy for 3D measurements with cameras?" *Proc. of ICPR'96*, A80.1, pp.354-358, 1996.

[11] W. Förstner, "10 Pros and cons against performance characterization of vision algorithms", *Workshop on performance characterization of vision algorithms*, Robin College, Cambridge, 1996.

[12] R. Haralick, "Performance characterization in computer vision", *Performance versus methodology in computer vision*, Haralick and Meer, editors, University of Washington, Seattle, 1994.

[13] R. Haralick, "Propagating Covariance in Computer Vision", *Workshop on performance characterization of vision algorithms*, Robin College, Cambridge, 1996.

[14] E. Healey and R. Kondepudy, "Radiometric CCD Camera Calibration and Noise Estimation, *IEEE Trans. on Pattern Analysis and Machine Intelligence*, 16(3), pp. 267-276, 1994.

[15] J. Hoggs, *The Microscope: its history, construction and application*, London, 1854.

[16] G. Holst, *CCD arrays, cameras, and displays* , Winter Park, FL : JCD Pub., 1996.

[17] B. Jähne, *Digital image processing: concepts, algorithms and scientific applications*, Springer-Verlag, 1993.

[18] J. Janesick et al, "The future scientific CCD", *Proc. of SPIE State-of-the-art imaging arrays and their applications, San Diego, California*, Vol. 501, pp. 2-33, 1984.

[19] B. Julesz, *Foundations of Cyclopean Perception*, Chicago, 1971.

[20] G. Kamberova and R. Bajcsy, "Precision of 3-D Points Reconstructed from Stereo", GRASP Lab, Technical Report, 1997, http://www.cis.upenn.edu/ kamberov/doc/papers.html

[21] G. Kamberova and M. Mintz, "Minimax Rules Under Zero-one Loss for a Restricted Location Parameter", *Journal of Statistical Planning and Inference*, accepted, 1997.

[22] G. Kamberova, "The Effect of Radiometric Correction on Multicamera Algorithms", Technical Report, MS-CIS-97-17, CIS Department, University of Pennsylvania, 1997

[23] R. Lenz and D. Fritsch, "Accuracy of videomery with CCD sensors", *ISPRS Journal of Photogrammetry and Remote Sensing*, Vol. 45, pp. 90-110, 1990.

[24] R. Mandelbaum, G. Kamberova and M. Mintz, "Stereo Depth Estimation: a Confidence Interval Approach", *Proc, Intl. Conf. Computer Vision ICCV98*, 1998.

[25] D. Marr and T. Poggio, "Cooperative Computation of Stero Disparity", *Science*, 194, pp 283-287, 1976.

[26] D. Marr and T. Poggio, "A Computational Theory of Human Stereo Vision", *Proc. of the Royal Soc. of London*, B204, pp 301-328, 1979.

[27] S. Maybank and O. Faugeras, "A Theory of Self-Calibration of a Moving Camera", *Intl. J. of Computer Vision*, 8(2), pp 123-151, 1992.

[28] I. Mclean, *Electronic and computer-aided astronomy: from the eyes to electronic sensors*, Ellis Horwood Limited, England, 1989.

[29] H. Moravec, "Robot Rover Visual Navigation", *Computer Science:Artificial Intelligence*, pp 13-15, 105-108, UMI Research Press 1980/1981.

[30] M. Okutomi and T. Kanade, "A Multiple-baseline Stereo", *IEEE Trans. on Pattern Analysis and Machine Intelligence*, 15(4), pp. 353-363, 1993.

[31] Photometrics Homepage, *Photometrics High performance CCD Imaging*, http://www.photomet.com, 1996.

[32] K. Price, "Anything you can do, I can do better (No you can't)...", *Computer Vision, Graphics, and Image processing*, Vol. 36, pp. 387-391, 1986.

[33] R. Šára and R. Bajcsy, On Occluding Contour Artifacts in Stereo Vision", *Proc. Int. Conf. Computer Vision and Pattern Recognition*, IEEE Computer Society, Puerto Rico, 1997.

[34] R. Šára, "Reconstruction of 3-D Geometry and Topology from Polynocular Stereo", http://cmp.felk.cvut.cz/ sara

[35] R. Šára and R. Bajcsy, "Fish-Scales: Representing Fuzzy Manifolds", *Proc, Intl. Conf. Computer Vision ICCV98*, 1998.

[36] H. Sahabi and A. Basu, "Analysis of Error in Depth with Vergence and Spatially Varying Sensing", *Computer Vision and Image Understanding*, 63(3), pp. 447-461, 1996.

[37] C. Slama, edt., *Manual of photogrammetry, IVth edition*, American Society of Photogrammetry, 1980.

[38] D. Snyder *et al.*, "Compensation for readout noise in CCD images", *Journal of the Optical Society of America, A*, Vol. 12, No. 2, pp. 272-283, 1995.

[39] A. Sripad and D. Snyder, "A necessary and sufficient condition for the quantization error to be uniform and white", *IEEE Trans. on acoustic, speech, and signal processing*, Vol. ASSP-25,No 5, pp. 442-448, 1977.

[40] A. Theuwissen, *Solid-State imaging with charge-coupled devices*, Kluwer Academic Publishers, 1995.

[41] A. Yuille and T. Poggio, "A Generalized Ordering Constraint for Stereo Correspondence", *MIT, Artificial Intelligence Laboratory Memo*, No. 777, 1984.

Fingerprint Image Enhancement: Algorithm and Performance Evaluation

Lin Hong, Yifei Wan, and Anil Jain
Pattern Recognition and Image Processing Laboratory
Department of Computer Science
Michigan State University
East Lansing, MI 48824
{honglin,wanyifei,jain}@cps.msu.edu

Abstract

A critical step in automatic fingerprint matching is to automatically and reliably extract minutiae from the input fingerprint images. However, the performance of a minutiae extraction algorithm relies heavily on the quality of the input fingerprint images. In order to ensure that the performance of an automatic fingerprint identification/verification system will be robust with respect to the quality of input fingerprint images, it is essential to incorporate a fingerprint enhancement algorithm in the minutiae extraction module. We present a fast fingerprint enhancement algorithm, which can adaptively improve the clarity of ridge and furrow structures of input fingerprint images based on the estimated local ridge orientation and frequency. We have evaluated the performance of the image enhancement algorithm using the goodness index of the extracted minutiae and the accuracy of an online fingerprint verification system. Experimental results show that incorporating the enhancement algorithm improves both the goodness index and the verification accuracy.

1: Introduction

Fingerprint identification is one of the most important biometric technologies which has drawn a substantial amount of attention recently [12, 14]. A fingerprint is the pattern of ridges and furrows on the surface of a fingertip. Each individual has unique fingerprints. The uniqueness of a fingerprint is exclusively determined by the local ridge characteristics and their relationships [12, 13]. A total of one hundred and fifty different local ridge characteristics, called minute details, have been identified [13]. These local ridge characteristics are not evenly distributed. Most of them depend heavily on the impression conditions and quality of fingerprints and are rarely observed in fingerprints. The two most prominent ridge characteristics, called minutiae, are (*i*) *ridge ending* and (*ii*) *ridge bifurcation*. A ridge ending is defined as the point where a ridge ends abruptly. A ridge bifurcation is defined as the point where a ridge forks or diverges into branch ridges. A good quality fingerprint typically contains about 40–100 minutiae. Examples of minutiae are shown in Figure 1.

Automatic fingerprint matching depends on the comparison of these local ridge characteristics and their relationships to make a personal identification [12]. A critical step in fingerprint matching is to automatically and reliably extract minutiae from the input

(a)

(b)

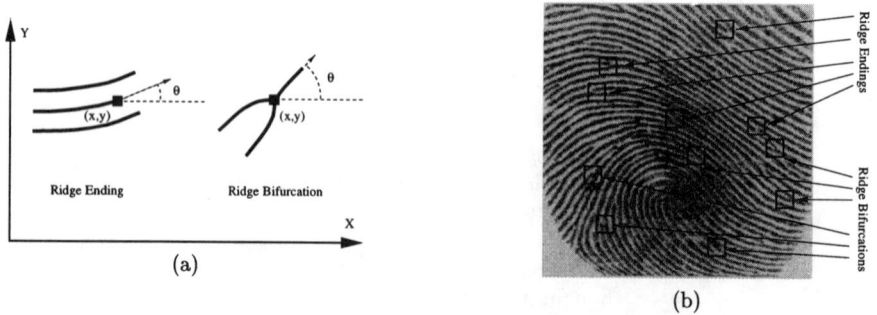

Figure 1. Examples of minutiae; (a) a minutiae can be characterized by its position and its orientation, (b) minutiae overlaid on a fingerprint image. ordinary paragraph.

fingerprint images, which is a difficult task. The performance of a minutiae extraction algorithm relies heavily on the quality of the input fingerprint images. In an ideal fingerprint image, ridges and furrows alternate and flow in a locally constant direction and minutiae are anomalies of ridges, *i.e.* ridge endings and ridge bifurcations. In such situations, the ridges can be easily detected and minutiae can be precisely located from the binary ridges. Figure 1(b) shows an example of good quality live-scan fingerprint image. However, in practice, due to variations in impression conditions, ridge configuration, skin conditions (aberrant formations of epidermal ridges of fingerprints, postnatal marks, occupational marks), acquisition devices, and non-cooperative attitude of subjects, *etc.* a significant percentage of acquired fingerprint images (approximately 10% according to our experience) is of poor quality. The ridge structures in poor-quality fingerprint images are not always well-defined and hence they can not be correctly detected. This leads to following problems: (*i*) a significant number of spurious minutiae may be created, (*ii*) a large percent of genuine minutiae may be ignored, and (*iii*) large errors in their localization (position and orientation) may be introduced. Examples of fingerprint images of very poor quality, in which ridge structures are completely corrupted, are shown in Figure 2. In order to ensure that the performance of the minutiae extraction algorithm will be robust with respect to the quality of input digital fingerprint images, an enhancement algorithm which can improve the clarity of the ridge structures is necessary.

A fingerprint expert is often able to correctly identify the minutiae by using various visual clues such as local ridge orientation, ridge continuity, ridge tendency, *etc.*, as long as the ridge and furrow structures are not corrupted completely. It is possible to develop an enhancement algorithm that exploits these visual clues to improve the clarity of ridge structures in corrupted fingerprint images. Generally, for a given digital fingerprint image, the region of interest can be divided into the following three categories (Figure 3):

- *Well-defined region*, where ridges and furrows are clearly differentiated from one another such that a minutiae extraction algorithm is able to operate reasonably.

- *Recoverable corrupted region*, where ridges and furrows are corrupted by a small amount of creases, smudges, *etc.* But, they are still visible and the neighboring regions provide sufficient information about the true ridge and furrow structures.

Figure 2. Fingerprint images of very poor quality.

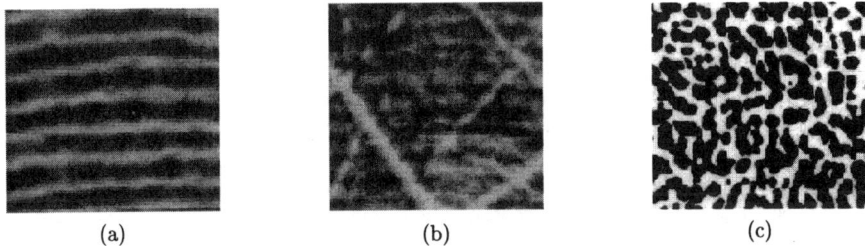

(a)   (b)   (c)

Figure 3. Fingerprint regions; (a) well-defined region; (b) recoverable corrupted region; (c) unrecoverable corrupted region.
ordinary paragraph.

- *Unrecoverable corrupted region*, where ridges and furrows are corrupted by such a severe amount of noise and distortion that no ridges and furrows are visible and the neighboring regions do not provide sufficient information about the true ridge and furrow structures either.

We refer to the first two categories of regions as *recoverable* and the last category as *unrecoverable*. The goal of an enhancement algorithm is to *improve the clarity of ridge structures* of fingerprint images in recoverable regions and to remove the unrecoverable regions. Since the objective of a fingerprint enhancement algorithm is to improve the clarity of ridge structures of input fingerprint images to facilitate the extraction of ridges and minutiae, a fingerprint enhancement algorithm should not result in any spurious ridge structures. This is very important because spurious ridge structure may change the individuality of input fingerprints.

Fingerprint enhancement can be conducted on either (i) *binary ridge images* or (ii) *gray-level images*. A binary ridge image is an image where all the ridge pixels are assigned a value 1 and non-ridge pixels are assigned a value 0. The binary image can be obtained by applying a ridge extraction algorithm on a gray-level fingerprint image [6]. Since ridges and furrows in a fingerprint image alternate and run parallel to each other in a local neighborhood, a number of simple heuristics can be used to differentiate the spurious ridge configurations from the true ridge configurations in a binary ridge image [5]. However, after applying a ridge extraction algorithm on the original gray-level images, information about the true ridge structures is often lost depending on the performance of the ridge extraction algorithm. Therefore, enhancement of binary ridge images has its inherent limitations.

In a gray-level fingerprint image, ridges and furrows in a local neighborhood form a sinusoidal-shaped plane wave which has a well-defined frequency and orientation. A number of techniques that take advantage of this information have been proposed to enhance gray-level fingerprint images [2, 15, 8, 18, 19]. However, they usually assume that the local ridge orientations can be reliably estimated. In practice, this assumption is not valid for fingerprint images of poor quality, which greatly restricts the applicability of these techniques. Hong *et al* [4] proposed a decomposition method to estimate the orientation field from a set of filtered images obtained by applying a bank of Gabor filters on the input fingerprint images. Although this algorithm can obtain a reliable orientation estimate even for corrupted images, it is computationally expensive which makes it unsuitable for an on-line verification system. We will present a fast enhancement algorithm which is able to adaptively enhance the ridge and furrow structures using both the local ridge orientation and local frequency information. Instead of using a computational expensive method to precisely estimate the local ridge orientation, a simple but efficient method is used. In addition, since this algorithm is designed to be integrated in an online system, a computationally efficient filtering technique is used.

In the following sections we will describe in detail our fast fingerprint enhancement algorithm. Section 2 addresses the main steps of our algorithm. A goal-directed performance evaluation of the implemented fingerprint enhancement algorithm on fingerprint databases is described in section 3. Section 4 contains the summary and discussion.

2: Fingerprint Enhancement

A fingerprint image enhancement algorithm receives an input fingerprint image, applies a set of intermediate steps on the input image, and finally outputs the enhanced image. In order to introduce our fingerprint image enhancement algorithm, a list of notations and some basic definitions are given below.

2.1 Notation

A *gray-level fingerprint image*, $\mathcal{I}$, is defined as a $N \times N$ matrix, where $\mathcal{I}(i,j)$ represents the intensity of the pixel at the *ith* row and *jth* column. We assume that all the images are scanned at a resolution of 500 *dots per inch* (*dpi*), which is the resolution recommended by FBI [12]. The *mean* and *variance* of a gray-level fingerprint image, $\mathcal{I}$, are defined as

$$M(\mathcal{I}) \;=\; \frac{1}{N^2} \sum_{i=0}^{N-1} \sum_{j=0}^{N-1} \mathcal{I}(i,j) \;\; and \tag{1}$$

$$VAR(\mathcal{I}) \;=\; \frac{1}{N^2} \sum_{i=0}^{N-1} \sum_{j=0}^{N-1} (\mathcal{I}(i,j) - M(\mathcal{I}))^2, \tag{2}$$

respectively.

An *orientation image*, $\mathcal{O}$, is define as a $N \times N$ image, where $\mathcal{O}(i,j)$ represents the *local ridge orientation* at pixel (i,j). Local ridge orientation is usually specified for a block

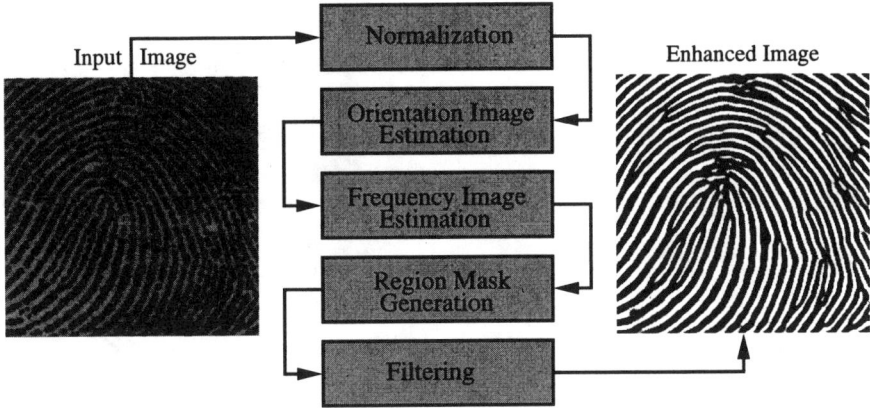

Figure 4. A flowchart of the proposed fingerprint enhancement algorithm.

rather than at every pixel; an image is divided into a set of $w \times w$ non-overlapping blocks and a single local ridge orientation is defined for each block. Note that in a fingerprint image, there is no difference between a local ridge orientation of 90° and 270°, since the ridges oriented at 90° and the ridges oriented at 270° in a local neighborhood can not be differentiated from each other.

A *frequency image*, $\mathcal{F}$, is a $N \times N$ image, where $\mathcal{F}(i, j)$ represents the *local ridge frequency*, which is defined as the frequency of the ridge and furrow structures in a local neighborhood along a direction normal to the local ridge orientation. The ridge and furrow structures in a local neighborhood where minutiae or *singular points* [9] appear do not form a well-defined sinusoidal-shaped wave. In such situations, the frequency is defined as the average frequency of its neighbors. Like orientation image, frequency image is specified block-wise.

The *region mask*, $\mathcal{R}$, is defined as a $N \times N$ image with $\mathcal{R}(i, j)$ indicating the category of the pixel. A pixel could be either (*i*) a *non-ridge-and-furrow* (unrecoverable) pixel (with value 0) or (*ii*) a *ridge-and-furrow* (recoverable) pixel (with value 1). Region mask is also specified block-wise.

2.2 Algorithm

The flowchart of the fingerprint enhancement algorithm is shown in Figure 4. The main steps of the algorithm include:

1. *Normalization:* an input fingerprint image is normalized so that it has a pre-specified mean and variance.

2. *Local orientation estimation:* the orientation image is estimated from the normalized input fingerprint image.

3. *Local frequency estimation:* the frequency image is computed from the normalized input fingerprint image and the estimated orientation image.

4. *Region mask estimation:* the region mask is obtained by classifying each block in the normalized input fingerprint image into a recoverable or a unrecoverable block.

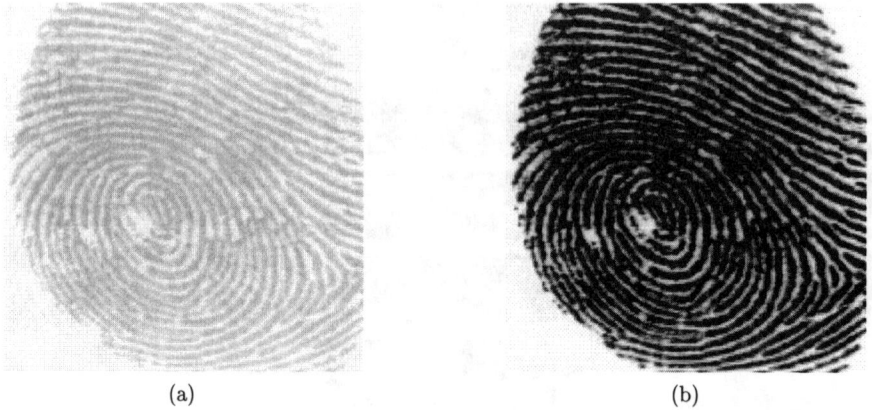

Figure 5. The result of normalization; (a) input image; (b) normalized image.

5. *Filtering*: A bank of *Gabor filters* which is tuned to local ridge orientation and ridge frequency is applied to the ridge-and-furrow pixels in the normalized input fingerprint image to obtain an enhanced fingerprint image.

2.3 Normalization

Let $\mathcal{I}(i,j)$ denote the gray-level value at pixel (i,j), M and VAR denote the estimated mean and variance of $\mathcal{I}$, respectively, and $\mathcal{G}(i,j)$ denote the normalized gray-level value at pixel (i,j). The normalized image is defined as follows:

$$\mathcal{G}(i,j) = \begin{cases} M_0 + \sqrt{\frac{VAR_0(\mathcal{I}(i,j)-M)^2}{VAR}}, & \text{if } \mathcal{I}(i,j) > M \\ M_0 - \sqrt{\frac{VAR_0(\mathcal{I}(i,j)-M)^2}{VAR}}, & \text{otherwise,} \end{cases} \tag{3}$$

$$\tag{4}$$

where M_0 and VAR_0 are the desired mean and variance values, respectively. Normalization is a pixel-wise operation. It does not change the clarity of the ridge and furrow structures. The main purpose of normalization is to reduce the variations in gray level values along ridges and furrows, which facilitates the subsequent processing steps. Figure 5 shows an example of image normalization.

2.4 Orientation Image

The orientation image represents an intrinsic property of the fingerprint images and defines invariant coordinates for ridges and furrows in a local neighborhood. By viewing a fingerprint image as an oriented texture, a number of methods have been proposed to estimate the orientation field of fingerprint images [11, 16, 10, 1]. We have developed a least mean square orientation estimation algorithm. Given a normalized image, $\mathcal{G}$, the main steps of the algorithm are as follows:

1. Divide $\mathcal{G}$ into blocks of size $w \times w$ (16×16).

2. Compute the gradients $\partial_x(i,j)$ and $\partial_y(i,j)$ at each pixel, (i,j). Depending on the computational requirement, the gradient operator may vary from the simple *Sobel* operator to the more complex *Marr-Hildreth* operator.

3. Estimate the local orientation of each block centered at pixel (i,j) using the following equations:

$$\mathcal{V}_x(i,j) \;=\; \sum_{u=i-\frac{w}{2}}^{i+\frac{w}{2}} \sum_{v=j-\frac{w}{2}}^{j+\frac{w}{2}} 2\partial_x(u,v)\partial_y(u,v), \tag{5}$$

$$\mathcal{V}_y(i,j) \;=\; \sum_{u=i-\frac{w}{2}}^{i+\frac{w}{2}} \sum_{v=j-\frac{w}{2}}^{j+\frac{w}{2}} (\partial_x^2(u,v) - \partial_y^2(u,v)), \tag{6}$$

$$\theta(i,j) \;=\; \frac{1}{2}tan^{-1}(\frac{\mathcal{V}_y(i,j)}{\mathcal{V}_x(i,j)}), \tag{7}$$

where $\theta(i,j)$ is the least square estimate of the local ridge orientation at the block centered at pixel (i,j). Mathematically, it represents the direction that is orthogonal to the dominant direction of the *Fourier spectrum* of the $w \times w$ window.

4. Due to the presence of noise, corrupted ridge and furrow structures, minutiae, *etc.* in the input image, the estimated local ridge orientation, $\theta(i,j)$, may not always be a correct estimate. Since local ridge orientation varies slowly in a local neighborhood where no singular points appear, a low-pass filter can be used to modify the incorrect local ridge orientation. In order to perform the low-pass filtering, the orientation image needs to be converted into a *continuous vector field*, which is defined as follows:

$$\Phi_x(i,j) \;=\; \cos(2\theta(i,j)), \;\; and \tag{8}$$
$$\Phi_y(i,j) \;=\; \sin(2\theta(i,j)), \tag{9}$$

where δ_x and δ_y, are the x and y components of the vector field, respectively. With the resulting vector field, the low-pass filtering can then be performed as follows:

$$\Phi'_x(i,j) \;=\; \sum_{u=-w_\Phi/2}^{w_\Phi/2} \sum_{v=-w_\Phi/2}^{w_\Phi/2} W(u,v)\Phi_x(i-uw, j-vw) \;\; and \tag{10}$$

$$\Phi'_y(i,j) \;=\; \sum_{u=-w_\Phi/2}^{w_\Phi/2} \sum_{v=-w_\Phi/2}^{w_\Phi/2} W(u,v)\Phi_y(i-uw, j-vw), \tag{11}$$

where W is a 2-dimensional low-pass filter with unit integral and $w_\Phi \times w_\Phi$ specifies the size of the filter. Note that the smoothing operation is performed at the block level. The default size of the filter is 5×5.

5. Compute the local ridge orientation at (i,j) using

$$\mathcal{O}(i,j) = \frac{1}{2}\tan(\frac{\Phi'_y(i,j)}{\Phi'_x(i,j)}). \tag{12}$$

With this algorithm, a fairly smooth orientation field estimate can be obtained. Figure 6 shows an example of the orientation image estimated with our algorithm.

(a) (b)

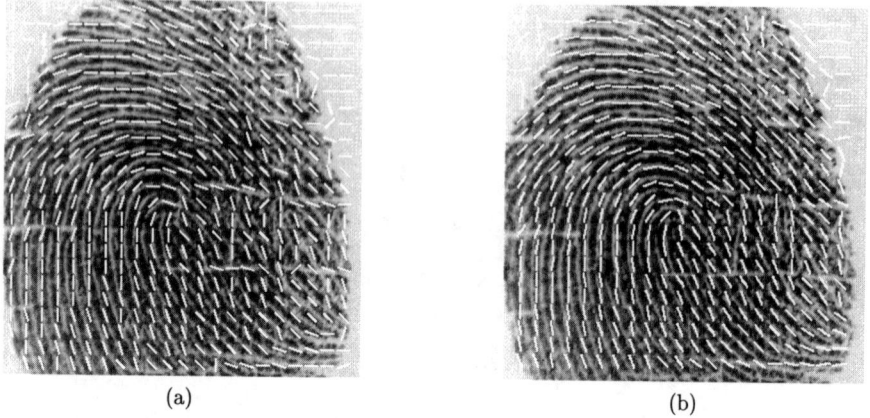

Figure 6. Comparison of orientation fields by the method proposed in [16] (a) and our method; $w = 16$ and $w_\Phi = 5$ (b).
ordinary paragraph.

2.5 Ridge Frequency Image

In a local neighborhood where no minutiae and singular points appear, the gray levels along ridges and furrows can be modeled as a sinusoidal-shaped wave along a direction normal to the local ridge orientation (see Figure 7). Therefore, local ridge frequency is another intrinsic property of a fingerprint image. Let $\mathcal{G}$ be the normalized image and $\mathcal{O}$ be the orientation image, then the steps involved in local ridge frequency estimation are as follows:

1. Divide $\mathcal{G}$ into blocks of size $w \times w$ (16×16).
2. For each block centered at pixel (i,j), compute an *oriented window* of size $l \times w$ (32×16) that is defined in the ridge coordinates system (Figure 7).
3. For each block centered at pixel (i,j), compute the *x-signature*, $X[0], X[1], ...X[l-1]$, of the ridges and furrows within the oriented window, where

$$X[k] \;=\; \frac{1}{w}\sum_{d=0}^{w-1}\mathcal{G}(u,v), \;\; k=0,1,...,l-1, \tag{13}$$

$$u \;=\; i + (d - \frac{w}{2})\cos\mathcal{O}(i,j) + (k - \frac{l}{2})\sin\mathcal{O}(i,j), \tag{14}$$

$$v \;=\; j + (d - \frac{w}{2})\sin\mathcal{O}(i,j) + (\frac{l}{2} - k)\cos\mathcal{O}(i,j). \tag{15}$$

If no minutiae and singular points appear in the oriented window, the x-signature forms a discrete sinusoidal-shape wave, which has the same frequency as that of the ridges and furrows in the oriented window. Therefore, the frequency of ridges and furrows can be estimated from the x-signature. Let $\mathcal{T}(i,j)$ be the average number of pixels between two consecutive peaks in the x-signature, then the frequency, $\Omega(i,j)$, is computed as: $\Omega(i,j) = 1/\mathcal{T}(i,j)$. If no consecutive peaks can be detected from the x-signature, then the frequency is assigned a value of -1 to differentiate it from the valid frequency values.

4. For a fingerprint image scanned at a fixed resolution, the value of the frequency of the ridges and furrows in a local neighborhood lies in a certain range. For a 500 *dpi* image, this range is $[1/3, 1/25]$. Therefore, if the estimated value of the frequency is out of this range, then the frequency is assigned a value of -1 to indicate that an valid frequency can not be obtained.

5. The blocks in which minutiae and/or singular points appear and/or ridges and furrows are corrupted do not form a well-defined sinusoidal-shaped wave. The frequency values for these blocks need to be interpolated from the frequency of the neighboring blocks which have a well-defined frequency. The interpolation is performed as follows:

(i) For each block centered at (i, j),

$$\Omega'(i, j) = \begin{cases} \Omega(i, j), & \text{if } \Omega(i, j) \neq -1 \\ \dfrac{\sum_{u=-w_\Omega/2}^{w_\Omega/2} \sum_{v=-w_\Omega/2}^{w_\Omega/2} W_g(u,v)\mu(\Omega(i-uw, j-vw))}{\sum_{u=-w_\Omega/2}^{w_\Omega/2} \sum_{v=-w_\Omega/2}^{w_\Omega/2} W_g(u,v)\delta(\Omega(i-uw, j-vw)+1)} & \text{otherwise,} \end{cases} \quad (16)$$

where

$$\mu(x) = \begin{cases} 0, & \text{if } x \leq 0 \\ x, & \text{otherwise,} \end{cases}$$

$$\delta(x) = \begin{cases} 0, & \text{if } x \leq 0 \\ 1, & \text{otherwise,} \end{cases}$$

W_g is a discrete *Gaussian kernel* where mean and variance is 0 and 9, respectively, and $w_\Omega = 7$ is the size of the kernel.

(ii) If there exists at least one block with the frequency value of -1, then swap Ω and Ω' and go to step (i).

6. Inter-ridges distances change slowly in a local neighborhood. A low-pass filter can be used to remove the outliers in f':

$$F(i, j) = \sum_{u=-w_\Omega/2}^{w_l/2} \sum_{v=-w_\Omega/2}^{w_l/2} W_l(u, v)\Omega'(i - uw, j - vw), \quad (17)$$

where W_l is a 2-dimensional low-pass filter with unit integral and $w_l = 7$ is the size of the filter.

2.6   Region Mask

As mentioned earlier, a pixel (or a block) in an input fingerprint image could be either in a recoverable region or an unrecoverable region. Classification of pixels into recoverable and unrecoverable categories can be performed based on the assessment of the shape of the wave formed by the local ridges and furrows. In our algorithm, three features are used to characterize the sinusoidal-shaped wave: amplitude (α), frequency (β), and variance (γ). Let $X[1], X[2], ..., X[l]$ be the x-signature of a block centered at (i, j). The three features corresponding to pixel (block) (i, j) are computed as follows:

1. α = (average height of the peaks - average depth of the valleys).

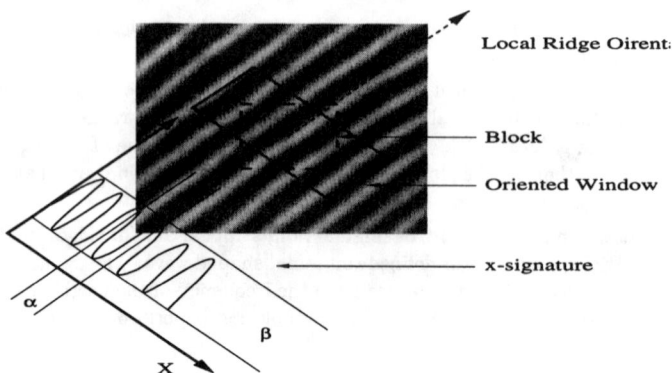

Figure 7. Oriented window and x-signature.

2. $\beta = 1/T(i,j)$, where $T(i,j)$ is the average number of pixels between two consecutive peaks.

3. $\gamma = \frac{1}{l}\sum_{i=1}^{l}(X[i] - (\frac{1}{l}\sum_{i=1}^{l}X[i]))^2$.

We selected several typical fingerprint images with both labeled recoverable and unrecoverable regions and computed these three features. A total of 2,000 3-dimensional patterns were obtained. In order to find representative patterns for the two classes, we fed the 2,000 patterns to a squared-error clustering algorithm and identified six clusters. Four of these clusters correspond to recoverable regions and the remaining two correspond to unrecoverable regions. The six prototypes (corresponding to cluster centers) were used in an one-nearest neighbor (1NN) classifier to classify each $w \times w$ block in an input fingerprint image into a recoverable or an unrecoverable block. If a block centered at (i,j) is recoverable, then $\mathcal{R}(i,j) = 1$, else $\mathcal{R}(i,j) = 0$. After the image $\mathcal{R}$ is obtained, the percentage of recoverable regions is computed. If the percentage of recoverable regions is smaller than a threshold, $\Gamma_{recoverable} = 40$, then the input fingerprint image is rejected. An accepted image is then passed through the filtering stage.

2.7 Filtering

The configurations of parallel ridges and furrows with well-defined frequency and orientation in a fingerprint image provide useful information which helps in removing undesired noise. The sinusoidal-shaped waves of ridges and furrows vary slowly in a local constant orientation. Therefore, a bandpass filter that is tuned to the corresponding frequency and orientation can efficiently remove the undesired noise and preserve the true ridge and furrow structures. Gabor filters have both frequency-selective and orientation-selective properties and have optimal joint resolution in both spatial and frequency domains [3, 7]. Therefore, it is appropriate to use Gabor filters as bandpass filters to remove the noise and preserve true ridge/valley structures.

The even-symmetric Gabor filter has the general form [7]

$$h(x,y:\phi,f) = exp\left\{-\frac{1}{2}\left[\frac{(x\cos\phi)^2}{\delta_x^2} + \frac{(y\sin\phi)^2}{\delta_y^2}\right]\right\}\cos(2\pi f x \cos\phi), \qquad (18)$$

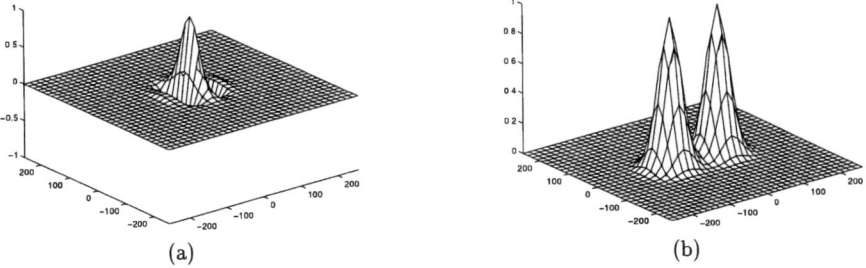

(a)   (b)

Figure 8. An even-symmetric Gabor filter: (a) the Gabor filter with F being 1/10 and $0°$ orientation; (b) the corresponding MTF.
ordinary paragraph.

where ϕ is the orientation of the Gabor filter, f is the frequency of a sinusoidal plane wave, and δ_x and δ_y are the space constants of the Gaussian envelope along x and y axes, respectively. The modulation transfer function (MTF) of the Gabor filter can be represented as

$$H(u,v:\phi,f) = 2\pi\delta_x\delta_y exp\left\{-\frac{1}{2}\left[\frac{[(u-2\pi/f)\sin\phi]^2}{\delta_u^2} + \frac{(v\cos\phi)^2}{\delta_v^2}\right]\right\} +$$
$$2\pi\delta_x\delta_y exp\left\{-\frac{1}{2}\left[\frac{[(u-2\pi/f)\sin\phi]^2}{\delta_u^2} + \frac{(v\cos\phi)^2}{\delta_v^2}\right]\right\} \tag{19}$$

where $\delta_u = 1/2\pi\delta_x$ and $\delta_v = 1/2\pi\delta_y$. Figure 8 shows an even-symmetric Gabor filter and its MTF.

To apply Gabor filters to an image, three parameters must be specified: (i) the frequency of the sinusoidal plane wave, u_0, (ii) the filter orientation, and (iii) the standard deviations of the Gaussian envelope, δ_x and δ_y. Obviously, the frequency characteristic of the filter, f, is completely determined by the local ridge frequency and the orientation is determined by the local ridge orientation. The selection of the values of δ_x and δ_y involves a trade-off. The larger the values, the more robust to noise the filters are but the more likely the filters will create spurious ridges and furrows. On the other hand, the smaller the values, the less likely the filters will create spurious ridges and furrows but then they will be less effective in removing the noise. The values of δ_x and δ_y were set to 4.0 and 4.0, respectively based on empirical data. Let $\mathcal{G}$ be the normalized fingerprint images, $\mathcal{O}$ be the orientation image, $\mathcal{F}$ be the frequency image, and $\mathcal{R}$ be the recoverable mask, the enhanced image $\mathcal{E}$ is obtained as follows:

$$\mathcal{E}(i,j) = \begin{cases} 255, & \text{if } \mathcal{R}(i,j) = 0, \\ \sum_{u=-w_g/2}^{w_g/2}\sum_{v=-w_g/2}^{w_g/2} h(u,v:\mathcal{O}(i,j),\mathcal{F}(i,j))\mathcal{G}(i-u,j-v), & otherwise, \end{cases} \tag{20}$$

where $w_g = 11$ specifies the size of the Gabor filters.

3: Experimental Results

The purpose of a fingerprint enhancement algorithm is to improve the clarity of ridges and furrows of input fingerprint images and make them more suitable for the minutiae extraction algorithm. The ultimate criterion for evaluating such an enhancement algorithm is the total amount of "quality" improvement when the algorithm is applied to the noisy input fingerprint images. Such an improvement can be assessed subjectively by a visual inspection of a number of typical enhancement results. However, a precise and consistent characterization of the quality improvement is beyond the capability of subjective evaluation. Examples of the enhancement results are shown in Figure 9. From these examples, we can see that our enhancement algorithm does improve the clarity of the ridge and furrow structures of input fingerprint images.

A goal-directed performance evaluation assesses the overall improvement in the system performance that incorporates the enhancement module as a component. Therefore, it is capable of providing a more reliable assessment of the performance benchmark and is directly associated with the ultimate goal of the system [20]. In the following, we present the results of the goal-directed performance evaluation of our enhancement algorithm.

3.1   Evaluation Using Goodness Index

We have used the *goodness index* (GI) of the extracted minutiae to quantitatively assess the performance of our fingerprint enhancement algorithm. Let $M_d = (f_d^1, f_d^2, ..., f_d^n)$ be the set of n minutiae detected by the minutiae extraction algorithm and $M_e = (f_e^1, f_e^2, ..., f_e^m)$ be the set of m minutiae identified by human expert in an input fingerprint image. We define the following terms:

- *Paired minutiae* (p): Minutiae f_d and f_e are said to be paired if f_d is located in a *tolerance box centered around* f_e. In this evaluation, the tolerance box size is 8×8.

- *Missing minutiae* (a): A minutiae that is not detected by the minutiae extraction algorithm.

- *Spurious minutiae* (b): A minutiae that is detected by the minutiae extraction algorithm, but which is not in the tolerance box of any minutiae, f_e.

The goodness index (GI) is defined as follows [17]:

$$GI = \frac{\sum_{i=1}^{r} q_i[p_i - a_i - b_i]}{\sum_{i=1}^{r} q_i t_i}, \tag{21}$$

where r is the number of 16×16 windows in the input fingerprint image, p_i represents the number of minutiae paired in the ith window, q_i represents the quality factor of the ith window (good=4, medium=2, poor=1), a_i represents the number of missing minutiae in the ith window, b_i represents the number of spurious minutiae in the ith window, and t_i represents the number of true minutiae in the ith window. GI penalizes both the missing minutiae and spurious minutiae. It is a reasonable measure of the quality of the extracted minutiae. The larger the value of GI, the better the minutiae extraction algorithm. The maximum value of GI equals 1, which means there are no missing and spurious minutiae.

Our fingerprint enhancement algorithm was tested on 50 typical poor fingerprint images obtained from IBM. First, we computed the goodness index of the extracted minutiae without applying the enhancement algorithm and then the goodness index of the extracted minutiae was computed with the enhancement algorithm applied to the input fingerprint images before the minutiae extraction was performed. Examples of minutiae extraction with/without enhancement are shown in Figure 10. Table 1 shows the GI values of 8 typical fingerprint images and the mean and standard deviation of GI values for all the 50 images. The GI values after applying the enhancement algorithm are always larger

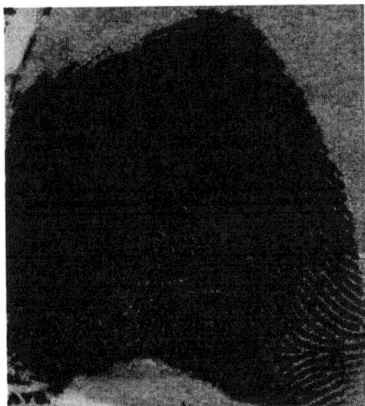

(a)

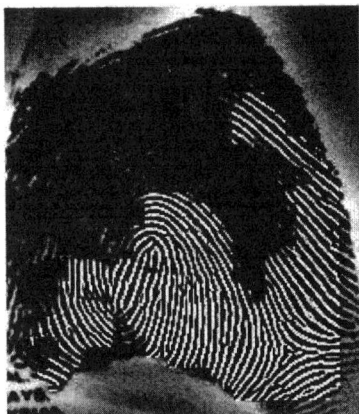

(b)

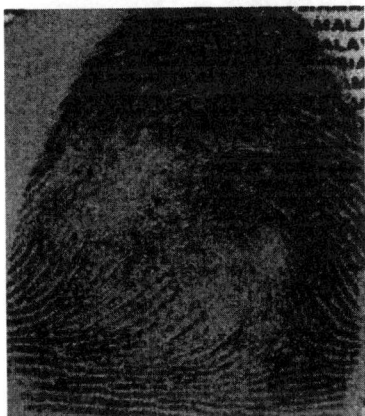

(c)

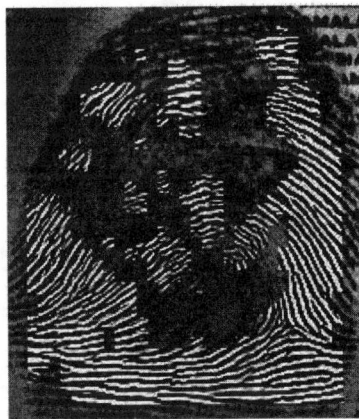

(d)

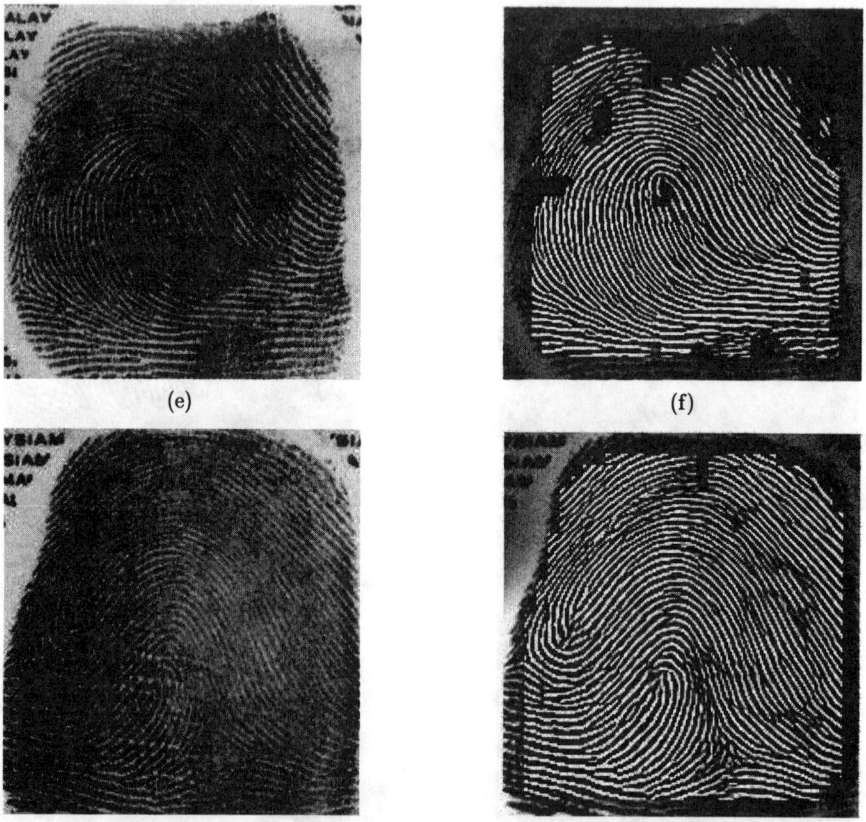

Figure 9. Examples of enhancement results; (a), (c), (e), and (g) are the input images; (b), (d), (f) and (h) show enhanced recoverable regions superimposed on the corresponding input images.
ordinary paragraph.

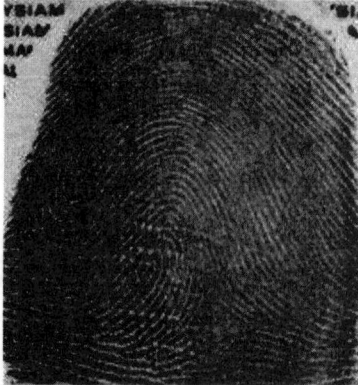

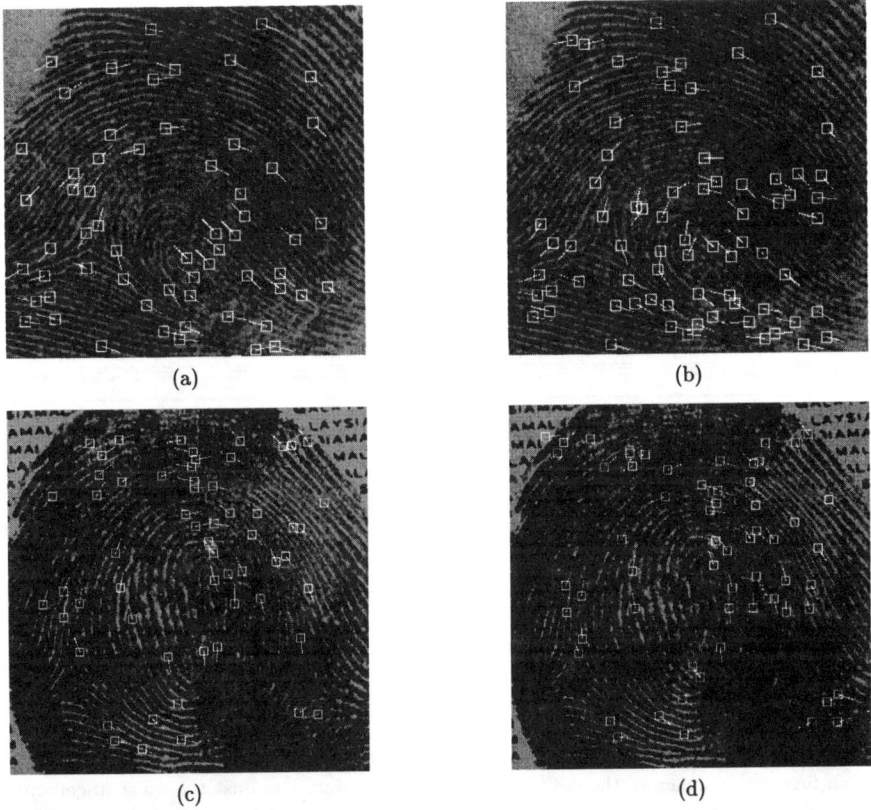

(a) (b)

(c) (d)

Figure 10. Examples of minutiae extraction with/without enhancement; (a) and (c) show the extracted minutiae without applying the enhancement algorithm; (b) and (d) show the extracted minutiae with the enhancement algorithm applied before the minutiae extraction.
ordinary paragraph.

Image #	Goodness Index (GI)	
	Without Enhancement	With Enhancement
1	0.46	0.55
2	0.38	0.52
3	0.29	0.42
4	0.26	0.39
5	0.21	0.35
6	0.12	0.31
7	0.11	0.26
8	0.10	0.29
mean	0.24	0.39
std	0.05	0.04

Table 1. The GI values of 8 typical fingerprint images and the mean and standard deviation of 50 IBM fingerprint images.
ordinary paragraph.

Normalization (seconds)	Orientation (seconds)	Frequency (seconds)	Region Mask (seconds)	Filtering (seconds)	Total (seconds)
0.11	0.14	0.09	0.07	2.08	2.49

Table 2. The wall time of the enhancement algorithm on a Pentium 200MHz PC

than that without the enhancement algorithm. Thus, we can conclude that our fingerprint enhancement algorithm does improve the quality of the fingerprint images, which, in turn, improves the accuracy and reliability of the extracted minutiae.

3.2 Evaluation Using Verification Performance

The performance of the enhancement algorithm was also assessed on the first volume of the MSU fingerprint database (700 live-scan images; 10 per individual) using the verification accuracy of an online fingerprint verification system [6]. We demonstrated that incorporating the enhancement algorithm in the fingerprint verification system improves the system performance. In the first test, the fingerprint enhancement algorithm was not applied. Each fingerprint image in the data set was directly matched against the other fingerprint images in the database. In the second test, the fingerprint enhancement algorithm was applied to each fingerprint image in the data set. Then, the verification was conducted on the enhanced fingerprint images. The receiver operating curves (ROC) resulting from these two tests are shown in Figure 11. From these experimental results, we can observe that the performance of the fingerprint verification system is significantly improved when our fingerprint enhancement algorithm is applied to the input fingerprint images. In particular, the enhancement algorithm substantially reduced the false reject rate while maintaining the same false acceptance rate.

In order to incorporate the enhancement algorithm into an online fingerprint verification/identification system, the whole enhancement process should take only a few seconds. Table 2 shows the wall time for different stages of the enhancement algorithm and the total time.

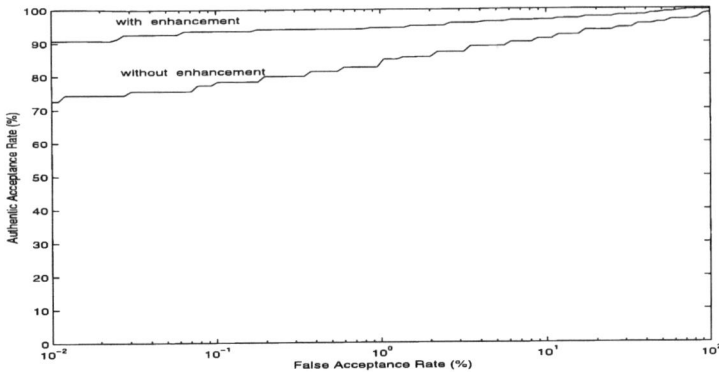

Figure 11. Receiver Operating Curves (ROC); the ROC shows the improvement in verification performance using the enhancement algorithm.
ordinary paragraph.

4: Summary and Conclusions

We have developed a fast fingerprint enhancement algorithm which can adaptively improve the clarity of ridge and furrow structures based on the local ridge orientation and ridge frequency estimated from the inputed images. The performance of the algorithm was evaluated using the goodness index of the extracted minutiae and the performance of an online fingerprint verification system which incorporates our fingerprint enhancement algorithm in its minutiae extraction module. Experimental results show that our enhancement algorithm is capable of improving both the goodness index and the verification performance. The algorithm also identifies the unrecoverable corrupted regions in the fingerprint and removes them from further processing. This is a very important property because such unrecoverable regions do appear in some of the corrupted fingerprint images and they are extremely harmful to minutiae extraction. These properties suggest that our enhancement algorithm should be integrated into an online fingerprint verification/identification system.

The global ridge and furrow configuration of fingerprint images presents a certain degree of regularity. A global model of the ridges and furrows that can be constructed from partial "valid" regions can be used to correct the errors in the estimated orientation images, which, in turn, will help the enhancement. Currently, we are investigating such a model-based enhancement algorithm.

The configurations of ridges and furrows within a local neighborhood vary with the quality of input fingerprint images, so a well defined sinusoidal-shaped waves of ridges and furrows may not always be observed. Global features are needed for a more precise region mask classification.

References

[1] T. Chang. Texture analysis of digitized fingerprints for singularity detection. In *Proc. 5th ICPR*, pages 478–480, 1980.

[2] P. E. Danielsson and Q. Z. Ye. Rotation-invariant operators applied to enhancement of fingerprints. In *Proc. 9th ICPR*, pages 329–333, Rome, 1988.

[3] J. G. Daugman. Uncertainty relation for resolution in space, spatial-frequency, and orientation optimized by two-dimensional visual cortical filters. *J. Opt. Soc. Am.*, 2:1160–1169, 1985.

[4] L. Hong, A. K. Jain, S. Pankanti, and R. Bolle. Fingerprint enhancement. In *Proc. 1st IEEE WACV*, pages 202–207, Sarasota, FL, 1996.

[5] D. C. Huang. Enhancement and feature purification of fingerprint images. *Pattern Recognition*, 26(11):1661–1671, 1993.

[6] A. Jain, L. Hong, and R. Bolle. On-line fingerprint verification. *IEEE Trans. Pattern Anal. and Machine Intell.*, 19(4):302–314, 1997.

[7] A. K. Jain and F. Farrokhnia. Unsupervised texture segmentation using Gabor filters. *Pattern Recognition*, 24(12):1167–1186, 1991.

[8] T. Kamei and M. Mizoguchi. Image filter design for fingerprint enhancement. In *Proc. ISCV' 95*, pages 109–114, Coral Gables, FL, 1995.

[9] K. Karu and A. K. Jain. Fingerprint classification. *Pattern Recognition*, 29(3):389–404, 1996.

[10] M. Kass and A. Witkin. Analyzing oriented patterns. *Comput. Vision Graphics Image Process.*, 37(4):362–385, 1987.

[11] M. Kawagoe and A. Tojo. Fingerprint pattern classification. *Pattern Recognition*, 17(3):295–303, 1984.

[12] H. C. Lee and R. E. Gaensslen. *Advances in Fingerprint Technology*. Elsevier, New York, 1991.

[13] A. Moenssens. *Fingerprint Techniques*. Chilton Book Company, London, 1971.

[14] E. Newham. *The Biometric Report*. SJB Services, New York, 1995.

[15] L. O'Gorman and J. V. Nickerson. An approach to fingerprint filter design. *Pattern Recognition*, 22(1):29–38, 1989.

[16] A. R. Rao. *A Taxonomy for Texture Description and Identification*. Springer-Verlag, New York, 1990.

[17] N. Ratha, S. Chen, and A. K. Jain. Adaptive flow orientation based feature extraction in fingerprint images. *Pattern Recognition*, 28(11):1657–1672, 1995.

[18] D. Sherlock, D. M. Monro, and K. Millard. Fingerprint enhancement by directional fourier filtering. *IEE Proc. Vis. Image Signal Processing*, 141(2):87–94, 1994.

[19] A. Sherstinsky and R. W. Picard. Restoration and enhancement of fingerprint images using m-lattice: A novel non-linear dynamical system. In *Proc. 12th ICPR-B*, pages 195–200, 1994.

[20] O. Trier and A. Jain. Goal-directed evaluation of binarization methods. *IEEE Trans. Pattern Anal. and Machine Intell.*, 17(12):1191–1201, 1996.

Empirical Evaluation of Laser Radar Recognition Algorithms Using Synthetic and Real Data

Sandor Der

Qinfen Zheng*

U.S. Army Research Lab
2800 Powder Mill Road
Adelphi, MD 20783-1197

Center for Automation Research
University of Maryland
College Park, MD 20742-3285

Abstract

Automatic target recognition (ATR) systems require a large data set to tune free parameters of the algorithm as well as for evaluating the algorithm. Synthetic data is frequently used when the quantity of real data is insufficient, or covers limited pose or articulation variations. This paper discusses a verification and validation methodology for synthetic data for the problem of automatic recognition of targets in laser radar (LADAR) imagery, and applies this methodology to a laser simulator. The simulator was used to generate data for training and evaluation of a LADAR target recognition algorithm. Performance results of the recognizer are presented for synthetic and real data. The range error in the synthetic range imagery corresponds closely with real range imagery, and the recognizer performs well on real imagery after being trained entirely on synthetic imagery, suggesting that the simulator is useful for the development and evaluation of recognition algorithms.

1: Introduction

Computer vision algorithm development efforts frequently suffer from a lack of adequate data for training and testing of algorithms. This is particularly true for algorithms that have a large number of free parameters, which require large data sets to ensure that the parameters are not over-fitted to the data. Data acquisition is often extremely expensive, forcing algorithm designers to develop algorithms that are too simple for the problem or to over-fit the data, or to use synthetic imagery.

Many computer vision researchers have used synthetic data to assist in the design and evaluation of their algorithms. Computer generated synthetic data is becoming more common as the processing capability of computers increases. Jones and Bhanu [1] have used computer generated synthetic aperture radar (SAR) imagery from the XPATCH [2] radar signature prediction code to identify features that are invariant to target articulation. These features are then used to identify targets in various articulations. Miller et al [3] used computer generated multisensor data to evaluate an algorithm for simultaneous detection, tracking, and recognition of objects.

---

*This work has been supported through collaborative participation in the Advanced Sensors Consortium sponsored by the U.S. Army Research Laboratory under the Federated Laboratory Program, Cooperative Agreement DAAL01-96-2-0001. The authors would like to thank Professor Rama Chellappa of the University of Maryland for many fruitful discussions.

Synthetic imagery generated from model boards has also been common. A number of authors [4, 5, 6] have used the model board derived TRIM2 [7] synthetic FLIR database. These algorithms are based primarily on the geometrical shape of targets, and therefore are not highly sensitive to the inadequacies of synthetic data. Liu, Kanungo, and Haralick [8] use both RADIUS [9] model board imagery and computer generated synthetic imagery to evaluate algorithms which simultaneously locate buildings and estimate their size and shape parameters. They also address the issue of validation of computer recognition software using Kolmogorov-Smirnov comparison of empirical and theoretical probability distributions.

Verification is used to ensure that an algorithm code performs as the designer intended. It does not serve to test the assumptions upon which the algorithm was designed, hence real data is not required to verify an algorithm. Validation is used to determine if a simulation accurately reflects the real world, and hence requires real data. Validation can serve to both verify that a simulation accurately encodes the assumptions of the designer, and that the design assumptions are reasonable. A separate verification step serves to eliminate the possibility of error due to mistakes in coding, so that remaining error between simulated and real data can safely be assumed to be due to the inadequacies of the assumptions used in the simulation. This maximizes the use of the limited real data.

This paper discusses verification and validation of a laser radar simulator. The simulator was used to help design a laser radar to be used with an automatic target recognizer (ATR) algorithm[10]. The simulator provided performance predictions of each potential laser radar design. The simulator was also used to create a set of synthetic images for the development and evaluation of a recognition algorithm. The performance of the algorithm on synthetic and real images is reported.

The rest of this paper is organized as follows. Section 2 briefly describes the laser simulation algorithm. Sections 3 and 4 describe the verification and validation of the simulation. Section 5 describes how the synthetic data was used to train and test a laser radar recognition algorithm. Section 6 discusses how the simulation could be expanded to include realistic portrayals of backgrounds.

2: Laser Simulation

A direct detection laser radar sends out a narrow pulse for each pixel, receives the returning signal energy, and measures the reflectivity by detecting the pulse in the return signal and integrating the energy in the pulse. The range to the pixel is calculated from the time elapsed between the transmitted and received pulse. The outputs of the laser radar is a reflectivity image and a range image. For most applications, the range image is more useful and less noisy than the reflectivity image.

The computer simulation described herein produces simulated range images for a direct detection laser radar. The simulation incorporates established theory at each stage. The simulation process is divided into six parts: component disassembly, speckle modulation, scintillation modulation, signal assembly, receiver noise, and pulse detection. The simulation is described in greater detail in [11].

In order to simulate the effects of target shape, speckle, and scintillation, we decomposed the laser pulse into components along the range and the two cross range dimensions. Thus the pulse can be expressed as $U(x, y, z)$ where x and y are the horizontal and vertical cross range dimensions, and z is the range dimension (the direction that the pulse is traveling). The beam's irradiance profile is assumed to be Gaussian, and the pulse shape in the z dimension is an input to the model and can be written as $V(z)$, or alternatively, as $V(ct)$ where c is the speed of light. So the outgoing pulse is decomposed as

$$U(x,y,z) = P_sG(x,y)V(z) \tag{1}$$

where P_s is the total pulse power, x, y, and z take on discrete values, $G(x,y)$ is the proportion of energy contained within a component located at (x,y) under the two dimensional Gaussian curve, and $V(z)$ is the discrete pulse shape in the range dimension. The distance that each component travels is determined by a geometric model, which is an input to the simulation.

The standard radar equation governs the return power received for each component sent[12]

$$P_R = P_c e^{-2\alpha R} \frac{d^2}{4R^2} \epsilon \rho, \tag{2}$$

where P_R is the received component power, P_c is the transmitted component power, α is the atmospheric extinction coefficient, R is the range to the target, d is the aperture diameter, ϵ is the receiver's optical efficiency, and ρ is the reflectivity of the target.

The power received from each component is modulated by speckle and atmospheric scintillation. Speckle is applied using the exponential distribution to modulate the received power. Similarly, scintillation is applied by modulating the power of each component with a lognormal random variable. Speckle distortion varies with optics size and the wavelength of the laser, and scintillation varies with optics size and the long term turbulence correlation structure constant.

The return signal is subject to background, shot, amplifier, and dark current noise, all of which are independent, identically distributed Gaussian. The detector noise, consisting of dark, shot, and background noise, is calculated as in [13]. The noise variables are added in quadrature to determine the overall noise figure for the detectors. The simulation generates a Gaussian random variable which is added to the signal, and then is filtered according to the electrical bandwidth of the sensor. The noisy signal is then passed to the pulse detector, which performs matched filtering to detect the return pulse. Quantization error of the system clock is simulated by shifting the transmitted and received pulse by a random variable that is uniformly distributed across $\pm$ one half of a clock cycle.

The simulation requires as input a range image of the object. This is usually obtained by raytracing BRL-CAD models of the targets. The background of the object is assumed to be a plane perpendicular to the line of sight of the laser radar and a few meters behind the object. Because the simulation was to be used to develop a recognition algorithm, the background was not important for training and testing. Clearly, a much more detailed background model would be necessary to train and test algorithms to detect targets in natural background scenes. Figures 1 and 2 show a CAD model range image and a corresponding synthetic range image.

3: Verification

The purpose of verification is to ensure that the simulation code performs as the designer intended. Verification does not attempt to test the assumptions of the design. Hence verification does not require real data. It does require that the simulation be run with a wide range of inputs, and that the outputs be compared to ensure that the input-output relationships perform as expected.

It was especially necessary to do this in our case because part of the purpose of the simulation was to aid in the design of a laser radar. Parameter studies were performed by running

Figure 1. Range image from CAD model

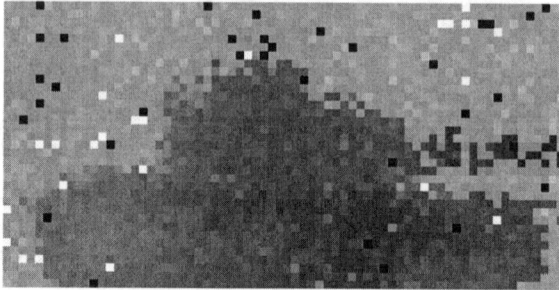

Figure 2. Synthetic image derived from CAD model

the simulation on a number of proposed preliminary designs. The synthetic images were input to an ATR to estimate recognition performance, to aid analysis of performance/cost tradeoffs. Figure 3 shows the procedure. Figure 4 gives an example of probability of recognition versus range curves for a fictitious laser radar. The only difference in the parameters is that the first figure represents a laser radar with a pulse energy of 5 millijoules, and the second a pulse energy of 8 millijoules. It can be seen that performance is improved with the increased pulse energy, as expected. The curves give recognition probability for a two class problem; thus probability of .5 corresponds to a completely random decision. Each point on the curve was generated by producing 216 synthetic images of each of the two targets, corresponding to 3 images of the target at each of 72 aspect angles (sampled every 5 degrees), and using the 432 images as input to an ATR which compares each synthetic range image to noise free range images obtained directly from CAD models. The probability of recognition was calculated as the number of times the best match corresponded to the correct target, but not necessarily the correct aspect angle. The method should be interpreted as a predictor of which laser radar will produce superior ATR performance rather than an accurate estimator of ATR performance. Accurate estimation of performance would require a more complex background which would allow the ATR to take target silhouette discontinuities into account, and would require more targets. As the method is implemented, estimated performance is highly dependent on the choice of two targets to be discriminated.

There were far too many simulation parameters to densely sample each parameter and run the simulation for every combination of parameter values. Instead, the parameter-

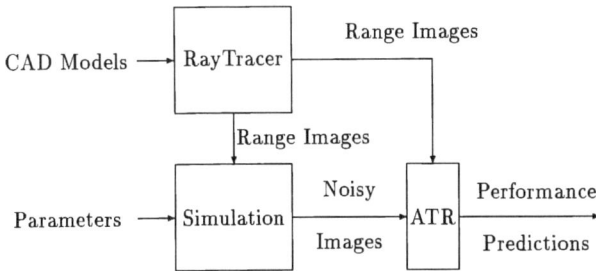

Figure 3. ATR performance predictions

performance relationships were verified systematically by varying each parameter in turn from a few canonical designs. Table 1 shows a partial list of parameters and the parameter values for a fictitious laser radar. The range of values a parameter could possibly take was determined by the designer, and a number of values in that range used as inputs to the simulation. The resulting range error versus parameter curve would be compared to expected *qualitative* performance. For example, the dark noise current density can take on values from zero to infinity. For the LADAR of table 1, the primary noise source is amplifier noise, so the expected behavior as the dark noise is increased is that the range error would be constant for small values of dark noise, increase steadily once the dark noise values were of the same order of magnitude as the amplifier noise, and then become unbounded when the dark noise drowns the signal. Figure 5 is a flow diagram of the procedure.

Knowledge of the physics behind the simulation suggests other parameter tests for verification. For example, increasing the optics size serves to increase the signal to noise ratio and also to reduce speckle effects by integrating over the speckle random field. These two effects can be isolated by simultaneously increasing the optics diameter and reducing the laser pulse energy so that signal to noise ratio is kept constant. Thus the dependence of speckle distortion on the optics size can be isolated. The expected result is that the range error will decrease sharply when the optics is approximately the same size as the speckle correlation length, but then become constant once the aperture is much larger than the speckle correlation length. Figure 6 shows the resulting simulated performance. The correlation length of the speckle in this case is .015 meters, which corresponds to the knee of the curve, as expected.

4: Validation

As a check on the quality of the simulation, we gathered a small amount of real data. We painted a 4x8 foot sheet of plywood carc green and positioned it perpendicular to the line of sight to the sensor. We captured laser radar images of the plywood, and extracted an empirical probability density function from the images. Figure 8 shows simulated range error as a function of range. The two x's mark the range's standard deviation for the real laser radar. Additional simulation runs showed that the LADAR is limited by quantization error of the clock at short ranges, and by signal to noise ratio at longer ranges. Many laser radar systems, including the Lockheed Martin Vought system used for this test, avoid the sharp increase in error at longer ranges by returning no range value for return pulses that do not exceed several times the ambient rms noise level. Such a pixel is called a dropout.

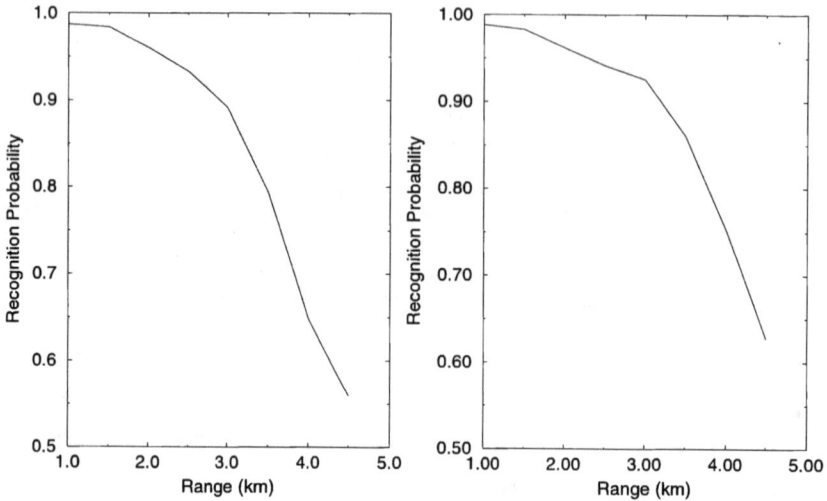

Figure 4. Estimated recognition performance with 5 and 8 millijoules.

The percentage of dropout pixels would then increase as range increases, while the range error on non-dropout pixels would increase much more slowly. Figure 7 is a laser radar range image of two plywood boards in the bottom right corner and an M1 tank in the lower left. The range image is in ambiguous greytone so that range differences are visible but brighter pixels do not necessarily correspond to greater ranges. The dynamic range of laser radar range images is too great for human interpretation using standard greytone.

The plywood target data that we collected has characteristics that limit its use for validation purposes. The Lockheed Martin Vought laser radar has a relatively large aperture, so aperture averaging makes the effects of speckle and scintillation almost negligible. The detector noises are dominated by amplifier noise, so background and shot noise effects are not adequately tested. While the comparison of real and simulated data was useful in validating the simulation, it is far from thorough. A thorough empirical validation would

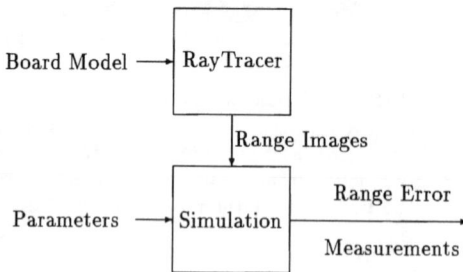

Figure 5. Range error predictions

Table 1. Partial list of simulation parameters

wavelength	1.5 microns
beam divergence	125 microradians
pulse width	15 nanoseconds
pulse energy	250 microjoules
optics diameter	.19 meters
pointing σ	.000042 radians
optics quality factor	1.1
receiver field of view	.00019 radians
dark noise current	5.e-8 amps
electrical bandwidth	40 Mhz
optical bandwidth	12.e-9 meters
receiver transmission	.75
noise factor	4.
load resistor	500 ohms
range to object	2.0 kilometers
atmospheric extinction	.0002731 km^{-1}
target reflectivity	.38
turbulence correlation constant	10e-13

require laser systems having at least several different values of each sensor parameter, to ensure that each sensor parameter is accurately modeled. Of course, this would be extremely expensive.

The plywood reflection boards only served to validate the simulation for the simplest possible geometry, in which the entire beam falls on a planar object perpendicular to the line of sight of the sensor. It would be quite useful to validate the simulation on more complex geometries, which would test the simulations ability to model the distortion of the pulse due to the different path lengths of the subpulses.

This experiment would require accurate measurement of the location of the beam footprint on the object, and a detailed geometric description of the object. Suppose that the object is two planar boards, perpendicular to the line of the sight of the sensor, offset so that one board is a half a meter closer than the other. The laser radar is positioned so that a portion of the beam falls on each of the boards, and the empirical pdf of the range returns is obtained by repeatedly measuring the range to the object. In order to calculate the estimated pdf using the simulation, the portion of the beam that falls on each board would need to be measured. One might be tempted to survey the location of the boards and of the laser radar, and then use a gimbal to measure the pointing angle of the laser radar. This is not practical because it would require a gimbal or theodylite with accuracy an order of magnitude smaller than the beamwidth of the laser. It would also require that the laser in the laser radar be aligned to similar accuracy, which is not generally a good assumption.

A more realistic option would be to use an infrared imager to view the boards and see the area of laser illumination. The infrared imager would not need to be colocated with the laser radar. It could be placed close to the boards to get a better image. Background noise could be reduced by placing a notch filter in front of the optics that matches the wavelength of the laser. Background noise could be further reduced by using an imager with short sample times. Unfortunately we were unable to perform these experiments due

Range Standard Deviation vs. Aperture Diameter

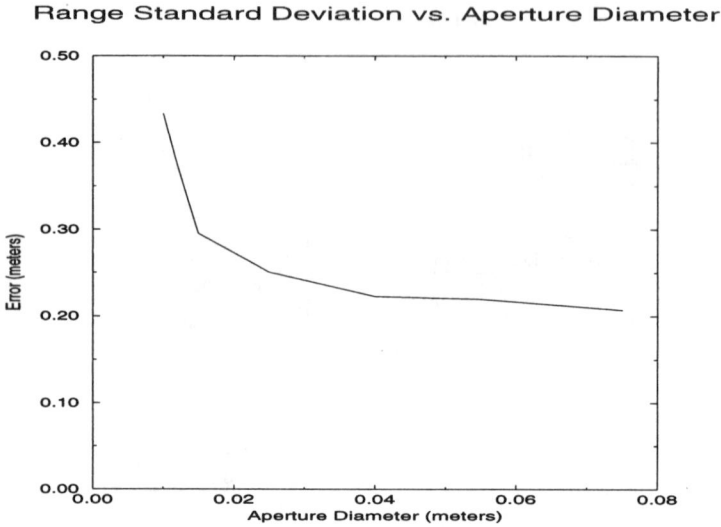

Figure 6. Range error versus aperture for fixed signal to noise ratio

to time and equipment constraints.

5: Training and Testing of a Laser Radar Recognition Algorithm

The laser radar ATR matches an input image to noise free range images obtained from CAD models. The matching is performed at two levels. A prescreener extracts one dimensional projections from the image in the horizontal, vertical, and range dimensions. These are compared to a set of stored one dimensional projections obtained from CAD models. Projections are obtained for each of seven targets and for every five degrees of azimuth, resulting in 72 poses per target and 504 total poses. If the correspondence between the image projection and stored projections is close, silhouette and surface matching are performed for that pose.

Figure 7. Ambiguous greytone range image of plywood boards.

Range Standard Deviation vs. Range

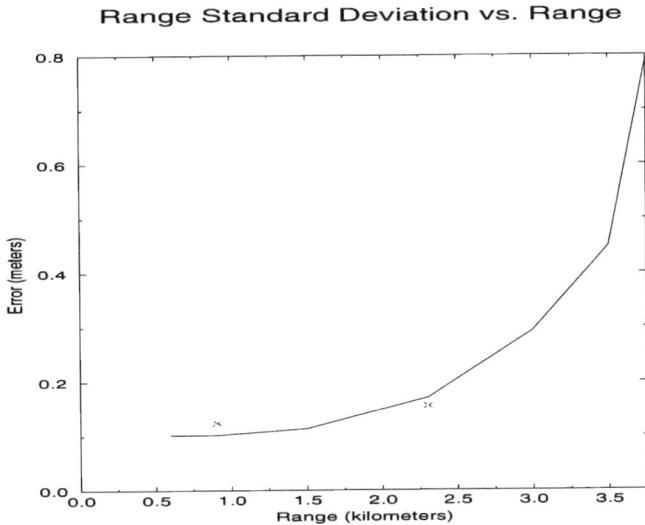

Figure 8. Estimated and measured range variance

Surface matching counts the number of pixels in the input image which can be matched to a noise free range image obtained from a CAD model for a particular pose, and then divides by the total number of pixels on the target. A pixel is matched if the absolute value of the range to the pixel minus the range to the pixel in the noise free range image is below a threshold. The ratio of the number of matched pixels (M) to the number of pixels on target (N) is a natural choice for LADAR imagery because range error density functions have large tails. A commonly used metric like mean square error would be dominated by outliers.

Silhouette matching is performed by projecting a silhouette extracted from a noise free range image onto the LADAR image. A silhouette pixel is matched if there is a range discontinuity between the two sides of the silhouette. The silhouette match score is the ratio between the number of matched pixels and the number of total silhouette pixels.

The surface and silhouette match scores are combined using a simple weighted average, with greater weight given to the surface match. The ATR is described in detail in [10]. The algorithm is primarily model based, so it has fewer free parameters than a learning algorithm such as a neural network, which must rely on a large set of imagery to train the free parameters. The few free parameters include the thresholds which determine whether a range discontinuity exists, how close a range value must be to the template value, and the weighting between the surface and silhouette match scores. This facilitates the use of synthetic data since there is less danger of overtraining a large number of free parameters to learn some anomaly of the synthetic data.

The ATR algorithm was written for a low resolution LADAR that would provide 200 pixel images covering a target size area of 4x8 meters with 10x20 pixels. The LADAR would automatically adjust the angular spacing of the pixels based on the range to the target so that the same physical area was covered regardless of range. The application for the LADAR was to serve as a recognition aid only, with another sensor acting as a cuer to

Table 2. Target Identifications

T1	T2	T3	T4	T5	T6	T7
m1	m113	btr60	bmp2	m60	zil131	m35

find potential targets.

A set of synthetic images was created using the simulation to set the free parameters. Figure 9 shows one example. One image was created at every five degrees of azimuth for each of seven targets, resulting in a set of 504 synthetic images. The parameters were set to give the best possible performance on the image set. Performance was defined as the probability that the closest matching template corresponded to the same target, but not necessarily the correct pose from the correct target. Tables 2 and 3 show the performance of the algorithm on the synthetic data.

Figure 9. Rendering, range image, and synthetic image of BTR60

The algorithm was tested on a set of real imagery. The imagery was from a high resolution LADAR, but was downsampled to 10x20 pixels to simulate the low resolution LADAR. Table 4 shows the performance on a set of 276 real images. Only two of the seven targets for which the ATR was trained were available. The performance is lower than for the synthetic imagery, as one might expect. Possible reasons for the decrease in performance include: 1) inadequacies of the simulation, 2) presence of partial occlusion of the targets in the real imagery, especially near the bottom of targets parked in tall grass, 3) discrepancy in pose (both azimuth and elevation angle) between the real images and the templates. The

Table 3. Confusion Matrix of Classification on Synthesized 10x20 LADAR Imagery

	T1	T2	T3	T4	T5	T6	T7	accuracy
T1	72	0	0	0	0	0	0	100%
T2	0	72	0	0	0	0	0	100%
T3	0	0	72	0	0	0	0	100%
T4	0	0	0	72	0	0	0	100%
T5	0	0	0	0	72	0	0	100%
T6	0	0	0	0	0	71	1	98.6%
T7	0	0	0	0	0	2	70	97.2%
total								99.4%

Table 4. Confusion Matrix of Classification on Real 10x20 LADAR imagery

	T1	T2	T3	T4	T5	T6	T7	accuracy
T1	110	4	0	2	6	1	0	89.4%
T2	6	142	1	1	0	0	3	92.8%
total								91.3%

third source of performance discrepancy could easily be removed by testing the algorithm on synthetic data with slightly different pose than the templates. There has not been adequate time to perform this experiment.

6: Realistic Portrayals of Background

The simulation as currently configured is not suitable for designing and evaluating detection algorithms because of its simplistic handling of the background. Realistic backgrounds are challenging to laser radar detection algorithms because of the presence of target competitive shapes in the form of natural objects such as foliage and rocks. Treelines in particular will frequently contain challenging shapes. Creating a realistic synthetic background requires accurate geometric modeling of terrain and foliage, and a knowledge of the reflectance properties of natural objects.

A great deal of effort has gone into the creation of synthetic imagery in natural backgrounds, e.g. [14, 15]. Much of the work relates to synthetic infrared imagery and consists of geometric models of the terrain and natural objects, and associated thermal models applied to the geometric models. The laser radar simulation requires only a highly detailed geometric model, so the infrared work is readily applicable to laser radar. In particular, highly detailed geometric tree models are available in [16, 17]. Detailed synthetic terrain geometry models are available in [18]. Random tree placement models that locate trees in biologically plausible configurations exist in [14, 15, 19]. CAD Target models are already being used in our simulation. Combining these models requires mostly software integration work: converting the geometric models into one CAD format, applying the tree placement models and using the results to insert trees into the CAD model of the terrain, ray tracing the CAD model, and then applying the simulation to the resulting range image.

The complex geometry of the background model would make it much more difficult to validate using the method, described earlier, using known geometries. Validation of the background imagery could be accomplished in a number of alternative ways. The false alarm rate of a detection algorithm on synthetic and real images could be compared. This would preferably be done with synthetic imagery that was designed specifically to match the geographic region where the real imagery was collected, since tree density and type and terrain roughness will clearly influence the false alarm rate of a detection algorithm. An alternative validation scheme might be to match a texture model to treelines, foliage canopies, and grass fields of both synthetic and real imagery and compare the associated texture parameters.

7: Conclusions and Future Work

A methodology used to verify and validate a simulation of a laser sensor was reported. The simulation was used to generate imagery to train and test a computer recognition algorithm, which was also tested on real data. The success of the recognition algorithm and the close correspondence between real and synthetic data on known geometries suggest that the simulation can produce imagery that is useful for the development and evaluation of computer vision algorithms.

The utility of the simulation would be enhanced if it could be made to produce realistic renderings of natural backgrounds. Accurate backgrounds would allow the synthetic imagery to assist in the development of detection algorithms, and would make evaluation of recognition algorithms more accurate by modeling target/background interactions such as partial occlusion.

References

[1] G. Jones and B. Bhanu, "Invariants for the Recognition of Articulated and Occluded Objects in SAR Images", Proceedings of the Image Understanding Workshop, pp. 1135-1144, January, 1996.

[2] D. Andersch, S. Lee, H. Ling, and C. Yu, "XPATCH: A High Frequency Electromagnetic Scattering Prediction Code using Shooting and Bouncing Rays", Proceedings of Ground Target Modeling and Validation Conference, pp. 498-507, August, 1994.

[3] M. Miller, U. Grenander, J. O'Sullivan, and D. Snyder, "Automatic Target Recognition Organized via Jump-Diffusion Algorithms", IEEE Transactions on Image Processing, Vol. 6, No. 1, pp. 157-174, January, 1997.

[4] J. Antoine, K. Bouyoucef, P. Vandergheyst, and R. Murenzi, "Target Detection and Recognition Using Two Dimensional Continuous Isotropic and Anisotropic Wavelets", Automatic Object Recognition V, SPIE-1995, Vol. 2485, pp. 20-31, April, 1995.

[5] S. Der and R. Chellappa, "Probe-Based Automatic Target Recognition in Infrared Imagery", IEEE Transactions on Image Processing, Vol. 6, No. 1, pp. 92-102, January, 1997.

[6] A. Kramer, D. Perschbacher, R. Johnston, and T. Kipp, "Relational Template Matching Algorithm for FLIR Automatic Target Recognition", Proceedings SPIE Automatic Target Recognition, April, 1993.

[7] "Training Collection at the Center for Night Vision and Electro-Optics", Proceedings of SPIE Signal and Image Processing Systems Performance Evaluation, Vol. 1483, April, 1991.

[8] X. Liu, T. Kanungo, and R. Haralick, "Statistical Validation of Computer Vision Software", Proceedings of the Image Understanding Workshop, pp. 1533-1540, January, 1996.

[9] V. Ramesh, K. Thornton, D. Nadadur, A. Bedekar, X. Liu, W. Hudson, X. Zhang, and R. Haralick, "Groundtruthing the RADIUS Model-Board Imagery", in *RADIUS: Image Understanding for Imagery Intelligence*, Morgan Kaufmann Publishers, San Francisco, pp. 469-480, 1997.

[10] Q. Zheng and S. Der, "Model-Based Target Recognition in LADAR Imagery", IEEE CVPR, June, 1998.

[11] S. Der, B. Redman, and R. Chellappa, "Simulation of Error in Optical Radar Range Measurements", Applied Optics - Lasers, Photonics, and Environmental Optics, Vol. 36, No. 27, pp. 6869-6874, September, 1997.

[12] J. Shapiro, B. Capron, and R. Harney, "Imaging and Target Detection with a Heterodyne-Reception Optical Radar", Applied Optics, Vol. 20, pp. 3292-3313, October, 1981.

[13] P. Webb, R. McIntyre, and J. Conradi, "Properties of Avalanche Photodiodes", RCA Review, Vol. 35, pp. 234-278, June 1974.

[14] S. Der, G. Dome, J. Horger, M. Lorenzo, and R. Moulton, "A Multi-Sensor Digital Terrain Board for Processor Test and Evaluation", IEEE MILCOM, San Diego, October, 1992.

[15] J. Penn, H. Nguyen, T. Kipp, C. Kohler, G. Huynh, and M. Sola, "The CREATION Scene Modeling Package Applied to Multispectral Missile Seekers and Sensors", Proceedings of Multispectral Missile Seekers Conference, Huntsville, Alabama, November, 1995.

[16] M. Aono and T. Kunii, "Botanical Tree Image Generation, IEEE Computer Graphics and Applications, Vol. 4, No. 5, pp. 10-34, May, 1984.

[17] J. Weber and J. Penn, "Creation and Rendering of Realistic Trees", Proceedings of SIGGRAPH, Los Angeles, California, August, 1995.

[18] H. Peitgen and D. Saupe, editors, *The Science of Fractal Images*, Springer-Verlag, New York, 1988.

[19] L. Balick, "A Forest Canopy Height Surface Model for Scene Simulation", Simulation, Vol. 49, pp. 5-12, 1987.

Disclaimer

The views and conclusions contained in this document are those of the authors and should not be interpreted as representing the official policies, either expressed or implied, of the Army Research Laboratory or the U.S. Government.

A WWW-Accessible Database for 3D Vision Research*

Patrick J. Flynn — Richard J. Campbell
School of Electrical Engineering and Computer Science
Washington State University
Pullman, WA 99164 USA
{flynn,rcampbel}@eecs.wsu.edu
http://www.eecs.wsu.edu/~flynn/3DDB/

Abstract

Reproducibility and thorough experimental evaluation are essential elements of the dissemination of systems, including the systems developed for object recognition by computer vision researchers. Systems tested on widely used image databases are more directly comparable than systems which are tested on customized data. This paper describes the components of an evolving database of 3D object models, 3D (range) images, and code to support research in and evaluation of model-based object recognition and range image analysis systems. Images and models are available in a variety of geometric formats (*e.g.*, VRML, GIF as well as documented neutral formats) and the entire database can be browsed and retrieved through the World Wide Web.

1 Introduction

Model-based three-dimensional object recognition continues to be a popular and productive area of research in the computer vision community. The different assumptions inherent in designing a complete system have spawned many original systems and the subsequent development of variants and improvements [3, 2]. Object recognizers employing dense range maps have been particularly popular systems for experimental computer vision researchers to construct. Recent drops in the prices of range sensors and dramatic improvements in the processing power and memory capacity of research workstations have combined to make experimentation with real (as well as synthetic) range data and historically-slow recognition strategies approachable, even in research labs with significantly limited resources.

At the 1991 NSF-sponsored workshop entitled "Future Directions in Computer Vision Research" [8], computer vision methodology (or lack of same) received focused attention, and a series of articles in *CVGIP: Image Understanding* [10] has hopefully prompted researchers to standardize those parts of their experiments, systems, and data where such standardization is warranted. While improvement of the situation requires activity on many fronts, one theme that has been sounded repeatedly in recent years is the value of comparisons against standard data sets. Such comparisons allow the strengths and weaknesses of competing techniques to be highlighted, as well as demonstrating the domain of applicability of the methods under study (promoting the 'where does it break?' question asked of researchers describing implementations and algorithms by reviewers and conference presentation attendees). There are signs that this need is beginning to be addressed by researchers, as evidenced by the use of standardized databases in applications as diverse as stereo matching [9], character recognition [7], and (most relevant to this work) range image segmentation [1].

In this article, we describe a database of range images and 3D objects constructed over an nine-year period at two institutions, which was designed to facilitate research in model-based 3D object recognition. Almost

*Support for the construction of the databases described in this article was provided by the National Science Foundation under grants CDA-8806599 (to Michigan State University), CDA-9121675, CDA-9422044, IRI-9209212, and IRI-9506414 (to Washington State University), by the Washington State University Research Grant-In-Aid program, by the Washington Technology Center, and by the Northrop Corporation.

by accident, this database (freely available over the Internet) has become a popular source of imagery for evaluation of range image segmentation algorithms. Dissemination and use of this database will allow for realistic comparative studies as well as a source of test data for development of new techniques for range image analysis and understanding.

The remainder of this paper is organized as follows. Section 2 describes the overall organization of the database, along with some commentary about the history of its construction. Section 4 describes two techniques for accessing the database and suggests a standard scheme for crediting the authors in publications employing the database. Section 5 presents some final comments.

2   Database Contents and Organization

At present, the database has two main components: an archive of over 250 3D object models, and an archive of over 400 real and synthetic range images. New items are added to the database regularly and documented on its home page. Access statistics for the database are incomplete but the available data indicates a minimum of 100 downloads of one or more items from the database each month.

2.1   Object Model Database

The object model database contains descriptions of 3D objects in a variety of formats. Different descriptions of the same object are employed for different purposes in our model-based object recognition system. For example, each object in the database is described as a polyhedron, which for objects with curved surfaces is merely an approximation to the true object shape. This polyhedral description is useful for synthesizing images of the object, both to provide data sets for system testing and to provide images of hypotheses for verification [5]. Many objects are described in terms of the types and parameters of their constituent surfaces; this description is useful during recognition as a source of surface primitives to correspond to surfaces extracted from range image segmentations. Each file format is described via example below.

There are presently three sub-archives in the model database:

1. The **MSU-Ideas** database contains descriptions of twenty 3D objects used in testing of the BONSAI and IFI object recognition systems developed by Flynn and Jain [5, 6]. These objects were originally designed in 1989 and 1990 with the I-deas solid modeler sold by Structural Dynamics Research Corporation. A subsequent redesign phase employed the IRIT public-domain solid modeler[1] written by Gershon Elber (of the University of Utah and Technion). Object shapes are generally simple and piecewise-planar, cylindrical, spherical, or conical. Objects in this database are described in four formats:

   (a) IRIT solid modeler input.

   (b) IGES 3.0 formatted data.

   (c) An ASCII polyhedral approximation.

   (d) An ASCII file of surface types and parameters.

2. The **USF** database contains descriptions of 81 polyhedral objects designed by students and faculty at the University of South Florida. A custom solid modeler developed at USF was used to fabricate these models. Each object is presented in two formats:

   (a) An ASCII polyhedral description.

   (b) An ASCII file of surface types and parameters (in this database, all object surfaces are planar).

3. The **NETLIB** database contains descriptions of 119 polyhedral objects obtained from the NETLIB scientific data repository. Each object is presented in two formats:

   (a) An ASCII polyhedral description.

   (b) An ASCII file of surface types and parameters (in this database, all object surfaces are planar).

---

[1] http://www.cs.technion.ac.il/~gershon/irit/home/irit_home.html

4. The **WSU-PRO/E** database contains descriptions of seven objects constructed from drawings in mechanical design texts. The models were designed using the Pro/Engineer CAD package sold by Parametric Technologies, Inc. and are currently presented only as polyhedral approximations. We are currently designing more models and making surface descriptions available.

Figure 1 shows a synthetic range image of an object from each of the four databases. Figure 2 shows excerpts from the four files used to describe the 'curvblock' object in the MSU-Ideas database. As mentioned above, the file formats most useful to our object recognizers are the polyhedral approximation (*i.e.*, the .poly format, Figure 2(c)) and the neutral surface description (*i.e.*, the .neutral format, Figure 2(d)).

2.2 Range Image Database

The range image archive consists of several sub-archives.

- The **Synthetic** sub-archive contains five synthetic range images of each of the models in the MSU-Ideas database described above (for a total of one hundred images). These images were generated from the polyhedral approximations (.poly files) accompanying each of the MSU-Ideas models. Software to generate these synthetic range images is also available in the archive and is described below.

- The **Isolated** sub-archive contains five real range images (taken with a Technical Arts 100x range sensor) of each of the twenty objects whose models are in the MSU-Ideas model database (for a total of 100 images).

- The **Cluttered** sub-archive contains ten real range images (taken with the 100x range sensor), each of a scene containing two of the objects in the WSU-Ideas object database.

- The **Miscellaneous** sub-archive contains 47 images taken with the 100x range sensor, containing one or more miscellaneous objects.

- The **seg-comp** sub-archive contains 80 images used in the segmentation comparison project conducted at the University of South Florida, Washington State University, the University of Bern, and the University of Edinburgh since 1993 [1]. Forty of the images were taken at Bern with a structured light scanner built by ABW Gmbh and forty images were taken at Oak Ridge National Laboratory with an Odetics Perceptron laser scanner.

- The **USF** sub-archive contains 19 range/reflectance image pairs taken with the Odetics Perceptron laser range finder at Oak Ridge.

2.3 Range Image Rendering Software

The archive also contains the source code and supporting libraries for a program that synthesizes range images from the polyhedral approximations (.poly files) mentioned above. The program currently available is somewhat limited in that it centers the object in the synthetic aperture, scales the coordinates appropriately to fill the image with the object, and specifies the object orientation by an index between 0 and 319. The index identifies one of 320 viewpoints which are drawn from the centers of the triangles comprising a 16-frequency subdivision of the icosahedron. The program requires the Mesa software library (which is a freely available implementation of most of the OpenGL graphics library) and a pointer to the home site for Mesa is provided in the database description.

3 Impact and Lessons Learned

The primary impact of a database such as the one described here on its research community is demonstrated by the its frequency of appearance in the literature. A brief examination of recent issues of *IEEE Trans. on PAMI* revealed several uses of database images, including these two uses.

- Dickinson *et al.* [4] used Technical Arts 100X images as input to a system to derive part models for the objects in the images.

- Sun and Sherrah [11] used images as input to a procedure that identifies symmetries in objects.

Images in the database have also been used in the literature to evaluate range image segmenters. The object models in the database have not been used to any great extent (this is perhaps due to the proliferation of free and inexpensive databases of CAD models developed for use in computer graphics and animation applications). Therefore, the record of success of this database is mixed. It is reasonably well-known and well-regarded as a source of range imagery, but has had little impact at a level above segmentation.

In retrospect, the design, updating and 'marketing' of the database might have been done differently to enhance its value to the community. Inexpensive CD-ROM writers and media as well as inexpensive high-capacity rewritable disks have impacted and are continuing to impact the for-free and for-profit approaches to selling data. More publicity about databases such as the 3D database also raises their profile in the community and the recent and continuing activities to improve the experimental rigor of computer vision research will help reinforce their value.

4   Access and Citation

The 'top' of the 3D model/image database is available at the following URL:

    http://www.eecs.wsu.edu/~flynn/3DDB/ .

The archive is organized as a tree; at the top level the visitor can elect to browse the object models, the range images, or the rendering software. Each leaf directory's contents (*e.g.* a group of related object models or images, or the entire rendering program source code) is available as a compressed UNIX `tar` file, and every individual file may also be browsed or downloaded separately. The several WWW pages associated with this database will be instrumented with programs to keep track of the database's usage; this will help to guide future development of the archive.

5   Conclusions

In this short article, we have described a nine-year-old archive of data used in model-based 3D object recognition. As the archive matures, additional items are begin added and its value to the computer vision research community should continue to increase. Ultimately, the success of this archive depends on the willingness of researchers to use it; this was the basis of our decision to make the data freely available over the Internet and World Wide Web.

References

[1] Adam Hoover and Gillian Jean-Baptiste and Xiaoyi Jiang and Patrick J. Flynn and Horst Bunke and Dmitry Goldgof and Kevin Bowyer and David Eggert amd Andrew Fitzgibbon and Robert Fisher. An Experimental Comparison of Range Image Segmentation Algorithms. *IEEE Trans. on Pattern Analysis and Machine Intelligence*, 18(7):673–689, 1996.

[2] Anil K. Jain and Patrick J. Flynn (eds.). *Three-Dimensional Object Recognition Systems*. Elsevier, Amsterdam, 1993.

[3] P. J. Besl and R. C. Jain. Three-Dimensional Object Recognition. *ACM Computing Surveys*, 17(1):75–145, March 1985.

[4] S. J. Dickinson, D. Metaxas, and A. Pentland. The Role of Model-Based Segmentation in the Recovery of Volumetric Parts from Range Data. *IEEE Trans. on Pattern Analysis and Machine Intelligence*, 19(3):259–267, 1997.

[5] P. J. Flynn and A. K. Jain. BONSAI: 3D Object Recognition Using Constrained Search. *IEEE Trans. on Pattern Analysis and Machine Intelligence*, pages 1066–1075, October 1991.

[6] P. J. Flynn and A. K. Jain. 3D Object Recognition Using Invariant Feature Indexing of Interpretation Tables. *Computer Vision, Graphics, and Image Processing: Image Understanding*, 55(2):119–129, March 1992.

[7] J. Hull. A Database for Handwritten Text Recognition Research. *IEEE Trans. on Pattern Analysis and Machine Intelligence*, 16(5):550–554, 1994.

[8] S. Negahdaripour and A. K. Jain. Challenges in Computer Vision: Future Research Directions. In *Proc. IEEE Computer Society Conference on Computer Vision and Pattern Recognition*, pages 189–199, June 1992.

[9] R. C. Bolles and H. H. Baker and M. J. Hannah. The JISCT Stereo Evaluation. In *Proc. ARPA Image Understanding Workshop*, pages 263–274, Washington, DC, 1993.

[10] Ramesh Jain and Thomas O. Binford. Ignorance, Myopia, and Naivete in Computer Vision Systems. *Computer Vision, Graphics, and Image Processing*, 53(1):112–117, 1991.

[11] C. Sun and J. Sherrah. 3D Symmetry Detection Using the Gaussian Image. *IEEE Trans. on Pattern Analysis and Machine Intelligence*, 19(2):164–168, 1997.

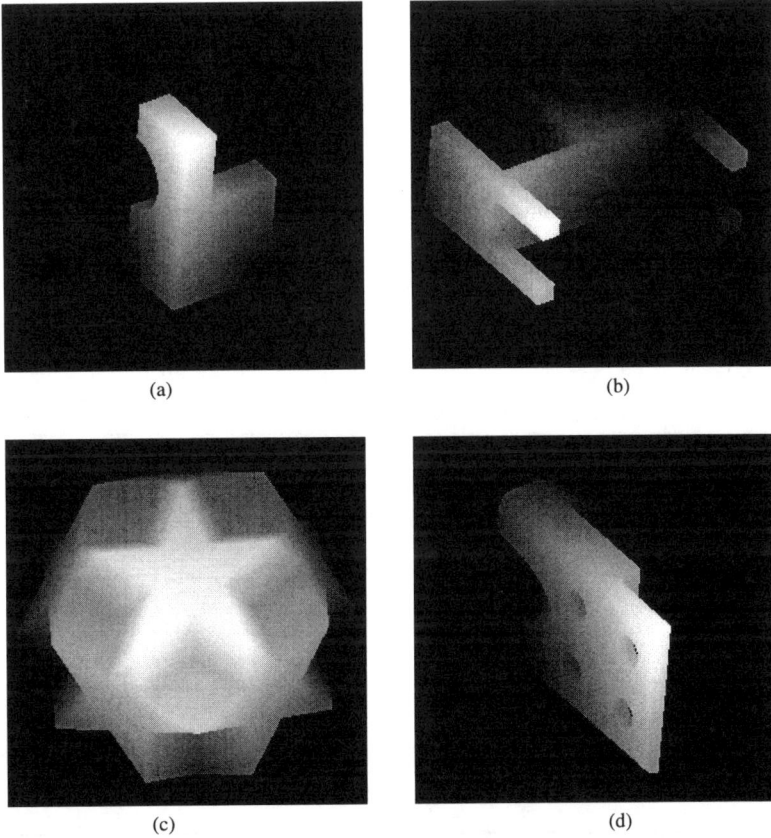

Figure 1. Synthetic range images of objects from the four databases in the object model database. (a): MSU-Ideas database. (b): USF database. (c): NETLIB database. (d): WSU-PRO/E database.

```
# the curvblock object
# 7/95 pat flynn

# x=0 polygon
i1 = poly(list(vector(0,0,0),vector(0,5.5,0),
                vector(0,5.5,2.75),vector(0,0,2.75)),FALSE);
# y=0 polygon
i2 = poly(list(vector(0,0,0),vector(6.75,0,0),
                vector(6.75,0,2.75), vector(0,0,2.75)),
          FALSE);

16 lines omitted

# z = 2.75 polygon (L-shaped. Note cylinder isn't
subtracted yet.)
i11 = poly(list(vector(0,0,2.75),vector(6.75,0,2.75),
                vector(6.75,1.25,2.75),
                vector(1.25,1.25,2.75),
                vector(1.25,5.5,2.75),
                vector(0,5.5,2.75)),
           FALSE);

# Note: negation needed because irit seems to create an
# inside-out object.
a = - mergepoly(list(i1,i2,i3,i4,i5,i6,i7,i8,i9,i10,i11));

# resolution=128;

c = cylin(vector(4,-1,2.75),vector(0,5,0),1.375);
b=a-c;

save("curvblock",convex(b));
```

(a)

```
GEOMETRY CREATED BY SDRC GEOMOD 4.0                              2    1
..,9HCURVBLOCK,9HCURVBLOCK, 3HHEDRC GEOMOD;V4.0 ,SUM   .TYPE-IGES3H.,32, G    1
8,24,8,56,9HCURVBLOCK,1.0,1,2HIN,,,13H900108.085616,1.0E-05,6.9E+00,,;   G    2
110       1         5         1         0         0         0  00000000D    1
110       0         5         1         0                     8D           2
110       3         1         1         0         0         0  00000000D    3
110       0         3         2         0                     8D           4
110       5         1         1         0         0         0  00000000D    5
110       0         5         2         0                     8D           6
110       7         5         1         0         0         0  00000000D    7
110       0         5         2         0                     8D           8

244 lines deleted

10S       26S        1         1         0         0         0  00000001D  253
10S        0         2         3         1                     8D          254
110, 0.1250000E+01, 0.5500000E+01, 0.2750000E+01,                    1P     1

0.1250000E+01, 0.5500000E+01, 0.0000000E+00, 0, 0;                   1P     2

261 lines deleted

110, 0.0000000E+00, 0.5500000E+01, 0.0000000E+00,                    249P  264
0.0900000E+00, 0.5500000E+01, 0.2750000E+01, 0, 0;                   245P  265
102,    4,    243,    245,    247,    249,                           251P  266
0,    0;                                                             151P  267
10S, 0.0000000E+00, 0.1000000E+01, 0.0000000E+00, 0.5500000E+01,     153P  268
251, 0.0000000E+00, 0.0000000E+00, 0.0000000E+00,                    253P  269
0.0000000E+00,0,0;                                                   153P  270
S     10      2D    154P    270                                        T    1
```

(b)

```
pnt 0: 4.000000 1.250000 1.250000
pnt 1: 4.000000 1.250000 0.000000
pnt 2: 6.750000 1.250000 0.000000
pnt 3: 6.750000 1.250000 1.250000
pnt 4: 1.250000 0.000000 0.000000

57 lines deleted

pnt 62: 2.692300 0.000000 2.325100
poly 0: [4]   25 26 44 45
poly 1: [4]   45 44 42 48

40 lines deleted

poly 42: [3]   62 61 59
poly 43: [4]   24 13 5 61
poly 44: [4]   14 15 7 6
surf 0: [10] 0
surf 1: [11] 10
surf 2: [2] 21
surf 3: [1] 23
surf 4: [1] 24
surf 5: [1] 25
surf 6: [2] 26
surf 7: [1] 28
surf 8: [3] 29
surf 9: [1] 32
surf 10: [1] 33
surf 11: [10] 34
surf 12: [1] 44
```

(c)

```
igesID: 1
type: cylinder     axis: 0 1 0      point: 4 0 3.75
length: 2.0        radius: 1.375000

igesID: 2
type: plane        coefficients: 0.000000 1.000000 0.000000 1.250000

igesID: 3
type: plane        coefficients: 1.000000 0.000000 0.000000 1.250000

igesID: 4
type: plane        coefficients: 0.000000 0.000000 1.000000 1.250000

igesID: 5
type: plane        coefficients: 0.000000 1.000000 0.000000 4.000000

igesID: 6
type: plane        coefficients: 1.0 0.0 0.0 4.0

igesID: 7
type: plane        coefficients: 0.000000 0.000000 1.000000 2.75

igesID: 8
type: plane        coefficients: 0.000000 0.000000 1.000000 2.75

igesID: 9
type: plane        coefficients: 0.000000 0.000000 -1.000000 0.000000

igesID: 10
type: plane        coefficients: -1.000000 0.000000 0.000000 0.000000

igesID: 11
type: plane        coefficients: 1.000000 0.000000 0.000000 6.750000

igesID: 12
type: plane        coefficients: 0 -1 0 0

igesID: 13
type: plane        coefficients: 0.000000 1.000000 0.000000 5.500000
```

(d)

Figure 2. Four descriptions of the MSU-Ideas curvblock **object. (a): IRIT (**.irt**) file. (b): IGES file (excerpts). (c): Polyhedral approximation (**.poly **file, excerpts). (d): surface descriptions (**.neutral **file).**

Shape of Motion and the Perception of Human Gaits

Jeffrey E. Boyd
Department of Electrical and Computer Engineering
University of California, San Diego, La Jolla CA 92093-0407

James J. Little
Department of Computer Science, University of British Columbia
Vancouver, B.C., Canada V6T 1Z4

Abstract

Researchers in computer vision have recently demonstrated several systems that interpret motion and optical flow without using a model of kinematic structure. These non-structural methods usually integrate a field of motion into a more compact representation. In this paper, we present the design of an experiment to investigate the relationship between the non-structural shape-of-motion algorithm and human perception of gait. We take the features used by the algorithm, and use them to synthesize gait-like optical flow. A group of subjects then views the flow stimuli and records their perceptions. The motion stimuli are designed to differ in structure but have similar shape-of-motion. We wish to show that we can vary the structure of a stimulus without altering its perception, so long as we maintain the shape-of-motion. Pilot study results illustrate the experiment design. By performing this experiment to relate gait perception with gait synthesis, we are able to probe the computer vision algorithm.

1: Introduction

Recent research in computer vision shows an interest in methods for perception of human locomotion and other activities. The methods fit into two broad categories: structural and non-structural. Structural methods use a model of human kinematic structure and possibly dynamics. In contrast, non-structural methods (sometimes referred to as *appearance-based* or *model-free*) avoid using such models. For example, Little and Boyd demonstrate non-structural gait recognition using *shape-of-motion* features [13, 6, 14] (described in Section 2). Polana and Nelson [15, 16, 17] look at global spatial distributions of motion for a figure engaged in some activity. They are able to recognize different activities by comparing motion statistics computed over a coarse mesh. Baumberg and Hogg [3] give a method to describe the shape of a walking human body as a function of time. In later work [4], they describe the variation in the shape over time as the changing shape of a vibrating plate. Bobick and Davis [10] describe another non-structural approach that analyzes the shape of a motion-energy image (MEI), a summation of optical flow over a sequences of images. Features that describe shapes in the MEI are used to recognize activities.

A common theme in these non-structural methods is the integration of a field of motion into a more compact representation. There is evidence in the psychophysics literature to suggest that spatial integration of motion is also important in human perception of motion. For example, Williams and Sekuler [20] show that perception of a field of randomly moving points is related to motion of the field as whole, i.e., spatial integral of the motion. If the motion of the points is randomly distributed over all directions, then there is no perceived large-scale motion. However, if they are distributed only over a smaller range of angles, then there is indeed a perception of the points moving *en masse.* Boyd and Little [6] sought to explain other psychophysical observations [2, 5, 11, 12, 18], based on moving light displays (MLDs), in terms of *shape-of-motion* features. The explanation was based on a comparison of psychophysical results and the properties of the *shape-of-motion* algorithm, indicating that there are many consistencies. The comparison leads to some conjecture about what one might observe in human perception if it is indeed related to *shape-of-motion.*

Our goal is to explore the relationship between non-structural properties of motion and optical flow. We form the hypothesis that motion stimuli may be treated as a field of optical flow that can be effectively characterized by a set of global *shape-of-motion* features. This paper presents the design of an experiment that tests this hypothesis, focusing on human locomotion.

The experiment uses the relationship that arises between perception and synthesis. We create a set of motion stimuli, all but one of which have non-structural *shape-of-motion* features like those for a gait, but varying in their underlying kinematic structure. If the hypothesis is true then we should see that perception is independent from the selection of stimulus. On the other hand, if the hypothesis is false and structure is critical, then the stimuli should all be perceived differently.

Although the experiment is primarily psychophysical in nature, it demonstrates a useful method to test a computer vision algorithm. We start with a computer vision algorithm that we know can perceive differences in human gaits under a set of controlled conditions. We then turn around and use that algorithm to synthesize human gaits. If the synthesis is successful, then we may assume that the algorithm is sensitive to the right things, at least for a gait. If not, then we know that there is more to the gait than what the algorithm measures, and perhaps we get a clue as to what else we should look for in the motion. The quality of the synthesis provides a means to evaluate the algorithm.

2: Background

2.1: Shape-of-Motion

Figure 1 illustrates the flow of data through the *shape-of-motion* system to create non-structural features that are used for recognition of individual gaits [14]. The system begins with an image sequence of $n + 1$ images featuring a single pedestrian walking in front of a static background, and then derives n dense optical flow images. For each of these optical flow images, the system computes m characteristics that describe the shape of the motion (i.e., the spatial distribution of the flow), for example, the centroid of the moving points, and various moments of the flow distribution. Some of these are pixel coordinates, but all are treated as time-varying scalar values. Table 1 summarizes the scalar values used. Rearranging the scalar values forms a time series for each scalar. A walking pedestrian undergoes periodic motion, returning to a standard position after a certain time period that depends on the frequency of the gait. The system analyzes the periodic structure of these time series and determines the fundamental frequency of the variation of each scalar.

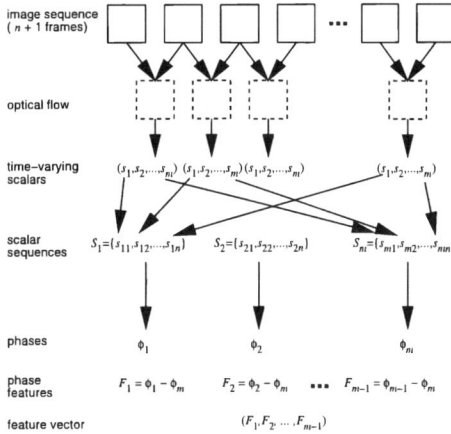

Figure 1. The data flow for *shape-of-motion* gait analysis. The spatial distribution of the optical flow computed for each frame of a video sequence is characterized by a set of scalars. Over the duration of the sequence, each scalar forms a time series that is cyclic. The phase relationships among the time series give features that are useful for gait recognition.

The set of time series for a view shares the same frequency, or simple multiples of the fundamental, but their phases vary. To make different sequences comparable, the system subtracts a reference phase, ϕ_m, derived from one of the scalars. Each image sequence is characterized by a vector, $F = (F_1, \ldots, F_{m-1})$, of $m - 1$ relative phase features. Little and Boyd showed that some of the features are sensitive to individual gaits, and that the sensitivity can be used for recognition.

At no point in *shape-of-motion* recognition is the structure of the walking subject recovered. The process operates entirely without a model of human kinematics.

2.2: Human Perception of Gait

Much of the psychophysics literature pertaining to motion perception refers to moving light displays (MLDs). MLDs are useful because they conceal the underlying structure of an image so that it cannot be perceived from a static image, only a moving sequence [11, 12]. We have suggested that gait perception from MLDs may not require identification of kinematic structure [6]. This is based on two observations:

1. the *shape-of-motion* algorithm applies equally well to both gray-scale images and MLDs, and

2. various observations concerning human perception of moving light displays can be explained in terms of the algorithm.

The first observation is based on a comparison of *shape-of-motion* features derived from gray-scale sequences and features derived from the equivalent MLD. The features are similar for the two types images. The second observation is based on the following evidence reported

Description	Label	Formula
x coordinate of centroid,	x_c	$\sum xT / \sum T$
y coordinate of centroid,	y_c	$\sum yT / \sum T$
x coordinate, centroid of $|(u,v)|$ distribution	x_{wc}	$\sum x|(u,v)|T / \sum |(u,v)|T$
y coordinate, centroid of $|(u,v)|$ distribution	y_{wc}	$\sum y|(u,v)|T / \sum |(u,v)|T$
x coordinate of difference of centroids	x_d	$x_{wc} - x_c$
y coordinate of difference of centroids	y_d	$y_{wc} - y_c$
aspect ratio (or elongation) – ratio of length of major axis to minor axis of an ellipse	a_c	$\lambda_{max}/\lambda_{min}$, where λs are eigenvalues of second moment matrix for motion distribution
elongation of weighted ellipse	a_{wc}	as in a_c, but for weighted distribution
difference of elongations,	a_d	$a_c - a_{wc}$
x coordinate, centroid of $|u|$ distribution	x_{uwc}	$\sum x|u|T / \sum |u|T$
y coordinate, centroid of $|u|$ distribution	y_{uwc}	$\sum y|u|T / \sum |u|T$
x coordinate, centroid of $|v|$ distribution	x_{vwc}	$\sum x|v|T / \sum |v|T$
y coordinate, centroid of $|v|$ distribution	y_{vwc}	$\sum y|v|T / \sum |v|T$

Table 1. Summary of scalar *shape-of-motion* descriptors. Summations are over the entire image. u and v are the x- and y-direction optical flow values respectively. The function T segments the image. $T = 1$ for pixels that are moving and $T = 0$ for stationary pixels.

in the psychophysics literature.

Sumi [18] looked at the effect of inverting the display of an MLD on recognition. His observations showed that while most people were often able to recognize a gait from an inverted MLD, they reported that the gait looked odd and failed to recognize that it was, in fact, a normal gait that was inverted (a perception that is contrary to the kinematic structure). Inverting the MLD reverses the phase of some, but not all of the *shape-of-motion* features. Furthermore, the relative phase of the features is preserved. Thus, the *shape-of-motion* is not expected to change much for the inverted image, making the observation of human perception consistent with the algorithm. Sumi's subjects often reported seeing gaits but thought that the gait was odd. Perhaps the oddness represents a failure to map human structure onto the stimulus once the subject believes that the stimulus represents a human.

Barclay *et al.* [2] showed that a detectable amount of gender recognition is possible from MLDs, although the recognition was only slightly better than chance. This evidence does not indicate the mechanism used for the recognition, but the success of the *shape-of-motion* algorithm suggests that kinematic structure is not necessary. In fact, Little and Boyd [14] suggest that although *shape-of-motion* features are not structurally based, they may be sensitive to variations in the build of the observed person, explaining the algorithm's ability to recognize individuals. The MLD evidence does not preclude a non-structural mechanism.

Bertenthal and Pinto identify the importance of phase in gait perception [5]. Their experiments use a stimuli that consists of an MLD for a walker, masked by dots moving in the background. Some of the stimuli are perturbed by altering the phase of oscillation of a limb. They observed that perception of gait from the unperturbed stimulus was better than for the perturbed one, but both were significantly better than chance. Bertenthal and Pinto assumed that kinematic structure was important and perturbed only the phase of the motion. Clearly, phase is important for perception of motion in humans, just as it is for the *shape-of-motion* algorithm.

2.3: Integrated Features of Cyclic Motion

Boyd and Little [6] offered the following conjecture to explain the psychophysical observations, based upon the *shape-of-motion* algorithm. Consider the image of a pedestrian to be a collection of moving points. As a simple approximation, the motion for each point in the pedestrian can be expressed as the sum of a linear motion and an oscillatory motion. For example, let the x-coordinate of an arbitrary point i be

$$x_i(t) = x_{i0} + v_x t + A_i \cos(\omega t + \phi_i), \tag{1}$$

where x_{i0} is a constant, v_x is the mean velocity of the person, and A_i, ω and ϕ_i are the amplitude, frequency and phase of the oscillation. $x_{i0} + v_x t$ is the linear part of the motion. All points share the same frequency, ω, but vary in A_i and ϕ_i depending on where they are in the body. The x-coordinate of the centroid is

$$\bar{x}(t) = \frac{1}{n} \sum_i x_i(t) = \frac{1}{n} \sum_i \{x_{i0} + v_x t + A_i \cos(\omega t + \phi_i)\}, \tag{2}$$

where n is the number of image points in the pedestrian. As part of extracting a phase feature we discard the linear portion of the motion leaving

$$\bar{x}(t) = \frac{1}{n} \sum_i A_i \cos(\omega t + \phi_i). \tag{3}$$

The summation in Equation (3) is the sum of a set of phase vectors, or phasors. Phase vectors are commonly used to perform computations with rotating vectors that share a common frequency, such as in electrical power systems. In short, the summation can be treated as the sum of a set of vectors, each vector having magnitude A_i and direction ϕ_i.

Several conclusions may be drawn from this conjecture:

1. Recognition of objects from MLDs does not necessarily happen because they capture the structure of objects in a scene, but because they adequately sample the motion of points in the entire object.

2. If it is only necessary for the dots in an MLD to sample the motion, then perception of gait should not require that the dots be at joints, since that is purely a structural concern.

3. Any MLD that has *shape-of-motion* features corresponding to gait should be perceived as gait.

The remainder of this paper describes the experiment we designed to to test the conjecture of Boyd and Little.

3: Experiment Design and Motion Stimuli

The conjecture in the previous section indicates how one should synthesize an optical flow field so that is has the same characteristics as the field for a human gait. We test the perception of flow synthesized this way, to see if it is indeed perceived as a human gait. The use of an algorithm for perception as a tool for synthesis can point to the specific part of the motion that the algorithm is sensitive to.

We use a factorial experiment to test the perception. Ideally, we should use the *posttest-only control group design* described by Campbell and Stanley [7], but the number of un-ordered combinations of factors is too numerous and forces us to use some repeated mea-sures. Section 4.1 describes the actual method we used in our pilot study. We discuss a final design in Section 5.

In the experiment, we present subjects with a motion stimulus and ask them to record, in not more than two or three words, what they perceive in the motion. We design the stimuli to have specific properties that are controlled by a set of factors. The factors, our independent variables, are:

1. **the motion:** how the motion was created for the stimulus and what its *shape-of-motion* features are,

2. **encoding:** how the motion was encoded into images in video sequence, and

3. **duration:** how long the subjects view the stimulus.

The following subsections describe these factors in detail.

3.1: Motion

The motion in the stimuli is a critical part of the experiment since we want to relate our results to the *shape-of-motion* algorithm. Our intention is to generate synthetic fields of optical flow with the *shape-of-motion* features of a human gait. Unfortunately, it is not a simple matter to create images to such a specification. We must also consider that it may not be necessary to reproduce all the *shape-of-motion* features in the stimulus. The reason for this stems from the difference between recognizing some motion as a gait, and recognizing and individual gait. In recognition tasks, *shape-of-motion* features that are consistent for individual gaits, but vary greatly over the population of gaits are desired. To simply recognize a gait requires a feature that remains consistent over all gaits, but varies greatly over the population of all motions. Sorting out what features are important for what perception is will not be a simple task. We make a start here by focusing on x_c and y_c.

In lieu of an algorithm to create synthetic flow fields to our specification, our strategy is to start with data for a real pedestrian, and then to distort it in a manner that maintains the *shape-of-motion* but may or may not be consistent with the underlying kinematic structure. Specifically, we generate the five motion patterns as follows:

On Joints (*on-joints*) The first stimulus is a simple MLD of a walking person. For this we use the synthetic canonical walker described by Cutting [9], which has the lights placed on major joints. We relied on a translation of Cutting's original Fortran program into C that was used by Bertenthal and Pinto [5].

Off Joints (*off-joints*) We perturb the *on-joints* stimulus by moving the points of light away from the joints and along the limbs. The displacement is random, sampled from a uniform distribution ranging from -0.75 to 0.75, and scaled by the length of the appropriate limb. For example, the point at the knee can end up anywhere from three quarters of the way up the thigh to three quarters of the way down the shin. Points for head, wrists and ankles can only move towards the shoulder, elbows and knees respectively, thus ensuring that all points are somewhere on the body, but not necessarily on the specific joint. Cutting's algorithm considers occlusions of points in an ad hoc manner. We ignore these occlusions to simplify the synthesis.

Off Body (*off-body*)   The *off-body* stimulus is also a perturbation of *on-joints*, but rather than keep all the points on the body, we move the points by a constant random displacement in a random direction. We select the magnitude of the displacement from a normal distribution, mean zero and standard deviation of 20 pixels (the torso of the walker is about 35 pixels). The direction of the displacement is selected from a uniform distribution ranging from −180° to 180°. We ignore the occlusions.

Oscillating Off Body (*osc-off-body*)   The *osc-off-body* stimulus is like *off-body*, with the exception that the displacement is not constant, but oscillates at the same frequency as the gait. The effect is one of the light orbiting about a fixed point on the joint of the walker. We ignore the occlusions here too.

Random Oscillations (*random-osc*)   The *random-osc* stimulus is the *straw man* in the set of motion. It consists of a set of randomly spaced points, placed such that the aspect ratio is about that of *on-joints*. The phases of the oscillations are randomly distributed over −180° to 180°. The amplitudes are normal with standard deviation of one quarter of the height of the distribution.

Selected frames from the five motions are shown in Figure 2 in MLD form. The perturbations of points in *off-joints*, *off-body* and *osc-off-body* are chosen such that the coordinates of the centroids should be nearly unchanged. For *off-body* and *osc-off-body*, this happens because the magnitude and direction of the displacement are unbiased. There is a bias to the direction of displacement for *off-joints* because the points are restricted to lie on the limbs, and because points at the extremities can only be displaced towards the center of the body. It is difficult to predict what will happen to the other features under these perturbations.

We encountered difficulty computing the *shape-of-motion* features from the synthetic stimuli. The first problem was with the motion of the walker with respect to the background. Previous work on *shape-of-motion* used sequences of pedestrians walking across a static background, allowing us to rely on motion to segment the moving figure. The Cutting synthetic walker data yields points for a static walker before a static background, giving the impression that the person is walking on a treadmill. This makes the torso nearly stationary. We were not able to resolve the tiny motions in the torso with the optical flow algorithm so they could not contribute to the distributions, and gave results that we could not interpret. We then resorted to computing an ideal optical flow given that we know the positions of the points and the frame-to-frame correspondences. Still, we could not see well-defined peaks in the spectra of the features. In the end we added a constant x velocity to all the points to simulate a walker moving across a static background. This failure of the synthetic data to behave like the real data that we had used often in the past caught us off guard and is a weakness in this experiment that we have yet to resolve. Clearly, the subjects will not see treadmill motion and we have yet to relate *shape-of-motion* to treadmill gaits.

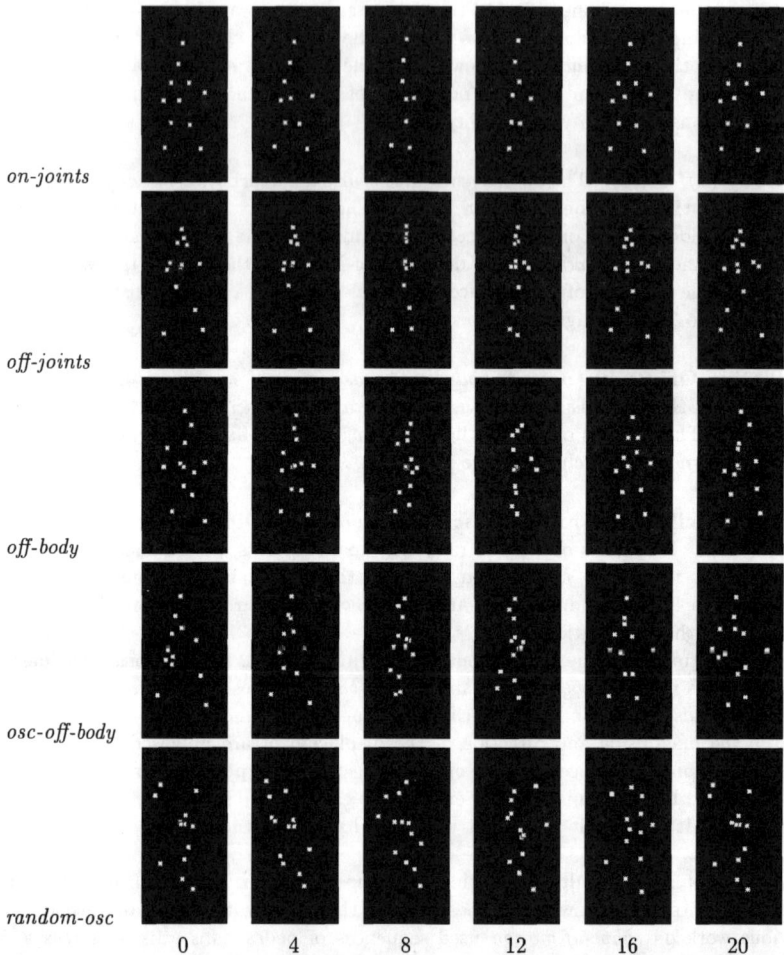

Figure 2. Sample images from motion stimuli: (*on-joints*) **the Cutting canonical walker MLD sequence,** (*off-joints*) **walker with dots on the body, but moved away from joints,** (*off-body*) **walker with dots randomly displaced from joints,** (*osc-off-body*) **walker with dots displaced from joints by oscillating distance of random amplitude and phase, and** (*random-osc*) **a set of random dots moving cyclically with random phase and amplitude, same frequency as** *on-joints*.

Table 2 summarizes the *shape-of-motion* features for the five stimuli, computed using point correspondences and a constant x velocity added. The right column of the table indicates the between-person standard deviations for the feature that was measured by Little and Boyd [14]. y_c is the reference phase and is, therefore, always zero. The table

Feature			Stimulus			
	on-joints	*off-joints*	*off-body*	*osc-off-body*	*random-osc*	σ
x_c	0.11	0.16(0.8)	0.14(0.6)	0.19(1.3)	0.30(2.8)	0.065
y_c	0.00	0.00	0.00	0.00	0.00	-
x_{wc}	0.05	0.18(2.9)	0.19(3.1)	0.18(2.8)	0.27(4.7)	0.045
y_{wc}	0.00	0.26(3.1)	0.23(2.7)	0.18(2.2)	-0.07(-0.9)	0.083
x_d	0.47	-0.31(6.9)	-0.25(8.8)	-0.34(5.9)	-0.37(5.0)	0.032
y_d	-0.28	-0.18(1.3)	-0.22(0.8)	-0.26(0.3)	0.41(-3.9)	0.077
a_c	-0.04	-0.06(-0.2)	-0.02(0.3)	-0.06(-0.3)	0.19(2.9)	0.077
a_{wc}	-0.04	-0.05(-0.2)	-0.04(0.0)	-0.07(-0.5)	0.27(5.1)	0.063
a_d	0.39	-0.14(-2.9)	0.12(-1.5)	0.01(-2.1)	-0.09(-2.6)	0.18
x_{uwc}	0.04	0.18(3.3)	0.22(4.1)	0.17(3.0)	0.26(5.1)	0.045
y_{uwc}	0.00	0.26(2.7)	0.23(2.5)	0.19(2.0)	-0.08(-0.9)	0.094
x_{vwc}	0.08	0.37(2.2)	0.09(0.0)	0.18(0.7)	0.27(1.4)	0.13
y_{vwc}	0.02	0.42(3.7)	-0.07(-0.8)	-0.09(-1.0)	0.16(1.3)	0.11

Table 2. Summary of *shape-of-motion* **features for the five motion stimuli. Values are phases scaled to lie in the range** $[-0.5, 0.5]$. **Values in the parenthesis are deviations from the feature value for** *on-joints* **scaled by** σ **(in the right column).** σ**s are the between-person standard deviation for the feature computed for earlier recognition experiments. These are used as a rough guide only for comparing features.**

verifies that x_c and a_c values for *off-joints*, *off-body* and *osc-off-body* are similar to that for *on-joints*. As expected, the weighted distribution and difference features vary from the *on-joints* value the most. *random-osc* is significantly different from *on-joints*.

3.2: Encoding

When creating the motion stimuli, we produce a sets of point coordinates (as described in Section 3.1; one set per time slice in the motion sequence. We encode those points into a motion stimulus in two ways. The first is to produce the well-know MLD. Each point coordinate gives an image coordinate for a point of light on a black background in the frame of a stimulus. The second type of encoding results from an effort to further hide the structure behind the motion. We create a static binary background image where each pixel is selected randomly from $\{black, white\}$, and $P(black) = P(white) = 0.5$. Then rather than represent the pixel coordinates with points of light, we use a small patch of binary pixels (generated using the same criteria as the background). Although we know from Johansson [11] that without *a priori* information regarding the contents of an image, a single frame of an MLD is difficult to interpret, if one knows that a frame represents a human, one could connect the dots and guess at the kinematic structure. With our random background and patch stimulus, a static frame cannot convey any information at all. All of the pixels are randomly generated. The sequence must be moving in order to see anything beyond uncorrelated pixels. MLDs have been adequate in the past for evaluating the perception of motion, and we have no particular reason to believe they would not suffice for our purposes. We derived this encoding scheme while trying to generate fields of optical flow, and this experiment offers an opportunity to test it. The encoding variable can have values M and R for MLDs and random dot patterns respectively. Figure 3 shows a sample

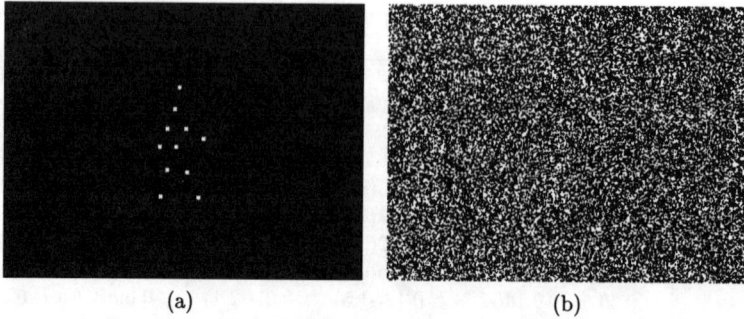

(a)   (b)

Figure 3. Examples of images in a motion stimulus illustrating the two encoding methods: (a) MLD and (b) random dot encoding. Nothing can be seen in the random dot image without any motion since there is no correlation between pixels.

frame for each of the two encoding methods.

3.3: Duration

In humans, the length of time a stimulus is presented can affect the perception. Presumably, a longer period of time allows slower mechanisms to come into play. Johansson [11], Barclay *et al.* [2], and Sumi [18] all cite variations in perception with the duration of the stimulus. Their observations suggest that a duration from about $1s$ up to about $4s$ should suffice. For the pilot study, we use four durations, $1s$, $2s$, $4s$, and $8s$.

4: Pilot Study

We collected data for a pilot study [8] to debug the design of the experiment. This section presents the experimental method and a sample analysis of the results.

4.1: Method

The five motion stimuli (*on-joints*, *off-joints*, *off-body*, *osc-off-body*, and *random-osc*), the two motion encodings (M and R) and four durations ($1s$, $2s$, $4s$, and $8s$) give a total of 40 separate motion sequences in the factorial experiment. We created digital movies at a resolution of $240 pixels$ by $320 pixels$, for each of these sequences using a Silicon Graphics workstation. We then transferred the sequences onto an SVHS video tape, in a random order, scaling the images to just fill the entire screen. We selected a group of subjects (students in a computer vision class) and showed every subject every stimulus in the same order. Before viewing the stimuli, we gave the subjects a sheet of paper with instructions saying that:

- they are about to see a series of 40 video clips,
- the clips vary in duration from about one second up to eight, and
- they should indicate what the believe is the cause of the motion in each clip.

They received no other information about the experiment. The paper had 40 numbered blanks with space for at most 3 or 4 words in which the subjects could record their responses. We review the responses and categorize them. This is a subjective process, but leaves the subjects free of expectations about the stimuli.

Once the responses are categorized, the raw data consists of a tuples containing the levels of the three independent variables, and the dependent variable, the categorized responses. There are 40 tuples for each subject, barring missing data. We form the data into a 4-way contingency table and test for dependencies among the variables [19, 8, 1]. Our analysis is described further in Section 4.2.

We recognize that by showing all subjects identical stimuli in the same order confounds validity due to the effects of multiple measures. With 40 stimuli it was simply convenient to show a single group all the stimuli in one shot. While the data in the pilot study may not lead to valid conclusions, it is sufficient for the purposes of the pilot study. In Section 5 we describe a better method for collecting data that will reduce the problems with multiple measures in the ultimate experiment.

4.2: Analysis

The first step in analysis of the data was to categorize the subject responses. After scanning the data, we decided that the responses fell naturally into three categories:

1. a person walking (W),

2. a person walking but in an unusual manner (WS), and

3. a some other motion not related to humans or human locomotion(X).

It was obvious when a subject perceived a normal human gait. Responses that we categorized as walking, but strange included

- the specific response that it was an odd human gait,

- a human gait, but the person is carrying something,

- a human gait, but the person is stepping back and forth, and

- a human running.

Explanations for the randomly oscillating points included:

- points swirling in the wind,

- rotation of some sort,

- random movement,

- a vortex,

- cellular motion, and

- rotating DNA helix.

Clearly the stimulus gave the impression of rotation. Although the process of categorizing the data is subjective, we can give the categories some validity by getting one or two other people to categorize the responses independently.

To analyze the categorized responses, we form a contingency table and look for dependencies among the variables. Tables 3 and 4 combined show the complete contingency table for the pilot study data.

encoding	duration	motion	response		
			W	WS	X
M	1*s*	*on-joints*	4	2	0
		off-joints	4	1	1
		off-body	0	3	3
		osc-off-body	0	3	2
		random-osc	0	0	5
	2*s*	*on-joints*	5	1	0
		off-joints	3	3	0
		off-body	0	4	2
		osc-off-body	0	4	2
		random-osc	0	0	6
	4*s*	*on-joints*	5	1	0
		off-joints	3	1	1
		off-body	1	4	1
		osc-off-body	1	4	1
		random-osc	0	0	6
	8*s*	*on-joints*	5	1	0
		off-joints	2	3	1
		off-body	0	4	2
		osc-off-body	1	4	1
		random-osc	0	0	6

Table 3. Pilot-study contingency table for MLD motion encoding only, showing frequencies split by all variables. Categories for responses are: (W) person walking, (WS) person walking, but strangely, and (X) explanations other than human locomotion. Duration values are: $1s$, $2s$, $4s$, and $8s$. Descriptions for motions are in Section 3.1.

The first thing we look at in the analysis is to determine if there is any dependency of the subject responses to the factors. For this we use a two-way contingency table analysis, comparing the responses against the joint distribution of encoding, duration and motion, i.e, the concatenation of Tables 3 and 4. We use the χ^2 statistic to test the hypothesis that the variables are independent. The G statistic [8] is a popular alternative, but for the pilot study data at least, there are several zero frequencies for which G is undefined. The result of the test for independence gives $\chi^2 = 202.6$ with degrees of freedom $df = 78$. The probability that such values occur if the variables are independent is $p < 0.0001$. We can safely conclude that responses depend jointly on encoding, duration and motion. This is not surprising since a quick scan of the data shows that subjects never mistook *random-osc* for a human and rarely thought that *on-joints* looked like anything other than a human walking.

To find the source of the dependency, we start by examining the conditional dependence of subject response on encoding. There are 20 two-way response versus encoding tables conditioned on the combinations of duration and motion. We compute χ^2 for each table and the sum over all tables[1]. This yields $\chi^2 = 32.6$ ($df = 29$) and $p = 0.29$. Thus

---

[1] A few of the tables had marginal sums of zero making it impossible to compute χ^2. In those cases we dropped the all-zero columns and compute χ^2 for the remaining data, adjusting df accordingly.

encoding	duration	motion	response		
			W	WS	X
R	1s	on-joints	4	2	0
		off-joints	4	1	1
		off-body	4	0	1
		osc-off-body	3	2	1
		random-osc	0	0	4
	2s	on-joints	4	1	0
		off-joints	5	1	0
		off-body	4	1	1
		osc-off-body	3	2	1
		random-osc	0	0	6
	4s	on-joints	2	3	1
		off-joints	2	1	3
		off-body	1	4	1
		osc-off-body	0	5	1
		random-osc	0	0	6
	8s	on-joints	4	1	1
		off-joints	4	2	0
		off-body	1	4	1
		osc-off-body	2	3	1
		random-osc	0	0	6

Table 4. Pilot study contingency table for random-dot motion encoding only, showing frequencies split by all variables. Categories for responses are: (*W*) person walking, (*WS*) person walking, but strangely, and (*X*) explanations other than human locomotion. Duration values are: 1*s*, 2*s*, 4*s*, and 8*s*. Descriptions for motions are in Section 3.1.

the encoding methods appear to be equivalent. However, an inspection of the individual conditioned tables suggest that, for the *off-body* motion at durations of only 1*s* or 2*s*, there is a dependency. A look at the data shows that subjects tended to interpret the *off-body* motion as a normal gait with random-dot encoding, but as a strange gait, or non-human, with MLD encoding. Given the methodological flaws in our data collection, we should still consider the random dot encoding for the ultimate experiment.

In a similar manner, we test for dependency of subject response on duration. There are 10 two-way response versus duration tables conditioned on the combinations of encoding and motion. The result is $\chi^2 = 37.7$ ($df = 45$) and $p = 0.77$, indicating that the responses were independent of duration. This comes as a surprise given the results of others, but can be attributed to the interference of repeated measures (see Section 5).

Finally, we want to know if there is a dependency between subject responses and the motion stimulus. This will indicate whether or not perception is tied to *shape-of-motion*. Again we produce a set of two-way response versus motion tables, one for every combination of encoding and duration. The result is $\chi^2 = 188.0$ ($df = 64$) and $p < 0.0001$, indicating a strong dependency. This is the major source of dependency in the data and as stated previously, because of the *random-osc* stimulus, that is to be expected. However, what we really want to know is whether or not maintaining *shape-of-motion* features while altering

motion	compare to			
	on-joints		random-osc	
off-joints	12.23(14)	$p = 0.59$	70.42(16)	$p < 0.0001$
off-body	38.19(16)	$p = 0.0014$	55.48(12)	$p < 0.0001$
osc-off-body	32.60(16)	$p = 0.0083$	59.81(13)	$p < 0.0001$

Table 5. Comparison of perception of *off-joints*, *off-body* **and** *osc-off-body* **to** *on-joints* **and** *random-osc*. χ^2 **values and the degrees of freedom in parentheses are the sums** χ^2**s for the two-way comparisons of subject response versus motion conditions on the joint distribution of encoding and duration.**

structure still leads to the same perception. To answer this we repeat the analysis, but for only two motions at a time. For example, when we compare *off-joints* with *on-joints* we get $\chi^2 = 12.23$ ($df = 14$) and $p = 0.59$. In other words, the perception does not appear to depend of whether or not the dots are on the joints. For completeness, we compare *off-joints* to *random-osc* and get $\chi^2 = 70.42$ ($df = 16$) and $p < 0.0001$. Comparisons of *off-joints*, *off-body* and *osc-off-body* to *on-joints* and *random-osc* are summarized in Table 5.

Table 5 shows that for the *on-joints* and *off-joints* stimulus, subject response is independent of the motion. This is the result we are looking for. However, we do not see the same independence for *on-joints* compared to *off-body* and *osc-off-body*.

5: Discussion

The purpose of the pilot study was to debug the experiment design. Therefore, we refrain from drawing conclusions from the data and analysis, and focus on what needs to be changed before collecting data in the final version.

Our data collection method is prone to multiple-treatment interference affecting the external validity (the ability to generalize the results) of the experiment [7]. For the full factorial experiment, each subject views 40 motion sequences, but there are only five different motions in all. A look at the randomized order in which the subjects view the stimuli shows that by the 12th sequence the subjects had viewed all five motions for at least two seconds. By the 19th sequence that rises to four seconds. It was clear from the responses that the subjects had identified the five motions and could quickly relate them to the previous sequences. This confounds the effect of duration in particular. Once a subject has seen the stimulus for four seconds or more, a brief exposure of one or two seconds is likely to produce the same response. In this light, it is not surprising to see that observations did not depend much on duration. Therefore, in the final experiment, we will reduce the number of levels of duration to three ($1s$, $2s$ and $4s$), then split the stimuli by duration, and show any individual subject only sequences of a single duration. This gives us the control we need over duration, but leaves interference within the motion factor. After subjects have seen what is obviously a human, it affects their subsequent responses. To address this we will present the stimuli to subjects in varying order, i.e., a counterbalanced design [7]. So long as there is not a complex interaction among the various motions, this method should be adequate.

Although we do not want to put too much emphasis on the pilot study data, it did suggest that the random dot stimulus was nearly equivalent to the moving light display.

For the final experiment we will use the random dot stimulus only. It is truer to our original intention of generating fields of optical flow, and there was a hint in the pilot data that the ability of the random dot stimulus to conceal structure may yet prove significant. In the worst case, it will make no difference at all. Using only the single encoding scheme also simplifies the analysis by removing a factor.

Another concern is that the *straw-man* stimulus, *random-osc*, is too feeble. In the small sample, no subject ever mistook it for a human. All descriptions suggested a perception of some sort of rotation. In the final experiment, we will replace the random elliptical oscillations with random pendular oscillations. Pendular oscillations should give us a *straw-man* stimulus that is not drastically different from the other stimuli. This will also give more credibility to one of our desired conclusions, that the failure to perceive the random pattern as human is because of *shape-of-motion*, and not for some other reason.

Finally, our goal is to learn about the *shape-of-motion* algorithm, and improve it by understanding its relationship with human perception, if there is one. To achieve this we need a better way to relate the stimuli to *shape-of-motion*. We have taken two steps to accomplish this:

1. perturb a known gait stimulus in such a way that we can predict the effect on at least some of the *shape-of-motion* features, and

2. compute (as best we could) the features of the the resulting randomly perturbed stimuli.

We did see that the x_c and a_c features for *on-joints*, *off-joints*, *off-body* and *osc-off-body* matched well, and did not match *random-osc*. However, we had no control over any of the weighted-distribution and difference features. Several questions remain to be answered about *shape-of-motion* features in this experiment, including:

1. how do we compute *shape-of-motion* not only for a walker moving across the field of view, but also when the field of view tracks the walker, and when the walker is on a treadmill,

2. why does the Cutting synthetic stimulus exhibit optical flow that is so differently from that observed for real stimuli, and

3. which *shape-of-motion* features are important?

This last item is particularly important since the list of *shape-of-motion* features is based purely on intuition about what is important for recognizing motion. Not everything on the list is likely to important, particularly for human observers, and almost certainly there are omissions form the list.

The experiment described tests a computer vision algorithm for recognition of gaits, by seeing whether or not it can be turned around and used to synthesize gait-like optical flow. It is unlikely that we will find, as a result of this experiment, that we can synthesize a completely realistic gait just from *shape-of-motion* features. It will be necessary to follow-up by refining the representation of the flow field used for synthesis, to get more gait-like perception. We can then proceed to repeat the cycle, producing a new algorithm that recognizes the features that were good for synthesis, and so on.

6: Conclusions

We present the design of an experiment that evaluates the quality of a computer vision algorithm by using the algorithm as a guideline for the synthesis of a stimulus, and then evaluating the quality of the resulting stimulus. In a factorial experiment, a group of subjects views a set of synthetic motion stimuli and records their perceptions. The factors that we control are actual motion, the length of time a subject views a stimulus, and how the motion is encoded into images. Four of the motions have the same *shape-of-motion* features as a walking person, but differ in structure, while the fifth contains random motion. We wish to show that we can vary the structure of a synthetic stimulus without altering its perception, so long as we maintain the *shape-of-motion*.

Pilot study results showed flaws in our method of data collection affecting the external validity of the experiment. To correct the flaws we propose the following changes for the final experiment:

- show each subject only sequences of a single duration, and
- show the stimuli to different subjects in different orders.

These changes address the interference of repeated measures that confounds our ability to generalize the results. We also encountered difficulty relating the stimuli to *shape-of-motion* because the algorithm behaved differently when applied to synthetic data than it did for real data.

References

[1] Abacus Concepts, Inc., Berkeley, CA. *StatView*, 1994.

[2] C. D. Barclay, J. E. Cutting, and Lynn T. Kozlowski. Temporal and spatial factors in gait perception that influence gender recognition. *Perception and Psychophysics*, 23(2):145–152, 1978.

[3] A. M. Baumberg and D. C. Hogg. Learning flexible models from image sequences. Technical Report 93.36, University of Leeds School of Computer Studies, October 1993.

[4] A. M. Baumberg and D. C. Hogg. Learning spatiotemporal models from training examples. Technical Report 95.9, University of Leeds School of Computer Studies, March 1995.

[5] B. I. Bertenthal and J. Pinto. Complementary processes in the perception and production of human movements. In L. B. Smith and E. Thelen, editors, *A Dynamic Systems Approach to Development: Applications*, pages 209–239. MIT Press, Cambridge, MA, 1993.

[6] J. E. Boyd and J. J. Little. Global versus structured interpreation of motion: moving light displays. In *Proceedings of IEEE Nonrigid and Articulated Motion Workshop*, pages 18–25, San Juan, Puerto Rico, June 1997.

[7] D. T. Campbell and J. C. Stanley. *Experimental and quasi-experimental designs for research*. Rand McNally, Chicago, 1963.

[8] P. R. Cohen. *Empirical methods for artificial intelligence*. The MIT Press, Cambridge, Massachusetts, 1995.

[9] J. E. Cutting. A program to generate synthetic walkers as dynamic point-light displays. *Behavior Research Methods and Instrumentation*, 10(1):91–94, 1978.

[10] J. W. Davis and A. F. Bobick. The representation and recognition of human movement using temporal templates. In *Proceedings of IEEE Computer Vision and Pattern Recognition*, pages 928–934, San Juan, Puerto Rico, June 1997.

[11] G. Johansson. Visual perception of biological motion and a model for its analysis. *Perception and Psychophysics*, 14(2):201–211, 1973.

[12] G. Johansson. Visual motion perception. *Scientific American*, pages 76–88, June 1975.

[13] J. J. Little and J. E. Boyd. Describing motion for recognition. In *IEEE Symposium on Computer Vision*, pages 235–240, Coral Gables, Florida, November 1995.

[14] J. J. Little and J. E. Boyd. Recognizing people by their gait: the shape of motion. *Videre*, 98. to appear.

[15] R. Polana and R. Nelson. Detecting activities. In *IEEE Conference on Computer Vision and Pattern Recognition*, pages 2–7, 1993.

[16] R. Polana and R. Nelson. Recognition of nonrigid motion. In *1994 DARPA Image Understanding Workshop*, pages 1219–1224, 1994.

[17] R. Polana and R. Nelson. Nonparametric recognition of nonrigid motion. Technical report, University of Rochester, 1995.

[18] S. Sumi. Upside-down presentation of the johansson moving light-spot pattern. *Perception*, 13:283–286, 1984.

[19] R. E. Walpole and R. H. Myers. *Probability and statistics for engineers*. MacMillan Publishing Co., Inc., New York, 1978.

[20] D. W. Williams and R. Sekuler. Coherent global motion percepts from stochastic local motions. *Vision Research*, 1984.

Empirical Evaluation of Automatically Extracted Road Axes

Christian Wiedemann, Christian Heipke, Helmut Mayer
Chair for Photogrammetry and Remote Sensing
Technische Universität München, D–80290 Munich, Germany
Tel: +49-89-2892-2676, Fax: +49-89-2809573
E-mail: {wied}{chris}{helmut}@photo.verm.tu-muenchen.de

Olivier Jamet
I.G.N., Laboratoire MATIS
2, avenue Pasteur, 94160 Saint Mandé, France
E-mail: Olivier.Jamet@ign.fr

Abstract

Internal self-diagnosis and external evaluation of the obtained results are of major importance for the relevance of any automatic system for practical applications. Obviously, this statement is also true for automatic image analysis in photogrammetry and remote sensing. However, so far only relatively little work has been carried out in this area. This is mostly due to the moderate results achieved. Only recently automatic systems reached a state in which a systematic evaluation of the results seems to be meaningful.

This paper deals with the external evaluation of automatic road extraction algorithms by comparison to manually plotted linear road axes used as reference data. The comparison is performed in two steps: (1) Matching of the extracted primitives to the reference network; (2) Calculation of quality measures. Each step depends on the other: the less tolerant is matching, the less exhaustive the extraction is considered to be, but the more accurate it looks. Therefore, matching is an important part of the evaluation process. The quality measures proposed for the automatically extracted road data comprise completeness, correctness, quality, redundancy, planimetric RMS differences, and gap statistics. They aim at evaluating exhaustivity as well as assessing geometrical accuracy. The evaluation methodology is presented and discussed in detail. Results of a comparative evaluation of three different automatic road extraction approaches are presented. They show the overall status of the road extractors, as well as the individual strengths and weaknesses of each individual approach. Thus, the applicability of the evaluation method is proven.

1: Introduction

Internal self-diagnosis and external evaluation of the obtained results are of major importance for the relevance of automatic systems for practical applications. Obviously, this statement is also true for automatic image analysis in photogrammetry and remote sensing. However, so far only relatively little work has been carried out in this area. This is mostly due to the moderate results achieved. Only recently automatic systems reached a state in which a systematic evaluation of the results seems to be meaningful.

Both, internal self-diagnosis and external evaluation should yield quantitative results which are independent of a human observer. A good description for the result of internal self-diagnosis is the traffic light paradigm [8]: a green light stands for a result found to be correct as far as the diagnosis tool is concerned, a red light means an incorrect result, and a yellow light implies that further probing is necessary. External evaluation needs some kind of reference data and compares them to the automatically obtained results. In this paper we deal with the external evaluation of automatic road extraction algorithms by means of comparison to manually plotted linear road axes used as reference data.

Only few approaches on evaluation of image analysis results are found in the literature. In [15] and [11] the evaluation of automated building extraction is reported. The results of the extraction are pixels (in image space) or voxels (in object space) which are classified as "building" or "non-building". The degree of overlap between the results of the automated extraction and a manually generated reference is determined by matching of the corresponding pixels or voxels, respectively. Subsequently, measures for quantifying exhaustivity and correctness of the extraction result are calculated. Road data from maps are analyzed with regard to distortions which are induced by the map production process in [10]. A data set of the French Topographic Database (BDTopo) is used as reference. The comparison is performed manually. The accuracy of the position of crossroads as well as the orientation of the connected roads, and their number and nature are investigated. Evaluation of the roads concentrates on measures for their geometrical accuracy. In [2] an evaluation methodology is proposed which is supposed to quantify the benefits of automatic and semi-automatic road extraction algorithms compared to manual data capture. The measures comprise geometric accuracy, success rate and in particular the time needed for data capture. [17] presents the evaluation of a multi-phase automatic road extraction. It points out the benefits of the different phases and quantifies the quality of the overall results. The reference data used is a data set of the BDTopo. Measures are geometric accuracy as well as exhaustivity of the extracted data. In [7] the evaluation is directed towards measuring the quality of (semi-)automatic road extraction with different levels of manual intervention. The reference data is generated by a procedure starting at manually selected positions, followed by automatic road tracking and manual editing. Roads are extracted as regions, and matching of the extracted data with the reference data is carried out using an intersection operation. Only the exhaustivity of the extracted data is further considered. [9] evaluates the effectiveness of different methods for the initialization of ribbon snakes as well as the geometric accuracy of the extracted road data. Manually generated road data serve as reference data. The evaluation focuses on the amount of effort needed by an operator which is measured by the number of necessary mouse actions. Measures for the geometric accuracy of the extracted road data are average and maximum deviation from the reference data.

This paper proposes and investigates a scheme for the evaluation of automatic road extraction. In this scheme various quality measures proposed in the literature are fused in a consistent manner. In the next Section three different road extraction schemes, for which the evaluation is carried out, are shortly reviewed. Section 3 is the main part of the paper. It presents the evaluation methodology in detail and discusses some implementation issues. In section 4 the results of the three different algorithms are presented and analyzed. The paper concludes with some final remarks and an outlook.

2: Road extraction

Basically, approaches for automatic road extraction use one or both of the following two properties: in low resolution imagery with pixel size of a few meters roads are modeled as lines, whereas in high resolution imagery (pixel size in the dm range) they are considered as homogeneous, elongated areas with parallel roadsides. In this paper three different approaches for the extraction of roads from digital imagery are evaluated: The first two combine line extraction in low resolution with a high resolution module based on grouping (TUM-G), and a snake-based technique, respectively (TUM-S). Both approaches were developed at Technische Universität München (TUM) and make use of the scale-space behavior of roads [14]. The third approach relies on homogeneity tracking in high resolution imagery and was developed at Institut Géographique National (IGN).

The **TUM-G** approach [4] is based on lines in an image of reduced resolution using the approach of [19] and edges in a high resolution image. By combining both resolutions and introducing explicit knowledge about roads, hypotheses for roadsides are generated. The roadsides are used to construct quadrilaterals representing road-parts and polygons representing intersections. Neighboring road-parts are chained into road-segments. Road-links, i.e., the roads between two intersections, are constructed by grouping of road-segments and closing of gaps between road-segments.

The **TUM-S** approach [13] is based on lines [19], too. In the high resolution imagery it uses so-called "snakes" [12] in the form of ribbon-snakes to extract roads and discriminate them from other line-type objects by means of the constancy of the width. What is more, the approach is able to bridge gaps between the lines, resulting, e.g., from shadows or partial occlusions. For the bridging, ziplock-snakes [16], i.e., snakes which are optimized starting at their end points, are employed.

The **IGN** road extractor is based on semi-automatic road plotting [1]. The core process is a road tracker that follows homogeneous elongated areas from a given seed point. This tracking is performed by searching for the continuation of the road through the construction of a local tree of possible paths. The best path is selected according to a quality function depending on the curvature of the path and its average homogeneity calculated from grey value variances of the branches. The stop criterion for the search relies on the computation of local parallel borders and of the dispersion of the search tree. For this experiment, the algorithm was seeded by a road seed detector based on parallel borders in a region-based segmentation of the image [18]. In order to perform road extraction in a fully automatic manner, seed detection was performed only within a buffer around the (approximately) known road position.

Global context and contextual information which describes so-called "outer characteristics" of roads, like "land cover area" and "bordered by" strongly influence the performance of road extraction [3, 5]. All three approaches cannot automatically extract roads in highly textured areas such as forests or urban areas due to their simplified models for roads. Therefore, the extraction is restricted to open areas in all cases. The delineation of the open areas is carried out automatically by texture classification.

3: Evaluation scheme

The evaluation of the extracted road data is carried out by comparing the automatically extracted road centerlines with manually plotted road axes used as reference data. Both data sets are given in vector representation. The evaluation is processed in two steps: (1) Matching of the extracted road primitives to the reference network and (2) Calculation of quality measures.

The proposed quality measures aim at assessing exhaustivity as well as geometrical accuracy. Each step depends on the other: the less tolerant is matching, the less exhaustive the extraction is considered to be, but the more accurate it looks. Therefore, matching is an important part of the evaluation process.

The quality measures address two questions: (1) How complete is the extracted road network, and (2) How correct is it. The completeness corresponds to the user's demands: ("how much is missing in the network"). The correctness, on the other hand is related to the probability of an extracted linear piece to be indeed a road. Thus, it is of high importance for a self-diagnosis module.

3.1: Matching procedure

The purpose of the matching is twofold: Firstly, it yields those parts of the extracted data which are supposed to be roads, i.e., which correspond to the reference road data. Secondly, it shows which parts of the reference data are explained by the extracted data, i.e., which correspond to the extracted road data.

There are various ways to perform the actual matching of two networks. Especially if the geometric distortions are large and not known beforehand, relational matching was used successfully [20, 6]. Special issues arise from the fact that the topologies of the reference and the extracted network can be different, and that the extraction can be redundant, i.e., extracted pieces overlap each other. The so called "buffer method", is a simple matching procedure in which every portion of one network within a given distance from the other network is considered as matched. The matching is not affected by different network topologies. The drawback of this procedure is that a highly redundant extraction will not be detected and that direction differences between parts of the two networks are not taken into account. Yet another method for matching consists in searching for a unique, i.e., bijective correspondence between the two networks. Such attempts have been made [21], however, it is not clear how to define such a correspondence for topologically different networks on a general basis.

In our case, there are very good approximations for position and orientation of the road data to be matched. As a consequence, matching is performed according to the buffer method and additional attention is paid to the problem of redundancy and direction differences.

In the first step both networks are split into short pieces of equal length. Then, a buffer of constant predefined width (buffer width) is constructed around the reference road data (see Fig. 1a). The parts of the extracted data within the buffer are considered as matched if the direction difference between the reference road data and the part to be matched does not exceed a given threshold. The direction difference can be derived directly from the vector representations of both networks. Following the notation of [15] and [7] the matched extracted data are denoted as *true positive* with length TP, emphasizing the fact that the extraction algorithm has indeed found a road. The unmatched extracted data is denoted as *false positive* with length FP, because the extracted road hypotheses are incorrect.

In the second step matching is performed the other way round. The buffer is now constructed around the extracted road data (see Fig. 1b), and the parts of the reference data lying in the buffer and fulfilling the direction constraint are considered as matched. In case of low redundancy their length can be approximated with TP. The unmatched reference data are denoted as *false negative* with length FN.

3.2: Quality measures

The quality measures for road extraction are intended to compare the results of different algorithms, rather than to evaluate the extraction and the matching results in an absolute way. Because these results are additionally quite different and still relatively far away from a perfect solution, a simplified set of measures is justified.

The definitions of the quality measures are presented in the following.

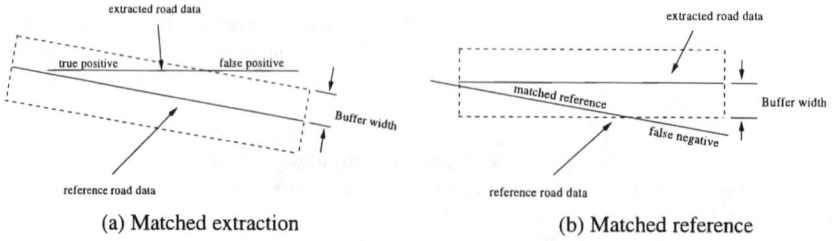

(a) Matched extraction   (b) Matched reference

Figure 1. Matching principle

- Completeness

$$completeness = \frac{length\ of\ matched\ reference}{length\ of\ reference}$$

$$\approx \frac{TP}{TP + FN}\ (for\ low\ redundancy)$$

$$completeness \in [0;1]$$

The completeness is the percentage of the reference data which is explained by the extracted data, i.e., the percentage of the reference network which lie within the buffer around the extracted data.

The optimum value for the completeness is 1.

- Correctness

$$correctness = \frac{length\ of\ matched\ extraction}{length\ of\ extraction}$$

$$= \frac{TP}{TP + FP}$$

$$correctness \in [0;1]$$

The correctness represents the percentage of correctly extracted road data, i.e., the percentage of the extracted data which lie within the buffer around the reference network.

The optimum value for the correctness is 1.

- Quality

$$quality = \frac{length\ of\ matched\ extraction}{length\ of\ extraction + length\ of\ unmatched\ reference}$$

$$= \frac{TP}{TP + FP + FN}$$

$$quality \in [0;1]$$

The quality is a more general measure of the final result combining completeness and correctness into a single measure:

$$quality = \frac{completeness * correctness}{completeness - completeness * correctness + correctness}$$

The optimum value for the quality is 1.

- Redundancy

$$redundancy \;=\; \frac{length\;of\;matched\;extraction - length\;of\;matched\;reference}{length\;of\;matched\;extraction}$$

$$redundancy \;\in\;]-\infty; +\infty[$$

The redundancy represents the percentage to which the correct (matched) extraction is redundant, i.e., it overlaps itself.

The optimum value for the redundancy is 0.

- RMS difference

$$RMS \;=\; \sqrt{\frac{\sum_{i=1}^{l}(d(extr_i; ref)^2)}{l}}$$

$$l \;=\; number\;of\;pieces\;of\;matched$$
$$extraction$$

$$d(extr_i; ref) \;=\; shortest\;distance\;between\;the$$
$$i-th\;piece\;of\;the\;matched$$
$$extraction\;and\;the\;reference$$
$$network$$

$$RMS \;\in\; [0; buffer\;width]$$

The RMS difference expresses the average distance between the matched extracted and the matched reference network, and thus the geometrical accuracy potential of the extracted road data. The value depends on the buffer width. If an equal distribution of the extracted road data within the buffer around the reference network is assumed, it can be shown that

$$RMS \;=\; \frac{1}{\sqrt{3}} * buffer\;width$$

The optimum value for RMS is 0.

- Gap statistics
  - Number of gaps per kilometer

$$No.\;of\;gaps\;per\;km \;=\; \frac{n}{length\;of\;reference\;[km]}$$

$$n \;=\; number\;of\;gaps$$

The number of gaps in the reference data, i.e., the number of connected *false negative* parts, is an indicator for the fragmentation of the extraction results.

The optimum value for the number of gaps per kilometer is 0.

  - Mean gap length

$$\mu_{gap\;length}\;[m] \;=\; \frac{\sum_{i=1}^{n}(gl_i)}{n}$$

$$gl_i \;=\; length\;of\;the\;i-th\;gap\;[m]$$

The optimum value for the mean gap length is 0.

Note that the completeness can be calculated from the number of gaps per kilometer and the mean gap length as follows:

$$completeness \;=\; 1 - (No.\;of\;gaps\;per\;km * \mu_{gap\;length}/1000)$$

3.3: Discussion of the evaluation scheme

There are several issues worth mentioning which may influence the significance of the proposed evaluation scheme and the accuracy of the results. They are mainly related to the employed matching strategy. Firstly, the two parameters which have to be chosen for the evaluation process as such, namely the buffer width and the maximum direction difference, are explained. Secondly, a property of extracted road data which may influence the result of the evaluation is discussed.

Buffer width: A suitable setting of the buffer width has to consider the expected internal accuracy of the road extraction algorithm. If the buffer width is set too large, false extractions close to the actual road will incorrectly be considered as roads. If it is chosen too small, correct road data which are only slightly geometrically inexact will be rejected.

For the evaluation of the results of the approaches described in Section 2, a buffer width of approximately half of the road width was chosen, i.e., it is assumed that a road is extracted correctly if the road axis lies between the roadsides.

Maximum direction difference: Some errors will not be detected, if the direction of the road axes is not taken into account during matching. Extracted road data can, e.g., result in a scenario similar to the one displayed in Figure 2. Without considering the direction constraint, the extracted road data within the buffer would be matched to the reference data, although this result is obviously incorrect. The directions of the extracted and the reference roads differ significantly which is an indicator for the error. This problem is solved by investigation of the direction difference of the two pieces to be matched. If it is larger than a threshold (maximum direction difference) no match between these pieces is established. Nevertheless, it is possible that matches exist to other pieces further away but with smaller direction differences.

The maximum direction difference should not be chosen too restrictive because of the direction uncertainty of short line segments, especially if they are an approximation of highly curved lines.

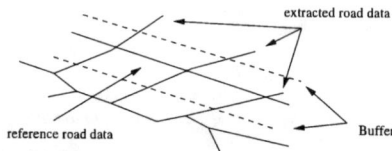

Figure 2. Incorrectly extracted road data

Shape: In some cases the results can wiggle around the reference data as depicted in Figure 3. A shape measure for the extracted data, e.g. based on curvature, can detect such problems. In the current implementation, however, no such measure has been included, because roads are implicitly or explicitly modeled as more or less straight segments in all three approaches. Consequently, no such effects have been observed in the extracted data. It should be pointed out that a shape measure becomes important as soon as generalized road axes are used as reference data.

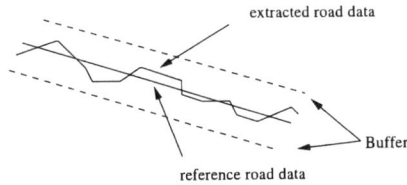

Figure 3. Extracted road data wiggling around the reference

3.4: Implementation issues

Both, the extracted data and the reference data are introduced into the evaluation procedure in vector representation, each as a set of polygons. Both networks are split into pieces of equal length (split length). Then, the shortest distance between each piece of one network and the other network data is calculated as explained below. The resulting distances are assigned to the respective pieces. Each piece is labeled as *matched* or *unmatched* based upon a check if its distance value is below the buffer width and if the direction constraint is fulfilled.

Because of the known and equal length of the pieces (split length) the length of the matched and unmatched network data can be easily computed by multiplication of the number of *matched/unmatched* pieces with the split length. This yields the length of the unmatched reference data (*FN*), the length of the matched extracted data (*TP*), and the length of the unmatched extracted data (*FP*). From these values the quality measures completeness, correctness, quality, and redundancy are computed using the formulas given in Section 3.2. The RMS difference is calculated from the distances assigned to the matched pieces of the extracted data. For the determination of the number of gaps per km and the mean gap length, the connectivity of the unmatched pieces of the reference data is analyzed.

There are three aspects to be discussed in more detail:

1. The definition of the shortest distance

2. The accuracy of the whole matching and the computed measures.

Ad 1: The definition of the shortest distance defines the outline of the buffer used for the matching. In the following, two possible definitions are described in detail (there are, of course, other options, they are not further pursued here).

Assume network **m** to be the one which will be labeled as *matched* or *unmatched*, depending on the overlap with the buffer around network **b**. For a particular point **P** on **m** the distance to **b** can be defined as the Euclidean distance between **P** and the intersection **I** between **b** and a straight line **l** through **P**, where **l** can be chosen either perpendicular to **m** in **P** (case 1, see Fig. 4a), or perpendicular to **b** in **I** (case 2, see Fig. 4b).

As **P** travels on **m** the shortest distance to **b** is calculated for every piece. In case 1 the calculation is self-evident, whereas it needs some further explanations in case 2: First, the correct **b**-piece has to be chosen. This is done by computing **l** to all **b**-pieces in a predefined vicinity of **P**, and subsequently selecting the piece with the smallest **l** fulfilling the direction constraint. There are pieces of **m** for which **I** lies outside any piece of **b** (see Fig. 5a). This problem is solved by allowing a non-perpendicular intersection of **l** with **b** if **I** is placed on a node of **b** which is connected to more than one other node (see Fig. 5b). In those cases, the direction of **b** is defined as the direction of the

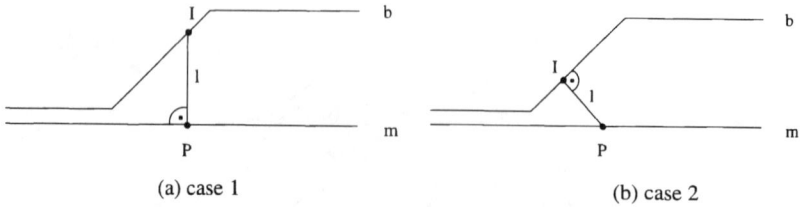

(a) case 1   (b) case 2

Figure 4. Distance definition

adjacent piece which yields the smallest direction difference.

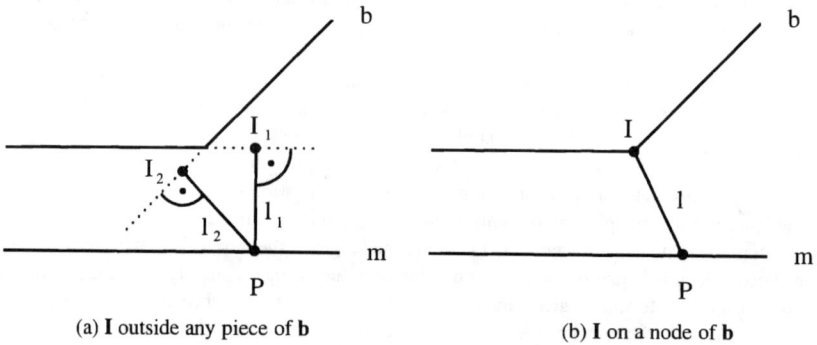

(a) **I** outside any piece of **b**   (b) **I** on a node of **b**

Figure 5. Special case of distance calculation in case 2

The value of the label (*matched/unmatched*) is determined for each piece of **m** based on a check if the distance value is below the buffer width, i.e., the parts of **m** which lie within the buffer around **b** are labeled as matched.

Case 1 of the distance definition yields a buffer width which depends on the direction difference between the two parts to be matched (see Fig. 6a). Problems which can be solved by an analysis of the direction difference (see Section 3.3) are treated implicitly by using a smaller buffer in case of larger direction difference.

In case 2 the resulting buffer width is constant and therefore independent of any direction difference (see Fig. 6b).

The evaluation results described in this paper are based on an implementation of case 2, because of the strict separation between distance and direction.

Ad 2: The accuracy of the matching and the quality measures is directly influenced by the split length used for splitting the networks. Obviously, a higher accuracy is achieved if a smaller split length is chosen. The split length used in this study was considerably smaller than the pixel size of the original grey value image (see also discussion on threshold parameters below). In this way, problems associated with the discretization are kept to an acceptable level.

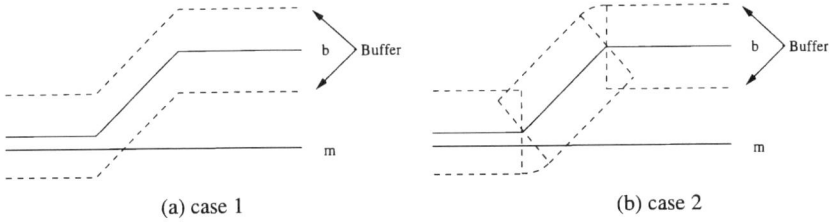

(a) case 1   (b) case 2

Figure 6. Resulting buffer

4: Experiments and results

The proposed scheme has been used to evaluate the results of the three described approaches on three different black and white images. A description of the test images and the evaluation procedure, the evaluation results and a detailed analysis thereof are presented in this Section.

4.1: Test images and test procedure

The test images are described in table 1 and are depicted below along with the extraction results. The image Marchetsreut with a groundel size (pixel size on the ground) of 0.225 m is a rather easy scene, Erquy (groundel size 0.45 m), and Montserrat (groundel size 0.225 m) are more difficult, because in some parts, the road models of the approaches are violated. It should be noted that all three images are rather large, their size amounts to approximately 1/16 of a photogrammetric aerial image.

Name of test site	description of image content	scale	pixel size in image space [μm]	groundel size in object space [m]	image size [pixel]	length of reference network [km]
Marchetsreut	flat, agricultural, easy	1:15,000	15	0.225	4000^2	3.84
Erquy	flat, agricultural, difficult	1:30,000	15	0.45	4500^2	24.10
Montserrat	hilly, agricultural, very difficult	1:15,000	15	0.225	4000^2	8.42

Table 1. Description of test images

For the test all three algorithms were run in a totally automatic fashion. As described in Section 2 road extraction was only carried out in the open areas. It should be mentioned that each algorithm requires a number of threshold parameters to be set prior to the computations. E.g., the line algorithm for the TUM approaches is based on low resolution images. They were generated by subsampling the test images to a pixel size of 3.6 m. As mentioned earlier, the aim of the study is to investigate the evaluation scheme. Information concerning the general applicability of each algorithm across different images is of secondary importance only. Therefore, it was thought acceptable to change threshold parameters from one image to the next according to visual inspection.

Also the matching procedure needs some parameters to be set, namely the buffer width, the maximum direction difference, and the split length. These parameters were chosen equal for all evaluations, and were set to 3 m for the buffer width (assuming a maximum road width of 6 m), 20° for the maximum direction difference, and 0.1 m for the split length.

The results are depicted in the Figures 7 through 9. For each test image the image superimposed with the mask for open area, the manually plotted reference road axes, the three results delivered by the extraction algorithms, and a table listing the quality measures are shown.

4.2: Discussion of results

First, the results of each algorithm are discussed separately. Then some comments valid for the whole investigation are given.

TUM-G algorithm: Due to the strong road model and the combination of low and high resolution image information, the TUM-G approach is able to deliver relatively stable road hypotheses which are also used to bridge small gaps. Consequently, the extracted road parts are well connected and thus quite long, the number of gaps is moderate, and the average gap length is not too high. Most important, the correctness is relatively high. The geometrical accuracy lies between 1 and 4 pixels. Although this result is acceptable, there is room for improvement.

Whereas for Marchetsreut most parts of the road network have been extracted, the algorithm has some problems with the more difficult scenes Erquy and especially Montserrat in which roads have a greatly varying appearance.

TUM-S algorithm: The TUM-S approach is especially aiming at bridging gaps due to shadows and occlusions. As desired TUM-S is bridging some of the gaps in the result of TUM-G. Therefore, the overall number of gaps is smaller than that of TUM-G which indicates that the connectivity of the extracted road parts is better. The geometrical accuracy again lies approximately between 1 and 4 pixels.

For Marchetsreut and Erquy the performance is a little worse than that of TUM-G (except the correctness), whereas better results were achieved on the Montserrat image.

IGN algorithm: The road model of the IGN approach depends mainly on the homogeneity of the road instead of the parallelity of the roadsides. Therefore, a performance different from the two TUM approaches can be expected. This is correctly expressed by the quality measures: the IGN approach performs good for the Marchetsreut image, rather poor for the Erquy image, but best for the Montserrat image. The geometrical accuracy lies between 2 and 4 pixels. The connectivity of the extracted road parts is relatively good.

The high correctness of the extracted data especially for the Montserrat image is partly due to the limited search space used for the road seed detection (see Section 2).

Additional comments: Looking at all results it becomes clear that the proposed quality measures adequately capture the impression obtained when visually inspecting the extracted road data. Thus, they can serve as a basis for the comparison of different automatic road extraction algorithms. Besides, the results are an indication for the applicability of the road extraction algorithms: they can serve as automatic extraction tools for easy scenes. It should be noted, however, that due to the effects mentioned in Section 3.3 and the discretization (see Section 3.4), the numerical accuracy of the quality measures is not extremely high.

The significance of these measures for a detailed comparison can be further improved by classifying the reference data into different local categories such as clearly visible road parts, roads in shadow, occluded roads etc. or regional categories like open area, urban area, forest, etc.

The number and mean length of the gaps needs some further discussion. Obviously the best result consists in having very few and short gaps. However, it is not clear whether a small number of long gaps is to be preferred to a large number of short gaps. The choice depends on the extra work necessary for closing the gaps. More detailed investigations are needed to clarify this issue.

5: Summary and conclusions

Automatic evaluation of the obtained results is an increasingly important topic in image analysis as results are approaching a point where they become useful for practice. In this paper a methodology for the evaluation of automatic road extraction algorithms based on the comparison to manually plotted reference data is presented. This methodology was tested using the results of three approaches across three different test images.

The obtained results are representative for the state-of-the-art of automatic road extraction from aerial images. In easy scenes a completely automatic extraction is possible. As the scenes become more difficult the obtained results start degrading. Low resolution images with a pixel size of a few meters can (and should) be used as a preprocessing step in the extraction. Reliable extraction only on the basis of these images, however, is not realistic. The key factor for improvement is a more detailed modeling of the roads and their surroundings. E.g., improvements of the TUM-G approach could be achieved firstly by analyzing the gaps in the extraction results, and finding and subsequently modeling the underlying reasons for these gaps such as shadows, occlusions etc. Secondly, other criteria like the homogeneity of the path employed in the IGN approach could complement the approach.

The proposed evaluation scheme adequately captures the characteristics of the individual extraction results and can thus serve as a basis for their comparison and integration. Depending on the application at hand some of the quality measures such as completeness in a semi-automatic environment may be more relevant than others. Additional measures could be thought of, e.g. an analysis of topological differences between extracted and reference data, especially in the vicinity of crossings, or the ratio between completeness and correctness, which should remain constant over different images, once a suitable ratio has been found. Also, the algorithmic complexity and thus the computational effort needed will become a criteria as automatic road extraction advances further towards practical applications.

The proposed evaluation scheme can also form the basis for automatic updating of geo-data, which is becoming an increasingly important issue. In this case, the reference data are substituted by existing, but out-dated geo-data, and these are compared to extracted data from an up-to-date image. It should be noted that in this scenario some of the proposed measures such as the correctness and the RMS difference loose their significance, and a sound uncertainty management is needed because the assumption that the given vector data be correct and complete is no longer valid. This topic will be further investigated in future research.

6: Acknowledgment

We would like to express our gratitude to Albert Baumgartner, Ivan Laptev, and Cecile Huet for contributing the software and the results of the three road extraction algorithms to this study.

References

[1] S. Airault and O. Jamet. Détection et réstitution automatique du réseau routier sur des images. *Traitement du signal*, 12(2):189–200, 1995.

[2] S. Airault, O. Jamet, and F. Leymarie. From Manual to Automatic Stereoplotting: Evaluation of Different Road Network Capture Processes. In *International Archives of Photogrammetry and Remote Sensing*, volume **31**(3), pages 14–18, 1996.

[3] A. Baumgartner, W. Eckstein, H. Mayer, C. Heipke, and H. Ebner. Context-Supported Road Extraction. In *Automatic Extraction of Man-Made Objects from Aerial and Space Images (II)*, pages 299–308, Basel, Switzerland, 1997. Birkhäuser Verlag.

[4] A. Baumgartner, C. Steger, H. Mayer, and W. Eckstein. Multi-Resolution, Semantic Objects, and Context for Road Extraction. In *Workshop on Semantic Modeling for the Acquisition of Topographic Information from Images and Maps*, pages 140–156, Basel, Switzerland, 1997. Birkhäuser Verlag.

[5] G. Bordes, G. Giraudon, and O. Jamet. Road Modeling Based on a Cartographic Database for Aerial Image Interpretation. In *Semantic Modeling for the Acquisition of Topographic Information from Images and Maps*, pages 123–139, Basel, Switzerland, 1997. Birkhäuser Verlag.

[6] W.J. Christmas, J. Kittler, and M. Petrou. Structural Matching in Computer Vision Using Probabilistic Relaxation. *IEEE Transactions on Pattern Analysis and Machine Intelligence*, 17(8):749–764, 1995.

[7] CMU. Performance Evaluation for Feature Extraction. Slides presented at the Terrain Week 1997 (http://www.cs.cmu.edu/afs/cs/usr/maps/www/rcvw/ terrainweek97/roads/tw97-roadeval.ROOT.html), 1997.

[8] W. Förstner. 10 Pros and Cons Against Performance Characterization of Vision Algorithms. In *European Conference on Computer Vision, Workshop "Performance Characteristics of Vision Algorithms"*, pages 13–29, 1996.

[9] P. Fua. *RADIUS: Image Understanding for Intelligence Imagery*, chapter Model-Based Optimization: An Approach to Fast, Accurate, and Consistent Site Modeling from Imagery. Morgan Kaufmann, 1997. O. Firschein and T.M. Strat, Eds. Available as Tech Note 570, Artificial Intelligence Center, SRI International.

[10] P. Guérin, O. Jamet, and H. Maître. Distortion Model in Road Networks from Topographic Maps: identification and Assessment. In *SPIE: Integrating Photogrammetric Techniques with Scene Analysis and Machine Vision II*, volume 2486, pages 232–243, April 1995.

[11] Y. Hsieh. Design and Evaluation of a Semi-Automated Site Modeling System. Technical Report CMU-CS-95-195, Computer Science Department, Carnegie Mellone University, 1995.

[12] M. Kass, A. Witkin, and D. Terzopoulos. Snakes: Active Contour Models. *International Journal of Computer Vision*, 1(4):321–331, 1987.

[13] H. Mayer, I. Laptev, A. Baumgartner, and C. Steger. Automatic Road Extraction Based on Multiscale Modeling, Context, and Snakes. In *International Archives of Photogrammetry and Remote Sensing*, volume **32**(3-2W3), pages 47–56, 1997.

[14] H. Mayer and C. Steger. A New Approach for Line Extraction and its Integration in a Multi-Scale, Multi-Abstraction-Level Road Extraction System. In *IAPR TC-7 Workshop: Mapping Buildings, Roads and other Man-Made Structures from Images*, pages 331–348, Vienna, Austria, 1996. Oldenbourg Verlag.

[15] C. McGlone and J. Shufelt. Projective and Object Space Geometry for Monocular Building Extraction. In *Computer Vision and Pattern Recognition*, pages 54–61, 1994.

[16] W. Neuenschwander, P. Fua, G. Székely, and O. Kübler. From Ziplock Snakes to Velcro[tm] Surfaces. In *Automatic Extraction of Man-Made Objects from Aerial and Space Images*, pages 105–114, Basel, Switzerland, 1995. Birkhäuser Verlag.

[17] R. Ruskoné and S. Airault. Toward an Automatic Extraction of the Road Network by Local Interpretation of the Scene. In *Photogrammetric Week '97*, pages 147–157, 1997.

[18] R. Ruskoné, S. Airault, and O. Jamet. Road Network Interpretation: A Topological Hypothesis Driven System. In *International Archives of Photogrammetry and Remote Sensing*, volume **30** (3/2), pages 711–717, 1994.

[19] C. Steger. Removing the Bias from Line Detection. In *Computer Vision and Pattern Recognition*, pages 116–122, 1997.

[20] G. Vosselman and N. Haala. Erkennung topographischer Paßpunkte durch relationale Zuordnung. *Zeitschrift für Photogrammetrie und Fernerkundung*, 6/92:170–176, 1992.

[21] V. Walter. *Zuordnung von raumbezogenen Daten - am Beispiel der Datenmodelle ATKIS und GDF*. PhD thesis, Fakultät für Bauingenieur- und Vermessungswesen, Universität Stuttgart, 1996.

a) Grey value image superimposed
with mask for open area

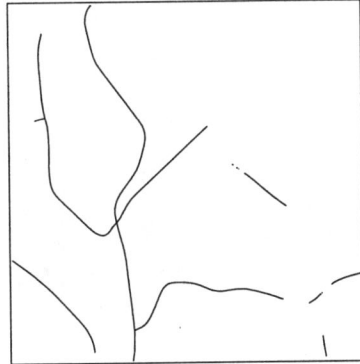

d) Results of TUM-G algorithm

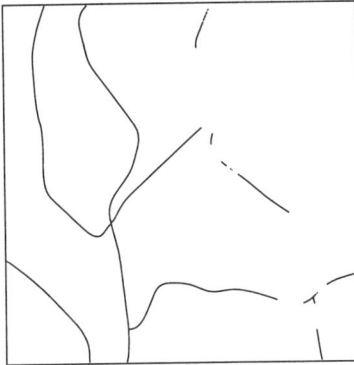

b) Manually plotted reference

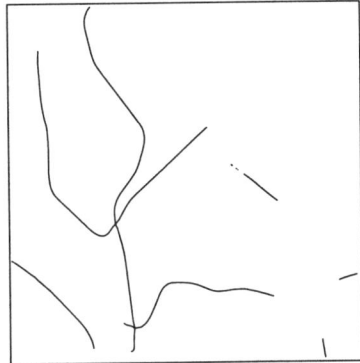

e) Results of TUM-S algorithm

	TUM-G	TUM-S	IGN
Completeness	0.90	0.85	0.80
Correctness	0.99	0.98	0.94
Quality	0.89	0.84	0.76
Redundancy	0.00	-0.01	0.00
RMS [m]	0.27	0.38	0.52
No. of gaps per km	7.06	5.97	7.06
$\mu_{gap\ length}$ [m]	13.58	24.80	27.84

c) Quality measures

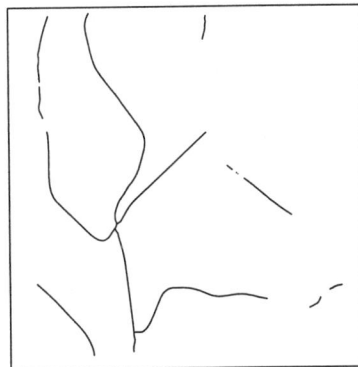

f) Results of IGN algorithm

Figure 7. Test image Marchetsreut

a) Grey value image superimposed
with mask for open area

d) Results of TUM-G extraction

b) Manually plotted reference

e) Results of TUM-S algorithm

	TUM-G	TUM-S	IGN
Completeness	0.77	0.72	0.43
Correctness	0.93	0.95	0.60
Quality	0.73	0.69	0.34
Redundancy	0.02	-0.01	0.00
RMS [m]	0.53	0.47	0.92
No. of gaps per km	5.11	4.85	4.94
$\mu_{gap\ length}$ [m]	44.35	58.00	115.05

c) Quality measures

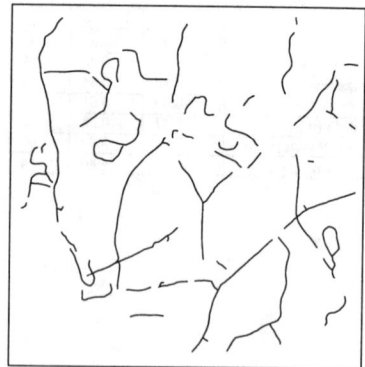

f) Results of IGN algorithm

Figure 8. Test image Erquy

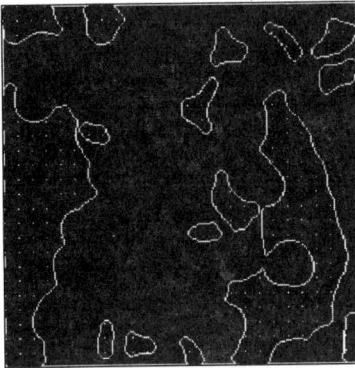

a) Grey value image superimposed
with mask for open area

d) Results of TUM-G extraction

b) Manually plotted reference

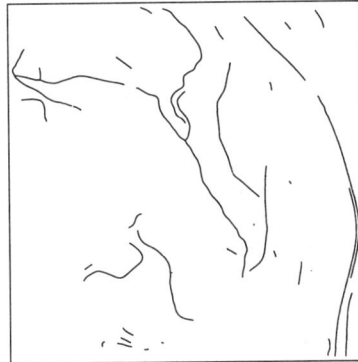

e) Results of TUM-S algorithm

	TUM-G	TUM-S	IGN
Completeness	0.31	0.45	0.46
Correctness	0.61	0.66	0.90
Quality	0.25	0.36	0.44
Redundancy	-0.01	-0.03	0.00
RMS [m]	0.91	0.66	1.05
No. of gaps per km	5.05	4.56	4.92
$\mu_{gap\ length}$ [m]	137.54	120.56	108.92

c) Quality measures

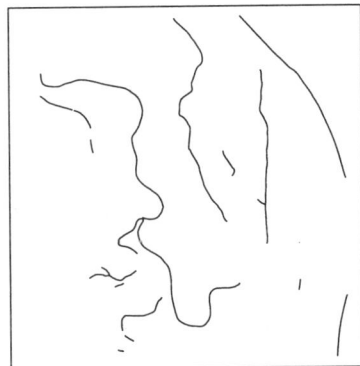

f) Results of IGN algorithm

Figure 9. Test image Montserrat

Analytical and Empirical Performance Evaluation of Subpixel Line and Edge Detection

Carsten Steger

Forschungsgruppe Bildverstehen (FG BV), Informatik IX

Technische Universität München, Orleansstr. 34, 81667 München

Phone: +49 (89) 48095-211, Fax: +49 (89) 48095-203

E-mail: stegerc@informatik.tu-muenchen.de

Abstract

An extensive analysis of the quality of the extraction results of a subpixel line detector and a subpixel edge detector is carried out. The localization quality of line and edge points in the presence of noise is analyzed analytically, and new formulas describing this relationship are derived. Tests on synthetic noisy images show the formulas to hold very well in practice. Experiments on real images show that subpixel accuracy better than one tenth of a pixel is possible in typical industrial inspection tasks.

1: Introduction

The analysis of the performance characteristics of a feature extraction algorithm is very important [8]. First, it makes an algorithm comparable to other algorithms, thus helping users in selecting the appropriate method for the task they have to solve. Second, it helps to identify breakdown points of the algorithm, i.e., areas where the algorithm cannot be used because some of the assumptions it makes are violated. Therefore, in this paper an attempt is made to characterize the performance of the subpixel line and edge detection algorithms proposed in [15, 16, 17]. The study is mainly concerned with the subpixel precision and accuracy of the extracted line and edge positions and line widths because of their importance for many applications, especially industrial inspection. Following [8], for the purpose of this paper precision refers to the repeatability of the extraction results, i.e., their variance, and accuracy refers to their absolute errors, including possible bias.

2: Subpixel line and edge detection algorithms

The line detector to be analyzed is discussed in detail in [15, 16, 17]. Therefore, only a brief overview is given here. The algorithm models lines as curves $s(t)$ that exhibit a characteristic 1D line profile in the direction perpendicular to the line. The most relevant type of line profile for applications is the asymmetrical bar-shaped profile given by

$$f_a(x) = \begin{cases} 0, & x < -w \\ 1, & |x| \leq w \\ a, & x > w \ , \end{cases} \tag{1}$$

where w is half the line width and $a \in [0, 1]$ is the asymmetry. General lines of contrast h can be obtained by considering a scaled asymmetrical profile.

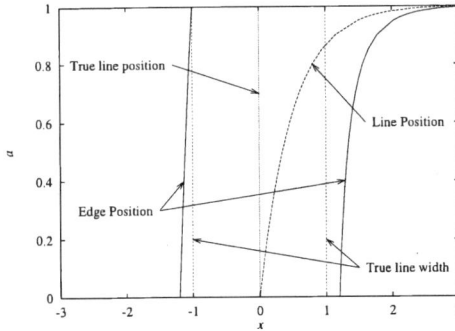

Figure 1. Position of an asymmetrical line and its corresponding edges with width $w = 1$, $\sigma = 1$, and $a \in [0, 1]$.

In 1D, line positions can be extracted by convolving $f_a(x)$ with the derivatives of a Gaussian smoothing kernel $g_\sigma(x)$ of width σ. This leads to a scale-space description of the model line profile:

$$r_a(x, \sigma, w, a) = \phi_\sigma(x + w) + (a - 1)\phi_\sigma(x - w) \tag{2}$$

$$r'_a(x, \sigma, w, a) = g_\sigma(x + w) + (a - 1)g_\sigma(x - w) \tag{3}$$

$$r''_a(x, \sigma, w, a) = g'_\sigma(x + w) + (a - 1)g'_\sigma(x - w) . \tag{4}$$

Line points are given by the points where $r'_a(x, \sigma, w, a) = 0$. Salient lines are selected based on the magnitude of the second derivative $r''_a(x, \sigma, w, a)$. For the extraction of the line width, the two edges to the left and right of the line position need to be extracted. They are given by the points where $r''_a(x, \sigma, w, s) = 0$ and $r'''_a(x, \sigma, w, a)r'_a(x, \sigma, w, a) < 0$. It can be shown that Gaussian smoothing necessarily leads to a bias of the extracted line positions if $a \neq 0$. The biased line position is given by

$$l = -\frac{\sigma^2}{2w} \ln(1 - a) . \tag{5}$$

Similarly, the edge position will generally be biased. An example of this behavior for $w = 1$, $\sigma = 1$, and $a \in [0, 1]$ is shown in Figure 1. It can be seen that as a gets larger the line and edge positions are pushed to the weak side, i.e., the side that possesses the smaller edge gradient. Furthermore, the line width is extracted too large for all a.

In [15, 16, 17] it is shown that the bias of the line position and widths can be predicted analytically. The true values of the line width and asymmetry are mapped to the line width and gradient ratio, which are extractable from the image. Since this map is invertible, the bias can be removed by plugging the extracted features into the inverted bias prediction.

In 2D, lines are modeled as curves $s(t)$ that exhibit the characteristic profile f_a in the direction perpendicular to the line, i.e., perpendicular to $s'(t) = n(t)$. This means that the first directional derivative in the direction $n(t)$ must vanish, while the second directional derivative should be of large absolute value. The necessary 1D zero crossing is obtained with subpixel accuracy by extrapolating it from a local second-order Taylor polynomial. The direction $n(t)$ is obtained from the eigenvector corresponding to the largest eigenvalue of the Hessian matrix of the image convolved with a Gaussian. To detect the width of the line, for each line point the closest points to the left

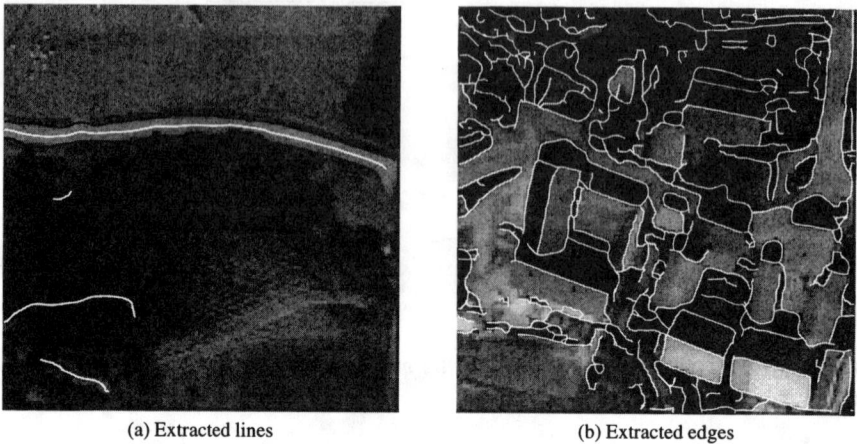

(a) Extracted lines   (b) Extracted edges

Figure 2. Lines and their width extracted in an aerial image (a). Edges extracted in an aerial image (b). Lines and edges are displayed in white, while the line width is displayed in black.

and right of the line point, i.e., along $-n$ and n, where the absolute value of the gradient takes on its maximum value, are determined. This is done by regarding the edges of the line as lines in the gradient image, and using the same techniques as mentioned above to achieve subpixel accuracy. The extracted line width and gradient ratio are plugged into the inverted bias function to achieve unbiased extraction results. The unbiasedness of the extracted line position, width, contrast, asymmetry, and orientation was established for noise-free synthetic images in [17]. Figure 2(a) shows an example of the lines extracted with this approach. Note that the lines are unbiased in the entire image, especially in the highly asymmetrical part in the center of the road.

The above mentioned approach to regard edges as lines in the gradient image leads quite naturally to a subpixel precise edge detector [17]. An example of the edges extracted with this approach is shown in Figure 2(b).

3: Quality of the localization of line and edge points

For real images, noise plays a significant role. Different "types" of noise can be distinguished: The first kind is characterized by random fluctuations of the image intensity. This is the "classical" notion of random noise. A second kind of "noise" is what might be called "clutter noise" or "structured noise," i.e., random structures that have the same appearance as the objects of interest. In this paper, only random noise in the classical sense will be considered for the following reasons: First, in the intended application area of industrial inspection one usually has excellent control over the lighting conditions. Often, objects are back-lit, and hence clutter noise can be eliminated à priori. Furthermore, the distinction between clutter noise and salient objects often involves a distinction on a semantical level, which cannot be decided in a low-level feature extractor. For example, in road extraction applications from low-resolution images roads can be detected as lines. However, house roofs often appear as short lines, and thus can be considered as clutter noise *in this particular application*, although for the low-level feature extractor they are perfectly good lines. Therefore, a meaningful performance evaluation of an algorithm in the presence of clutter noise seems only possible using the semantics of a particular application. For example, an evaluation of the line extractor discussed here and in [15, 16, 17] for the purpose of road extraction can be found in [10].

Random noise plays a significant role for the precision and accuracy of the extracted features, and therefore is of major importance for industrial inspection, where objects often have to be measured with subpixel precision and accuracy because of the limited resulution of the imaging sensor. This kind of noise enters the image in different stages of the imaging process, e.g., through the photon flux on the sensor or the quantization of the image intensity to a finite number of gray levels. The noise $n(x)$ can be modeled as a random process in one dimension and as a random field in higher dimensions. Thus, the observed version of the image is given by

$$i(x,y) = f(x,y) + n(x,y) \ .$$ (6)

The noise component n is assumed to be wide-sense stationary, statistically independent of the image content f, and white, i.e., its expected value $E(n) = 0$ and its autocorrelation $R(\tau) = \sigma_n^2 \delta(\tau)$, where σ_n^2 is the variance of the noise [13].

Obviously, if noise is added to an image the extracted line or edge positions, and in general any extraction results, such as the line widths, will not be the correct values that would have been obtained from the uncorrupted version of the image. Therefore, one of the major performance criteria of an algorithm is the dependence of the extracted features on the input noise level. Put into different terms, what needs to be determined is how the variance of the line position and width and the variance of the edge position varies as the variance of the noise in the image varies. Ideally, if this dependency has been determined the algorithm could return an estimate of the variance of the respective feature, e.g., the line position, if given an estimate of the variance of the image noise. The variance σ_n of the image noise can, for example, be determined by automatic procedures from an input image [2, 7]. Such a variance estimate can serve as an aid for self-diagnosis [8], e.g., to determine if certain requirements of the output data can be met. For example, if an application requires one tenth of a pixel accuracy in the line or edge position, and the returned standard deviation of the line position is larger than some fraction of this amount, an algorithm can alert the user to the fact that the lines cannot be extracted with the required precision. The user could then try to fix the problem, e.g., by changing the lighting conditions or by using a different sensor to reduce the noise level in the images.

3.1: Previous work on localization quality

Over the years, several attempts have been made to characterize the quality of the extracted feature positions. Most of them were made for edge detectors. Often, these quality measures are made in connection with deriving an "optimal" operator with respect to several criteria, one of which often is a term that tries to measure the goodness of edge locations. One of the earliest approaches to this problem is introduced in [3], where an optimal 1D edge detector with respect to three criteria is derived: good detection, i.e., the algorithm should have a high probability of extracting true edge points, while having a low probability of extracting false edge points, good localization, i.e., the extracted edge points should be as close as possible to the true edge points, and finally having only a single response to an edge. The optimal operator is described by a convolution of the image with a kernel of finite width $2w$. It is assumed to yield a maximum response at the edge location, i.e., the first derivative of the filter response at the edge is zero. With this, the localization criterion can be derived as follows: Assume the response of convolving the image i with the edge operator e to be given by

$$r(x) = h(x) * i(x) = h(x) * f(x) + h(x) * n(x) = r_f(x) + r_n(x) \ . \tag{7}$$

Then, due to the image noise n, the extracted edge location will lie at a point x_e, where $r'(x_e) = 0$. Thus, $r'_n(x_e) + r'_f(x_e) = 0$. If this equation is expanded into a Taylor series of order 1 about the origin, one obtains $r'(x_e) = r'(0) + r''(0)x_e + O(x_e^2)$. Since the model edge is located at 0, by assumption $r'(0) = 0$, and hence $x_e \approx -r'_n(x_e)/r''(0)$. In [3] it is now claimed that $r'_n(x_e)$ is a Gaussian random variable with variance

$$E(r'_n(x_e)^2) = \sigma_n^2 \int_{-w}^{w} f'(x)^2 \, dx \ . \tag{8}$$

Thus, the variance of the edge position is given by

$$E(x_e^2) = \frac{\sigma_n^2 \displaystyle\int_{-w}^{w} f'(x)^2 \, dx}{\left(\displaystyle\int_{-w}^{w} h'(-x)f'(x) \, dx \right)^2} \ . \tag{9}$$

These continuous criteria for an optimal edge detector are adapted for pixel-precise discrete edge operators in [4], where mainly the formulation of the single response criterion changes.

This definition of the localization quality measure was later criticized for being incorrect [19]. The major objection put forth there is that (8) only holds if x_e is constant for all realizations, which, of course, is not the case. Another objection is that the localization criterion in [3] only takes into account the first zero crossing of $r'(x)$, whereas it should take into account all zero crossings of the response. Therefore, the physical density of the zero crossings, i.e., the expected number of zero crossings per unit time, of the edge detector applied to a noisy step edge is derived in [19]. This density is given by

$$\mu(x) = \frac{1}{2\pi} \sqrt{-\frac{R''_{n'}(0)}{R_{n'}(0)}} \, \exp(-f(x)^2/2\sigma_{n'}^2) \ . \tag{10}$$

The authors note that $1 - \mu(x)$ is a measure for the suppression of the density of zero crossings by the edge operator, and go on to prove that the first derivative of the Gaussian kernel is the unique function that optimizes this criterion. With this criterion, in effect the localization and single response criteria used in [3] are combined into a single criterion, yielding a better optimality criterion for edge operators. Of course, $\mu(x)$ is not very useful to characterize the dependency of the localization of edges on the image noise, precisely for the fact that all zero crossings are taken into account.

For this reason, an attempt is made in [12] to derive the true variance of the edge position. The authors note that in addition to the problems mentioned in [19], i.e., the wrong derivation of $E(r'(x)^2)$, the Taylor expansion of the edge location should be done differently. They claim that the correct Taylor expansion is given by $r'(x_e) = r'_n(0) + (r''_f(0) + r''_n(0))x_e$, and hence $x_e \approx -r'_n(0)/(r''_f(0) + r''_n(0))$. They then go on to derive the probability density of x_e and arrive at the result that it is given by a sum of two Cauchy densities. Therefore, they note, the variance of x_e does not exist. Of course, in this case the mean of x_e also would not exist. What this obviously erroneous result implies is that if even the tiniest amount of noise is added to an image, the edge locations would fluctuate wildly all over the image, which is, of course, not supported by experimental evidence. If this were true, edges could in effect not be extracted at all because their positions would be meaningless.

A different approach is given in [1], where the imaging process is modeled in several steps. First, the ideal continuous image is assumed to be blurred by the imaging device with a Gaussian kernel with standard deviation σ_b. Then, the image is sampled and noise with standard deviation σ_n is added to the image. Finally, edge extraction is done by smoothing the image with a Gaussian kernel with standard deviation σ_e and calculating the gradient of the resulting image. The authors give the following formula for the variance of an edge of height h, if the edge extraction is performed in a direction forming an angle α with the direction perpendicular to the edge:

$$E(x_e^2) = \frac{3\sigma_n^2(\sigma_b^2 + \sigma_e^2)^3}{8h^2\sigma_e^6 \cos^6 \alpha} \quad . \tag{11}$$

Unfortunately, almost no details of the derivation of (11) are given, and hence it is hard to judge whether (11) is correct.

Contrary to this, edge detection is regarded as template matching in [8]. In this case, the variance of the edge position is given by

$$\sigma_e^2 = \frac{\sigma_n^2}{\displaystyle\sum_{r=-w}^{w}\sum_{c=-w}^{w} f_u^2(r, c)} \quad , \tag{12}$$

where the template is of size $(2w + 1) \times (2w + 1)$ and $f_u(r, c)$ is the derivative of the template in the direction perpendicular to the edge. Unfortunately, since the edge and line detection algorithms discussed in Section 2 do not use template matching, this result does not apply to them.

3.2: Variance of the zero crossing of an arbitrary function

Because none of the above derivations of the variance of the edge position are applicable to the problem of determining the variance of the line position and width, and because some of them are erroneous even for the problem of determining the variance of the edge position, there is a need for an, at least qualitatively, correct formula that also is supported by experimental evidence. The algorithm of determining the line position in the 1D continuous case can be broken down into two steps: the input signal is first convolved with the derivative of a Gaussian kernel, and then the zero crossings of the resulting response function $r'(x)$ are determined, where additionally $r''(x) < 0$ is required. For edges, the only change is that the zero crossings of the second derivative are determined. Since convolution with a Gaussian kernel is a linear operation, the response $r(x)$ of the line detection operator to the corrupted signal $i(x)$ is given by

$$r(x) = g_\sigma(x) * i(x) = g_\sigma(x) * f(x) + g_\sigma(x) * n(x) = r_f(x) + r_n(x) \tag{13}$$

and analogously for the corresponding derivatives. Thus, the response of the line detection operator to a model line profile $f(x)$ corrupted by wide-sense stationary white noise $n(x)$ can be broken down into a deterministic part $r_f(x)$ and a stochastic process $r_n(x)$. The deterministic part can, of course, be determined analytically, e.g., by (2)–(4) for the asymmetrical bar-shaped profile. Note that the deterministic part of the response has exactly one zero crossing of $r'(x)$ in $[-\infty, \infty]$. The random part of the response shifts this zero crossing and may introduce additional zero crossings of $r'(x)$. In order to determine the localization quality of the line detection algorithm, it is therefore useful to consider the variance of the zero crossing of $r'(x)$ closest to the zero crossing of the deterministic part $r'_f(x)$ of the response.

Since the noise $n(x)$ is assumed to be wide-sense stationary and white, the statistics of the response of the line detector to the noise can be calculated easily [13, 18]. Because the mean value $E(n(x))$ is assumed to be zero, convolution of $n(x)$ with an arbitrary derivative of the Gaussian kernel is also a stationary stochastic process with zero mean. Furthermore, the autocorrelation function $R_r(\tau)$ of the noise smoothed with a Gaussian kernel is given by [13]

$$R_r(\tau) = \sigma_n^2 g_{\sqrt{2}\sigma}(\tau) \ , \tag{14}$$

while the autocorrelation function of the noise convolved with the first derivative of a Gaussian kernel is

$$R_{r'}(\tau) = -R_r''(\tau) = -\sigma_n^2 g_{\sqrt{2}\sigma}''(\tau) \ . \tag{15}$$

Thus, the variances of the respective stochastic processes are

$$\sigma_r^2 \ = \ R_r(0) = \frac{\sigma_n^2}{2\sqrt{\pi}\sigma} \tag{16}$$

$$\sigma_{r'}^2 \ = \ R_{r'}(0) = \frac{\sigma_n^2}{4\sqrt{\pi}\sigma^3} \ . \tag{17}$$

With this, the problem of determining the variance of the line position can be transformed to the equivalent problem of determining the variance of the zero crossing of an arbitrary function with exactly one zero crossing, e.g., (3), corrupted by a the addition of a stationary stochastic process with zero mean and a given autocorrelation function, e.g., (17). Without loss of generality, the zero crossing of the function can be assumed to be at the origin.

Before an approximation of the variance of a zero crossing with added noise will be derived, it is useful to look at this problem in a purely geometrical manner to get an intuitive notion of the dependency. Assume for the moment that the uncorrupted signal is simply the linear function $f(x) = hx$, and that the distribution of the noise $n(x)$ is uniform in some interval $[-m, m]$ for all x. Then, conceptually one can think of the corrupt signal $i(x)$ to be contained entirely within a "tube" of diameter $dy = 2m$ around $f(x)$, as shown in Figure 3. Therefore, the zero crossings must all lie in an interval of diameter dx around 0. Obviously, $dy/dx = h$, and therefore the width of this interval is proportional to $1/h = 1/f'(0)$. Although the distribution of the zero crossings within the interval is unknown, the variance of the zero crossings can be expected to be proportional to $1/f'(0)^2$. Thus, it can be conjectured that for general functions the variance of the zero crossings is proportional to $\sigma_n^2/f'(0)^2$.

In order to prove this conjecture, a Taylor series expansion of order one of (13) can be performed, where for the moment it should be assumed that $r_f(x)$ and $r_n(x)$ are not obtained by convolution with a Gaussian, but are an arbitrary function and noise with a given autocorrelation function, respectively. Thus, $r(x_0) = 0$ for some x_0, and hence

$$r(x_0) \approx r(0) + r'(0)x_0 = r_f(0) + r_n(0) + (r'_f(0) + r'_n(0))x_0 \ . \tag{18}$$

By assumption, $r_f(0) = 0$, and hence

$$x_0 = -\frac{r_n(0)}{r'_f(0) + r'_n(0)} \ . \tag{19}$$

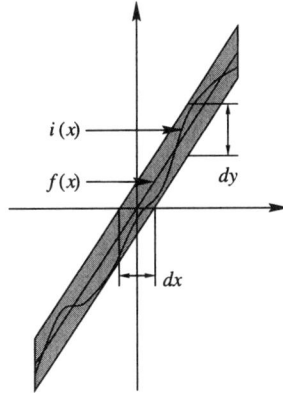

Figure 3. If the noise *n(x)* is assumed to be uniformly distributed in [−*m*, *m*] for all *x* and *f(x)* = *hx*, the zero crossings of *i(x)* = *f(x)* + *n(x)* must be contained in an interval of width *dx* = *dy*/*h*.

The calculation of the variance of x_0 from (19) is extremely difficult. However, if one assumes that the variance $\sigma_{r'}^2$ of $r'_n(0)$ is small with respect to the first derivative of the signal $r'_f(0)$, the term $r'_n(0)$ can be omitted from the denominator of (19):

$$x_0 \approx -\frac{r_n(0)}{r'_f(0)} \ . \tag{20}$$

With this simplification it is a simple matter to calculate the variance of x_0. The denominator is a known deterministic function, while the nominator is a stationary stochastic process with known autocorrelation $R_r(\tau)$, and therefore with known variance $\sigma_r^2 = R_r(0)$. Thus, the variance σ_0^2 of the location x_0 of the zero crossing is given by

$$\sigma_0^2 = \frac{\sigma_r^2}{r'_f(0)^2} \ . \tag{21}$$

It is useful to check whether this result also holds in practice. To do this, various functions $r_f(x)$ with exactly one zero crossing were sampled at discrete points x_i, $i = -32, \ldots, 32$. For the tests the following functions were used with varying signal levels h: $f(x) = hx$, $f(x) = h \arctan x$, $f(x) = h\sqrt{2\pi}(\phi(x) - 0.5)$, and $h \tanh(x)$. Note that all these functions have $r'_f(0) = h$. The noise $r_n(x)$ was generated by convolving Gaussian white noise of standard deviation 1 with a Gaussian smoothing kernel of width σ. Thus, it has the autocorrelation $R_r(\tau) = g_{\sqrt{2}\sigma}(\tau)$, i.e., $\sigma_r = 1/(2\sqrt{\pi}\sigma)$. For the experiments, $\sigma = 2$ was used. From the sampled values $r_i = r_f(x_i) + r_n(x_i)$, zero crossings are extracted to subpixel accuracy by linear interpolation. Since the relation $r'_f(0) = h$ does not hold for the sampled r_i, care is taken to choose h in such a way that the discrete approximation of the first derivative has the value h, i.e., $(r_1 - r_{-1})/2 = h$. For each function, 100000 experiments were performed, and the variance of the closest zero crossing to 0 was calculated from these experiments. The results are shown in Figure 4. As can be seen, for high signal to noise ratios, i.e., for large h, the variance of the zero crossings predicted by (21) and the experimentally determined variance match almost perfectly. If the signal level h is less than

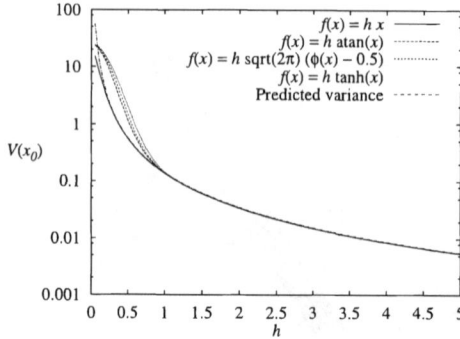

Figure 4. Variance of the zero crossing of different test functions when corrupted with noise of known autocorrelation and analytically predicted zero crossing variance. Note the logarithmic scale of the *y*-axis.

1, i.e., if the standard deviation of the noise is larger than approximately 40% of the amplitude of the signal, the term $r'_n(0)$ neglected in (20) becomes important. As can be seen, in this range of signal to noise ratios, the variance of the zero crossings is estimated too small by (21). Fortunately, such low signal to noise ratios occur extremely rarely in real images. One notable exception are Synthetic Aperture Radar (SAR) images, where the speckle effect can create very low signal to noise ratios [11]. Thus, (21) in most cases gives a very accurate estimate of the true variance of the zero crossings of a function corrupted by noise. If a more precise estimate of the variance is needed, one should note that this problem is intimately related to the problem of determining the first-passage density of a stochastic process to an arbitrary boundary, for which a solution was presented in [5]. However, since the first-passage density derived in [5] is rather complex and can only be calculated numerically, it is not applied here.

3.3: Variance of line and edge positions

With these preliminaries, the predicted variance of the line and edge positions can be determined easily. Since the edge model is much simpler than the line model it is considered first. To derive the variance of the edge locations in the 2D case, without loss of generality it can be assumed that the edge is straight and lies on the *y*-axis. This can be done because the independence of the unbiasedness of the edge and line positions on the orientation of the edge or line was shown for synthetic noise-free images in [17] and because the noise is assumed to be wide-sense stationary white noise, i.e., it is rotationally invariant. Therefore, the edge position is given by the zero crossings of the model edge convolved with the second directional derivative along the *x*-axis of a Gaussian kernel. Thus, the edge location is given by

$$r_{xx}(x,y) = r_{f,xx}(x,y) + r_{n,xx}(x,y) = 0 \ . \tag{22}$$

The 2D autocorrelation function $R_{r_{n,xx}}$ of $r_{n,xx}(x,y)$ is given by [18]

$$R_{r_{n,xx}}(\tau_x, \tau_y) = \sigma_n^2 g''''_{\sqrt{2}\sigma}(\tau_x) g_{\sqrt{2}\sigma}(\tau_y) \ , \tag{23}$$

where σ_n^2 is the variance of the noise in the original image. Thus, the variance of the random noise field is

$$\sigma_{r_{n,xx}}^2 = \frac{3\sigma_n^2}{16\pi\sigma^6} \ , \tag{24}$$

where σ is the standard deviation of the Gaussian smoothing kernel. For the response of the filter to the model step edge needed in (21), obviously

$$r_{f,xxx}(x,y) = hg_\sigma''(x) \ , \tag{25}$$

where h is the contrast of the edge. Thus,

$$r_{f,xxx}(0,0)^2 = \frac{h^2}{2\pi\sigma^6} \ . \tag{26}$$

Therefore, the variance of the location of an edge in a 2D image is given by

$$\sigma_e^2 = \frac{\sigma_{r_{n,xx}}^2}{r_{f,xxx}(0,0)^2} = \frac{\dfrac{3\sigma_n^2}{16\pi\sigma^6}}{\dfrac{h^2}{2\pi\sigma^6}} = \frac{3}{8}\frac{\sigma_n^2}{h^2} \ . \tag{27}$$

This result is quite surprising since it signifies that the variance σ_e^2 of the edge locations is independent of the choice of the smoothing parameter of the Gaussian kernel used to calculate the gradient of the image. This counterintuitive result should be verified by experiments. To do so, test images of size 32×32 were generated with exactly one edge in the center of the image. Note that the test images thus generated are mainly intended to verify (27) and to test whether it holds for discrete images. Real images will be considered in Section 4.3. From these images, edges were extracted with Gaussian derivatives of varying standard deviation σ. The input images were corrupted with Gaussian noise of varying standard deviation σ_n. For each combination of σ and σ_n, 1000 experiments were performed, resulting in roughly 30000 edge points from which the mean value and variance of the edge positions were computed. Figure 5 shows that the mean value of the edge positions is unbiased. The maximum deviation of the edge positions in this series of experiments is roughly one hundredth of a pixel. More importantly, Figure 6(a) shows the calculated variances of the edge positions. As can be seen, the results do not depend on the degree of smoothing. To check the quality of the edge position variance predicted by (27), Figure 6(b) displays the ratio of the predicted and extracted edge position variances. This ratio is a measure of the bias of (27). As can be seen, it is very close to 1 for most combinations of σ and σ_n. The maximum error occurs if σ and σ_n are close to zero. Overall, (27) gives an excellent estimate of the edge position variance.

In order to predict the variance of the line position in 2D images if the bias removal, i.e., the width and position correction, is not applied, the same technique can be used. As above, without loss of generality the line can be assumed to lie on the y axis. Thus, the line position is given by the zero crossings of the model line convolved with the first directional derivative along the x-axis of a Gaussian kernel. Hence, the line location is given by

$$r_x(x,y) = r_{f,x}(x,y) + r_{n,x}(x,y) = 0 \ . \tag{28}$$

The 2D autocorrelation function $R_{r_{n,x}}$ of $r_{n,x}(x,y)$ is

$$R_{r_{n,x}}(\tau_x,\tau_y) = -\sigma_n^2 g_{\sqrt{2}\sigma}''(\tau_x)g_{\sqrt{2}\sigma}(\tau_y) \ , \tag{29}$$

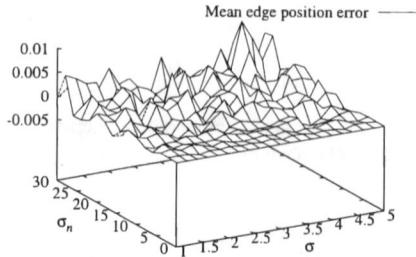

Figure 5. Mean value of the edge position as a function of the standard deviation σ of the Gaussian smoothing kernel and the standard deviation σ_n of the input noise.

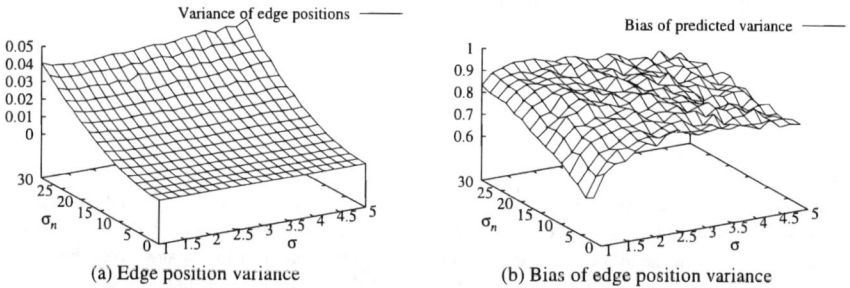

(a) Edge position variance   (b) Bias of edge position variance

Figure 6. Variance of the edge position as a function of the standard deviation σ of the Gaussian smoothing kernel and the standard deviation σ_n of the input noise (a). Ratio of the predicted and extracted edge position variances, i.e., bias of the prediction.

and hence the variance of the random noise field is

$$\sigma_{r_{n,x}}^2 = \frac{\sigma_n^2}{8\pi\sigma^4} \ . \tag{30}$$

Since the asymmetry of a line causes the line position to shift, the denominator of (21) has to be evaluated at the biased line position l given by (5). With this, the variance of the position of a 2D line is given by

$$\sigma_l^2 = \frac{\sigma_{r_{n,x}}^2}{r_{f,xx}(0,0)^2} = \frac{\sigma_n^2}{8\pi\sigma^4} \frac{1}{h^2 r_a''(l,\sigma,w,a)^2} \ , \tag{31}$$

where $r_a''(l,\sigma,w,a)$ is given by (4). Unfortunately, this expression cannot be simplified very much. Therefore, to get an impression of the dependency of the line position variance on different parameters, Figure 7(a) shows the predicted variance as a function of the standard deviation σ of the Gaussian smoothing kernel, while Figure 7(b) shows the predicted variance as a function of the line width w. As can be seen, the line position variance is smallest for $\sigma = w$. Furthermore, if σ is chosen much too small for a given line width w, the line position variance increases rapidly.

Again, it has to be checked whether (31) holds for synthetic test images. To do so, images of size 32×32 were generated with exactly one line lying in the center of the image. Note again that the

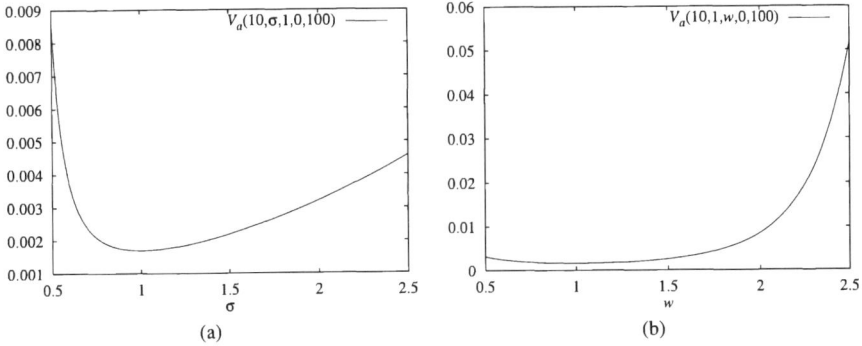

Figure 7. Predicted variance of the line position as a function of the standard deviation σ of the Gaussian smoothing kernel for a line of width $w = 1$, contrast $h = 100$, and noise of standard deviation $\sigma_n = 10$ (a), and as a function of the line width w for $\sigma = 1$, $h = 100$, and $\sigma_n = 10$ (b).

test images thus generated are mainly intended to verify (31) and to test whether it holds for discrete images. Real images will be considered in Section 4.2. To simulate the typical application case, the σ used for smoothing was kept constant, while the total width $2w$ of the line was varied in the interval $[\sigma, 5\sigma]$ in subpixel increments. To obtain subpixel line widths in the image, it was assumed that the sensor has a square aperture function, i.e., integrates the incoming light intensity over the area of each pixel. Therefore, the intensity of a pixel is proportional to the area of the pixel covered by the line. Furthermore, the asymmetry a of the line was varied in the interval $[0, 0.75]$. The input images were corrupted with Gaussian noise of varying standard deviation σ_n. For each combination of a, w, and σ_n, 1000 experiments were performed, resulting in roughly 30000 edge points from which the mean value and variance of the edge positions were computed. For the experiments, lines of contrast $h = 100$ were extracted with $\sigma = 2$. Figure 8 displays the mean line position error, i.e., the difference between the mean value of the line positions extracted from the image and the true line positions for $a = 0$. As can be seen, for all integer line widths the mean error is very close to zero. For non-integer line widths the mean may lie up to 0.07 pixels from the true line position. This effect is caused by extrapolation errors in the subpixel extraction, and not by noise, as is discussed in [17]. Thus, noise introduces no new bias of the line positions, and therefore the line detector can be regarded as unbiased in the presence of noise. Figure 9(a) shows the extracted line position variance. As can be seen, the cross sections of the surface in this figure closely resemble the shape of the predicted variance displayed in Figure 7(b). To check the validity of (31), Figure 9(b) shows the ratio of the predicted and extracted line position variances. This measure of the bias of (31) is very close to 1 for almost all combinations of w and σ_n. For integer line widths and small noise variances σ_n^2, the extracted line position variance is larger than the predicted variance by up to a factor of five because the line positions lie at the borders of a pixel, where the extrapolation error is largest, and thus one should expect the variance to be larger in these cases. Similar results are also obtained for all other values of the asymmetry a. Figure 10 exemplifies this by showing the line position variance and its bias for $a = 0.5$. As can be seen, the measure of the bias is again very close to 1 for almost all combinations of w and σ_n. Overall, (31) gives an excellent estimation of the line position variance for uncorrected line extraction results.

Another very important performance measure for lines is the variance of the extracted line

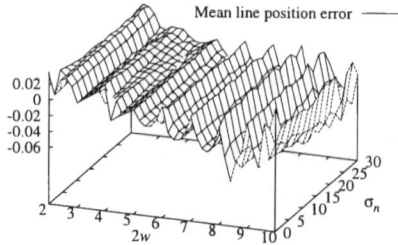

Figure 8. Mean value of the line position as a function of the total line width 2w and the standard deviation σ_n of the input noise.

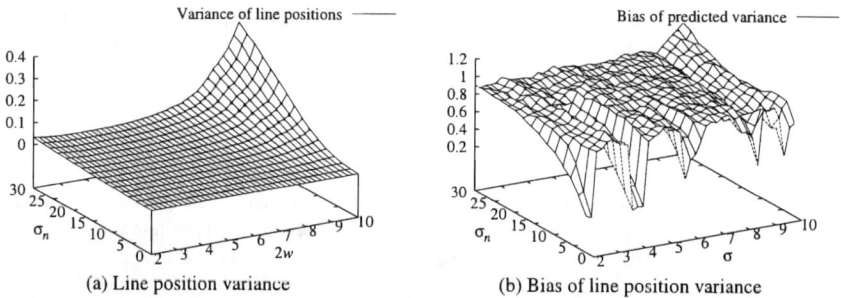

(a) Line position variance

(b) Bias of line position variance

Figure 9. Variance of the line position as a function of the total line width 2w and the standard deviation σ_n of the input noise (a). Ratio of the predicted and extracted line position variances, i.e., bias of the prediction (b).

widths. Since the total line width in case the width correction is not applied is given by the distance of the two edges of the line, in principle it seems possible to describe the variance of the line width using the approach introduced in this section. For example, one could calculate the variances of the two edge positions, assume they are independent, and obtain the variance of the line width as the sum of the two edge position variances. However, this has several problems. First, the edge locations at which (21) must be evaluated can only be calculated numerically. Therefore, no analytical result is possible. Instead, the width variance would have to be tabulated. More importantly, however, the edge positions, and thus their variances, are not independent of one another, as can be seen from the discussion in Section 2. Finally, since the first edge point found on each side of the line is used to calculate the width of the line without taking the edge strength into account, it can be expected that for high noise levels the edge position is much more affected by noise than predicted by an analytical model because then the first edge point found may be caused by noise instead of the true edge point. For these reasons, the approach of adding the variances of the two edge locations cannot return very good results, and this was verified on the synthetic examples used above for the line position variance. From this discussion it follows that a good analytical prediction of the variance of the line width seems very hard to obtain. Therefore, this topic has not been pursued further in this paper. Instead, only the experimentally obtained means and variances of the line width will be discussed. These are shown for $a = 0$ in Figures 11(a) and (b), respectively.

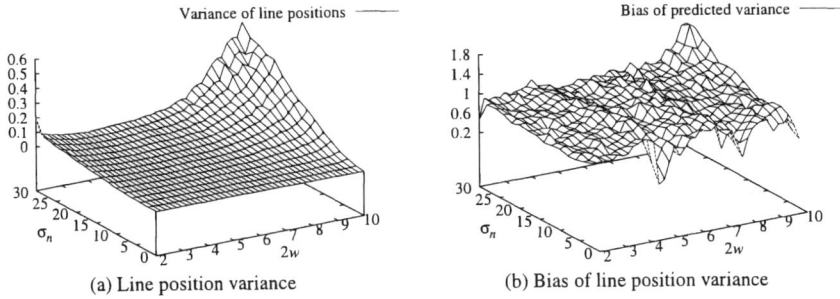

(a) Line position variance

(b) Bias of line position variance

Figure 10. Variance of the line position for $a = 0.5$ (a). Bias of the variance prediction for $a = 0.5$ (b).

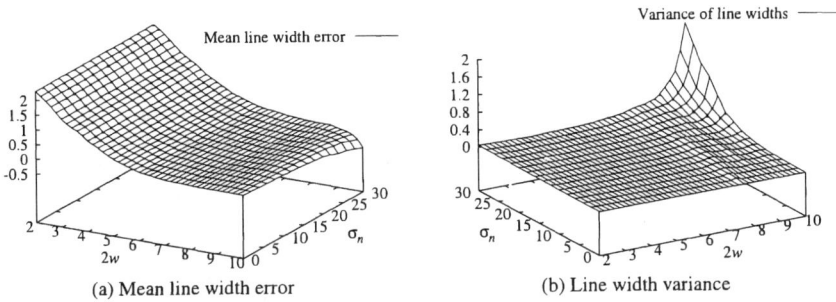

(a) Mean line width error

(b) Line width variance

Figure 11. Mean (a) and variance (b) of the line width as a function of the total line width $2w$ and the standard deviation σ_n of the input noise.

As can be seen from Figure 11(a), the noise has almost no influence on the extracted mean width error, i.e., the difference of the mean line width and the true line width. As is to be expected from the discussion in Section 2, the mean line width is biased. Figure 11(b) displays the experimentally obtained variance σ_w^2 of the line widths. As can be seen, it again increases proportionally to σ_n^2. Furthermore, the width variance is relatively small for small w. It increases significantly only for large w. Again, similar results hold for all asymmetries a.

The final important question is how the bias removal, i.e., the position and width correction, influences the variances of the line position and width. Conceptually, the bias removal is given by a function $f : (v, r) \mapsto (w, a)$, where v is the width extracted from the image, r is the gradient ratio at the two extracted edge points, w is the true line width, and a is the true asymmetry. Thus, standard error propagation schemes could be used to propagate the variances σ_l^2 and σ_w^2 through f. If the covariance $\sigma_{l,w}$ of the line position and width were known, the covariance matrix Σ_u of the uncorrected could be calculated, and the covariance matrix Σ_c of the corrected line position and width would simply be given by [9]

$$\Sigma_c = Df \Sigma_u Df^T , \tag{32}$$

where Df is the Jacobian matrix, i.e., the first derivative, of f. Since σ_w^2 is only known empirically, since an explicit estimation for $\sigma_{l,w}$ is very hard to obtain, and since f can only given in tabulated

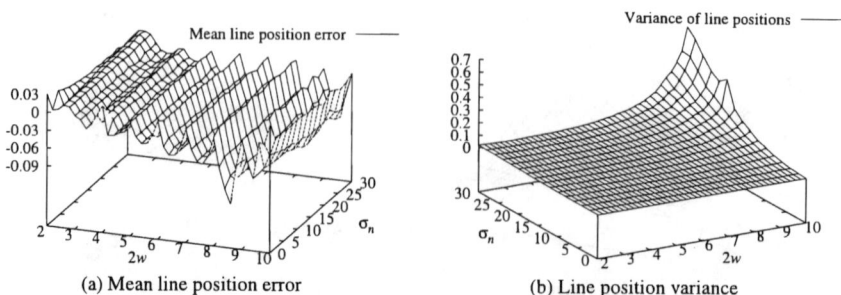

(a) Mean line position error   (b) Line position variance

Figure 12. Mean (a) and variance (b) of the line position as a function of the total line width 2w and the standard deviation σ_n of the input noise if the line width and position correction is applied.

form, no attempt has been made to derive an explicit formula for Σ_c. However, from the shape of the bias inversion function displayed in [15, 16, 17] it can be estimated that the variances of the corrected line position and width will be slightly larger than the uncorrected variances because the partial derivatives of f are larger than 1 almost everywhere. Figure 12 displays the mean and variance of the corrected line position, while Figure 13 shows the mean and variance of the corrected line width. From Figure 12(a) it can be seen that the line position correction slightly increases the mean error in areas of non-integer line width, i.e., in areas where the line position does not lie close to the center of a pixel. The maximum error increases from approximately 0.07 to 0.09 pixels. Furthermore, from Figure 13(a) it can be seen that the corrected line widths are much closer to the true line widths. The corrected line widths exhibit a small bias only if the line width is very small compared to σ. This fact was also observed for noiseless images in [17], and thus comes as no surprise. The variances of the line position are slightly decreased for small w, while they are slightly increased for large w, as can be seen when Figure 12(b) is compared to Figure 9(a). On the other hand, the line width variance increases by a large amount for small w, while the increase is small for large w, as can be seen by comparing Figures 13(b) and 11(b). Essentially, for small w the correction seems to trade better line position variance for worse line width variance. As above, qualitatively similar results hold for all other values of the asymmetry a.

In summary, from the discussion in this section it can be concluded that the extracted line positions and widths are unbiased in the presence of noise, and that the variances of the line positions and widths are very small. Thus, subpixel accuracy can definitely be achieved for noisy synthetic images, while it seems very likely that this kind of accuracy may also be achievable in real images.

To conclude this section, it should be noted that sometimes other performance measures apart from the localization quality also play an important role. The most often used measure is the detectability of a feature in the presence of noise. Essentially, this can be modeled by the signal to noise ratio, i.e., the ratio of squared magnitude of the derivative of the model feature in which the thresholding is done and the variance of the noise in this derivative. For edge detection, the relevant derivative is the first derivative, while for line detection it is the second derivative. Thus, for edge detection the signal to noise ratio r is given by

$$r = \frac{r'_e(0, \sigma, h)^2}{\sigma^2_{r_{n,x}}} = \frac{4h^2\sigma^2}{\sigma^2_n} \,, \tag{33}$$

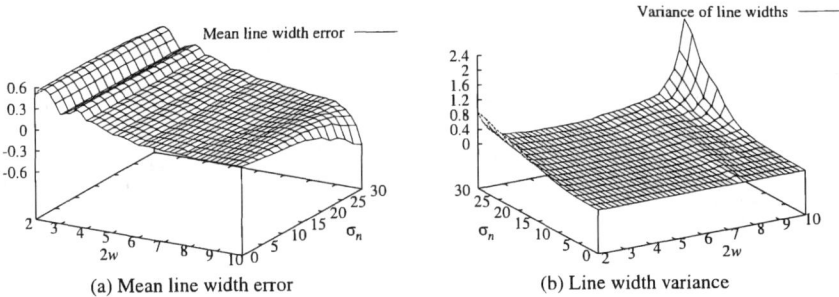

(a) Mean line width error   (b) Line width variance

Figure 13. Mean (a) and variance (b) of the line width as a function of the total line width 2*w* and the standard deviation σ_n of the input noise if the line width and position correction is applied.

while for lines it is given by

$$r = \frac{h^2 r_a''(l, \sigma, w, a)^2}{\sigma_{r_{n,xx}}^2} = \frac{16\pi\sigma^6 h^2 r_a''(l, \sigma, w, a)^2}{3\sigma_n^2} \ . \tag{34}$$

This leads to the fairly obvious conclusion that features become easier to detect as the standard deviation σ of the Gaussian used to smooth the image is increased. Of course, this only holds if there are no other nearby features complicating the extraction.

4: Experiments on real images

All the experiments so far have been done on synthetic test images. Since for real imaging conditions various assumptions made in the test so far may not hold, it is important to perform experiments on real images. For example, the camera used to acquire an image may not have an ideal square aperture function or the framegrabber may not exhibit a linear increase in gray values as the scene intensity increases linearly.

4.1: Experimental set-up

In order to test the subpixel accuracy of the line detection algorithms the following strategy was adopted: A test set containing five lines of widths 2 mm–10 mm were printed on a high-quality laser printer. The resulting print was mounted on a table which can be shifted in one direction in 10 μm increments. A camera with a 12 mm lens was mounted perpendicular to the table at a distance of approximately 50 cm. A standard analog framegrabber was used to acquire the images, since this is by far the most common setup in industrial applications, where subpixel measurements are extremely important. For edge detection, the test set contained a single edge in the center of the print. These test images were chosen because they are representative for many industrial inspection tasks, where often manufactured workpieces having straight or moderately curved edges or lines have to be measured. Of course, this test set-up is by no means intended to be exhaustive. Figure 14 shows an image of the line and edge test sets. Note that, although the lens used in this test is fairly good for computer vision standards, there is nevertheless a significant radial distortion. To test the subpixel accuracy, each of the test sets was shifted by 1 mm in increments of 50 μm, resulting in

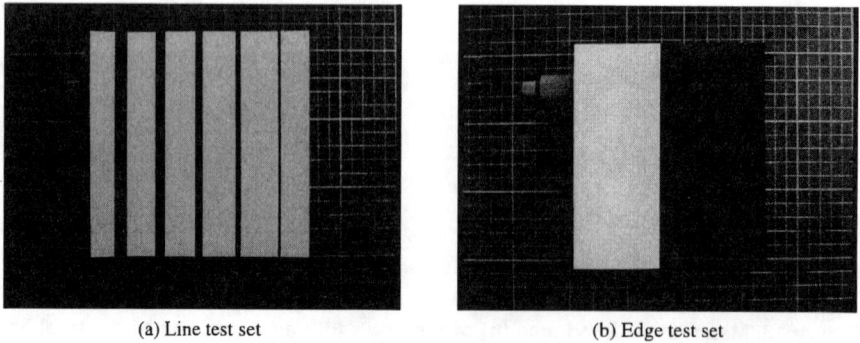

(a) Line test set   (b) Edge test set

Figure 14. Test sets used to check the subpixel accuracy of the line and edge detection algorithms.

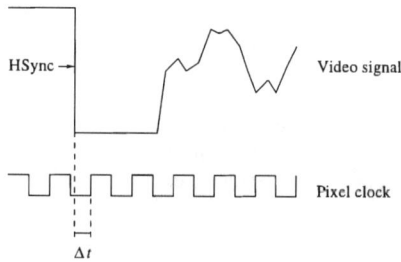

Figure 15. Scheme to start the digitization of one line of a video signal in current framegrabber architectures: Digitization is started at the first rising flank of the unsynchronized pixel clock after the horizontal sync signal has occurred, resulting in an offset of Δt.

a shift of approximately 2 pixels in total in the image. In each position of the measurement table, 20 images were taken. Thus, for each test set a total of 420 images were taken. The goal of the test is to determine the precision, i.e., the variance, of the line and edge position measurements and the absolute position shift of the line and edge positions in order to decide whether it is possible to detect subpixel shifts of 50 μm or, equivalently, approximately one tenth of a pixel.

Before this can be done, a few words on the architecture of standard analog framegrabbers are necessary since the technology used for them will influence the quality of the acquired images. The main implementation issue for framegrabbers is when to start digitizing a line of the video signal. To do this, most analog framegrabbers have a clock chip running at several times the pixel resolution, typically 4–10 times. They start the digitization process on the first upward flank of the pixel clock signal after the falling flank of the horizontal synchronization (HSync) signal of the video signal has occurred, as shown in Figure 15. Thus, for each line there is a random offset of Δt by which the line is shifted with respect to the true HSync signal, resulting in a random shift by up to one fourth to one tenth of a pixel, depending on the frequency of the pixel clock. This seems to prevent subpixel accuracy of better than one tenth of a pixel. Fortunately, however, the

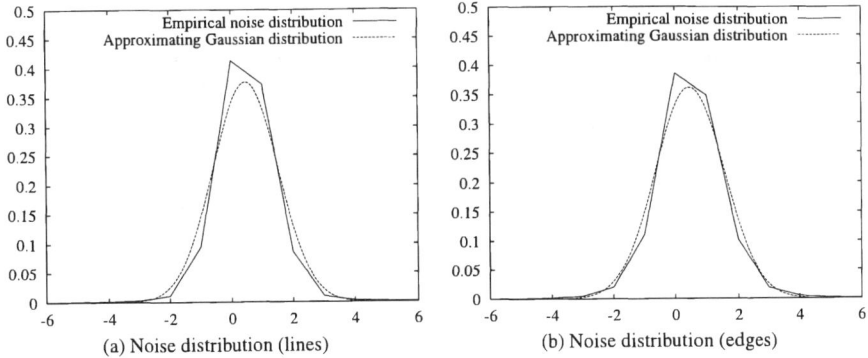

(a) Noise distribution (lines)   (b) Noise distribution (edges)

Figure 16. Experimentally determined noise distributions and Gaussian distributions with the same mean and variance for the line test set (a) and the edge test set (b).

random offset is independent for each line of the video signal. Thus, one can expect to achieve better accuracy by the smoothing done in the line or edge detector. In effect, the random shifts in the digitized video signal can be modeled as an additional noise source. From this discussion it follows that the noise should be larger at the border of the features of interest since the subpixel shifts result in larger gray value variations there, while they have very little influence in regions of approximately constant gray value. Of course, this problem does not occur if the framegrabber is synchronized with the pixel clock of the camera or if a digital camera is used.

In order to test the hypothesis that the random shifts in each line of the video signal lead to increased noise at edge positions and to obtain the noise statistics of the test images, the average of all 20 images within each shift can be calculated. Then, the individual images can be subtracted from the mean image to obtain the noise distribution for that particular image. If the distributions over all 420 images are averaged, a very good approximation of the noise distribution is obtained. Figure 16 shows the noise distributions thus obtained for the line and edge test set, superimposed by a Gaussian distribution of the same mean and variance. As can be seen, the Gaussian distribution yields a very good approximation of the noise distribution in both cases. The standard deviations of the noise were estimated as approximately 1.06 and 1.10, respectively, which is very small. Figure 17 shows small parts of test images subtracted from their corresponding mean image. As can be seen, the random shifts lead to significantly increased noise at the edges of the line in Figure 17(a) and at the edge in Figure 17(b). Therefore, the line and edge locations may not be as accurate as one would expect from the discussion in the previous section.

4.2: Subpixel accuracy of line position and width

To test the subpixel accuracy of the line detection algorithm, only the center line of the test set of width 6 mm was used since it turned out that the lens distortion was too large to obtain meaningful results for the other four lines. Although the center line was carefully aligned to be as vertical as possible, even for this line the lens distortion influences the extraction results, if statistics are calculated over the entire line. Therefore, small windows of height 25 were used to calculate the mean line position and width, and their corresponding variances, because for these small windows

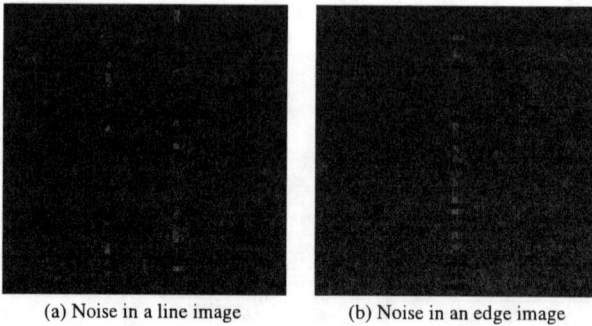

(a) Noise in a line image   (b) Noise in an edge image

Figure 17. Examples of a test image subtracted from its corresponding mean image for the line test set (a) and the edge test set (b). The gray values in the images have been scaled by a factor of 5 for clarity.

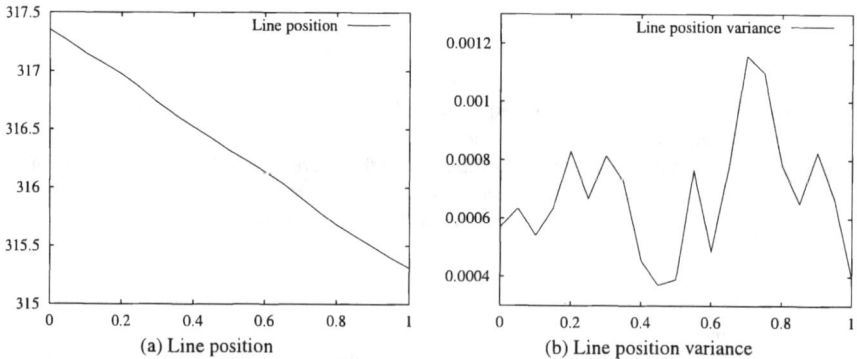

(a) Line position   (b) Line position variance

Figure 18. Line position and variance as a function of the shift of the line in mm.

the lens distortion played only a small role. Since the means and variances are calculated over 20 images, a total of 500 points are used to obtain the statistics. The experiments showed that the resulting statistics are independent of the position of the window. Because the center line of the test images has a total width of approximately 12 pixels, the optimal $\sigma = 6/\sqrt{3}$ [16] was used for the extraction. Figure 18(a) shows the extracted line position as a function of the shift of the line in mm. The resulting curve is almost a perfect straight line. The precision of the line position, i.e., its variance, is displayed in Figure 18(b), from which it can be seen that the maximum variance is approximately 0.00115, i.e., the maximum standard deviation is roughly one thirtieth of a pixel. From (31) the line position variance can be estimated as $1.84 \cdot 10^{-5}$. Thus, the extracted position variance is larger than the predicted variance by a factor of 62.5 mainly due to the random line shifts induced by the framegrabber, the still noticeable lens distortion in this small part of the image, and the fact that the line is not perfectly aligned with the vertical axis. However, since the position variance is still very small it can be expected that the goal of these experiments to detect subpixel shifts of one tenth of a pixel can be achieved. This can easily be verified by testing the hypothesis

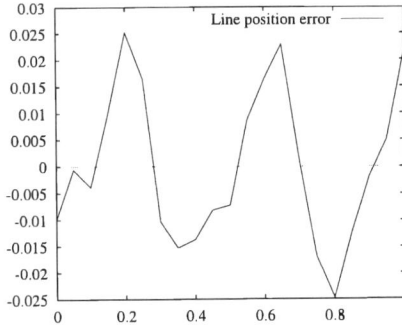

Figure 19. Absolute error of the line position calculated as the difference of the extracted line positions and their regression line as a function of the shift of the line in mm.

that the line positions corresponding to two adjacent shifts are equal [6]. These tests have been performed and the hypothesis that the line positions are equal can be rejected with a probability greater than 99.9% for all line positions. Therefore, relative shifts of one tenth of a pixel can definitively be detected in real images.

Another interesting point is the absolute position accuracy of the line. Since the camera was not calibrated, this is very hard to obtain. If, however, it is assumed that the linear shift of the object in the real world corresponds to a linear shift of the line in the image, a straight line can be fitted through the line positions extracted from the image. The fitting should, of course, take the extracted position variances into account. The equation of the line obtained for the experiment shown in Figure 18 is $-2.07633x + 317.364$. Therefore, it is likely that 1 mm in the real world corresponds to 2.07633 pixels in this part of the image. With this, the absolute position error of the line can be calculated as the difference of the extracted line position and the regression line. The results are shown in Figure 19. As can be seen, the absolute position errors are less than one fortieth of a pixel. The remaining errors show a systematic sinusoidal component, which may be caused by the mapping of the scene intensity to the gray values in the image performed by the framegrabber. Since they are very small, the line extraction algorithm yields very good absolute accuracy on real images as well. Of course, to get meaningful positions in real world coordinates the camera must be calibrated.

The final question for the line detection algorithm is the accuracy of the extracted line widths. The extracted line width and its variance are shown in Figure 20. The line width seems to be centered around 12 pixels. If the above discussion on the correspondence between 1 mm to 2.07633 pixels is taken into account, this would imply that the width of the 6 mm wide line is underestimated by approximately 3.7%. Therefore, the extracted line width is very close to the true line width. Furthermore, it can be seen by the regression line superimposed in Figure 20(a), that the line width increases as the shift increases. This can be attributed to the fact that for increasing shifts the line moves towards the center of the image where lens distortions are not as severe. Finally, Figure 20(b) displays the line width variance obtained from this experiment. As can be seen, the variance is less than 0.0005 almost everywhere, i.e., the standard deviation of the extracted line widths is less than one fortieth of a pixel. Thus, line widths can be extracted with very high subpixel accuracy.

(a) Line width

(b) Line width variance

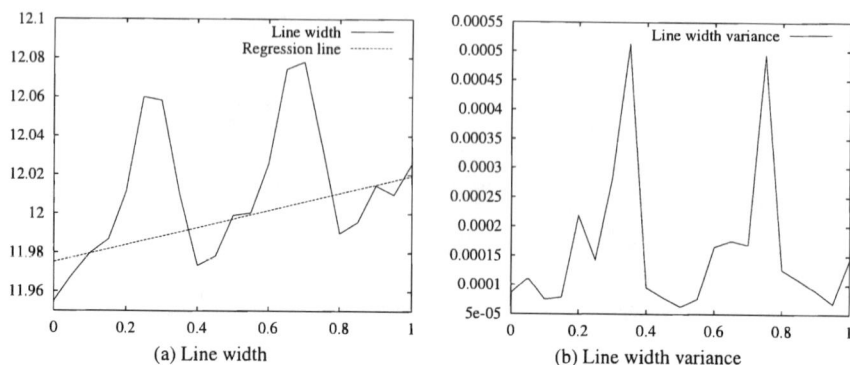

Figure 20. Line width and variance as a function of the shift of the line in mm.

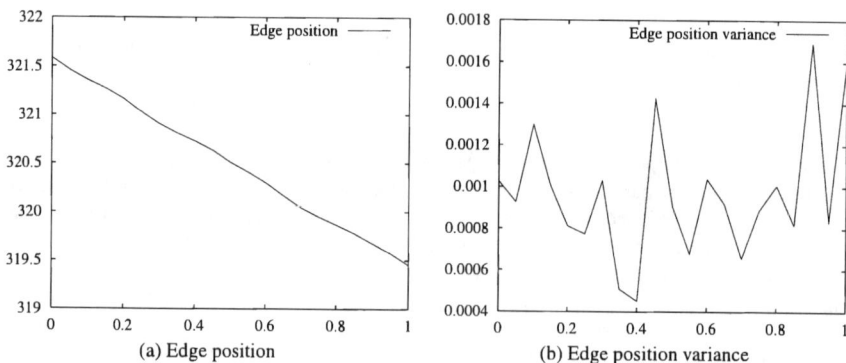

(a) Edge position

(b) Edge position variance

Figure 21. Edge position and variance as a function of the shift of the edge in mm.

4.3: Subpixel accuracy of edge position

To conclude this section, the results of performing the same kind of experiment for the subpixel edge detection algorithm are discussed. The edges were extracted using Gaussian derivatives with $\sigma = 1$. The resulting edge positions are exhibited in Figure 21(a). As was the case for lines, the extracted edge positions lie very close to a perfect straight line. Figure 21(b) displays the corresponding variances of the edge positions. The maximum variance is approximately 0.0017. The variance predicted by (27) is $1.57 \cdot 10^{-5}$, i.e., the extracted variance is larger than the predicted variance by a factor of 108. Again, this is caused by the random shifts of the discretized lines of the video signal, the lens distortion, and the fact that the edge is not perfectly aligned with the vertical axis. However, the standard deviation of the edge positions is still very small, being approximately one twenty-fifth of a pixel. Therefore, with the same hypothesis test as used above, it can be shown that edge shifts of one tenth of a pixel can be detected with better than 99.9% probability. Finally, as above, the absolute position error can be estimated by fitting a straight line through the extracted line positions. In this case, the equation of the line is $-2.12664x + 321.568$.

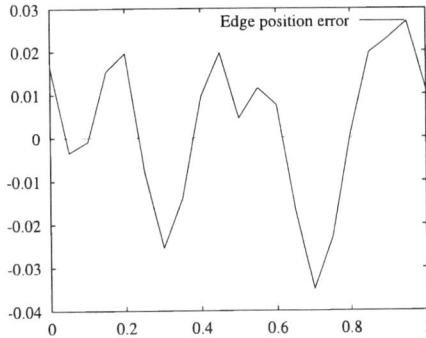

Figure 22. Absolute error of the edge position calculated as the difference of the extracted edge positions and their regression line.

The resulting absolute position errors are depicted in Figure 22. The maximum absolute error is approximately one thirtieth of a pixel. Therefore, edges can be extracted with very good absolute subpixel accuracy. If the remaining errors of up to one thirtieth of a pixel are too large for an application, a scheme as the one given in [14] could be used to estimate and remove the remaining errors. However, since the remaining errors are not generic, i.e., depend on the camera, lens, and framegrabber used, they are not modeled in this paper.

5: Conclusions

A thorough performance analysis is carried out for the line and edge extraction algorithms proposed in [15, 16, 17]. The quality of the line extraction results is evaluated for noisy images. Analytical formulas predicting the variance of the line and edge positions in case the variance of the image noise is known are derived and shown to correspond very well to experimentally determined line and edge position variances. Additionally, the variance of the line width in the presence of noise is determined experimentally. Finally, the quality of the line and edge positions and of the line widths is analyzed for real images. The results of this performance analysis indicate that the extraction results are unbiased for noisy synthetic and real images, that the variances of the extracted features are very small, i.e., the sub-pixel precision is very high, and that the absolute positions and widths can be determined to a very high degree of sub-pixel accuracy. For real images it is shown that position shifts of one tenth of a pixel can be detected with a probability of more than 99.9%, indicating that much better sub-pixel accuracy than one tenth of a pixel can be achieved for real images. Thus, it is shown that the line and edge extraction algorithms not only achieve sub pixel resolution, but also sub-pixel precision and accuracy.

References

[1] Kalle Åström and Anders Heyden. Stochastic modelling and analysis of sub-pixel edge detection. In *13th International Conference on Pattern Recognition*, volume II, pages 86–90, 1996.

[2] Regine Brügelmann and Wolfgang Förstner. Noise estimation for color edge extraction. In W. Förstner and S. Ruwiedel, editors, *Robust Computer Vision: Quality of Vision Algorithms*, pages 90–106, Karlsruhe, 1992. Wichmann.

[3] John Canny. A computational approach to edge detection. *IEEE Transactions on Pattern Analysis and Machine Intelligence*, 8(6):679–698, June 1986.

[4] Didier Demigny and Tawfik Kamlé. A discrete expression of Canny's criteria for step edge detector performances evaluation. *IEEE Transactions on Pattern Analysis and Machine Intelligence*, 19(11):1199–1211, November 1997.

[5] J. Durbin. The first-passage density of a continuous Gaussian process to a general boundary. *J. Appl. Prob.*, 22:99–122, 1985.

[6] Marek Fisz. *Wahrscheinlichkeitstheorie und mathematische Statistik*. VEB Deutscher Verlag der Wissenschaften, Berlin, 11th edition, 1989.

[7] Wolfgang Förstner. *Statistische Verfahren für die automatische Bildanalyse und ihre Bewertung bei der Objekterkennung und -vermessung*. Number 370 in Reihe C. Deutsche Geodätische Kommission, München, 1991.

[8] Wolfgang Förstner. 10 pros and cons against performance characterization of vision algorithms. In Henrik I. Christensen, Wolfgang Förstner, and Claus B. Madsen, editors, *Workshop on Performance Characteristics of Vision Algorithms*, pages 13–29, April 1996.

[9] Robert M. Haralick. Propagating covariance in computer vision. In Henrik I. Christensen, Wolfgang Förstner, and Claus B. Madsen, editors, *Workshop on Performance Characteristics of Vision Algorithms*, pages 1–12, April 1996.

[10] C. Heipke, H. Mayer, C. Wiedemann, and O. Jamet. Evaluation of automatic road extraction. In *International Archives of Photogrammetry and Remote Sensing*, volume XXXII, part 3–4W2, pages 151–160, 1997.

[11] Olaf Hellwich, Helmut Mayer, and Gerhard Winkler. Detection of lines in synthetic aperture radar (SAR) scenes. In *International Archives of Photogrammetry and Remote Sensing*, volume XXXI, part B3, pages 312–320, 1996.

[12] Jack Koplowitz and Vito Greco. On the edge location error for local maximum and zero-crossing edge detectors. *IEEE Transactions on Pattern Analysis and Machine Intelligence*, 16(12):1207–1212, December 1994.

[13] Athanasios Papoulis. *Probability, Random Variables, and Stochastic Processes*. McGraw-Hill, Inc., New York, NY, 3rd edition, 1991.

[14] Frederico Pedersini, Augusto Sarti, and Stefano Tubaro. Estimation and compensation of subpixel edge localization error. *IEEE Transactions on Pattern Analysis and Machine Intelligence*, 19(11):1278–1284, November 1997.

[15] Carsten Steger. Removing the bias from line detection. In *Computer Vision and Pattern Recognition*, pages 116–122, 1997.

[16] Carsten Steger. An unbiased detector of curvilinear structures. *IEEE Transactions on Pattern Analysis and Machine Intelligence*, 20(2), February 1998.

[17] Carsten Steger. *Unbiased Extraction of Curvilinear Structures from 2D and 3D Images*. PhD thesis, Fakultät für Informatik, Technische Universität München, 1998. Submitted.

[18] A. A. Sveshnikov. *Applied Methods of the Theory of Random Functions*. Pergamon Press, Oxford, 1966.

[19] Hemant D. Tagare and Rui J. P. deFigueiredo. On the localization performance measure and optimal edge detection. *IEEE Transactions on Pattern Analysis and Machine Intelligence*, 12(12):1186–1190, December 1990.

Objective Evaluation of Edge Detectors
Using a Formally Defined Framework

Sean Dougherty and Kevin W. Bowyer
Department of Computer Science & Engineering
University of South Florida, Tampa, Florida 33620-5399
doughert or kwb @bigpine.csee.usf.edu

Abstract

Edge detection is one of the most-studied problems in the field of computer vision. However, there is no standard method of objectively and accurately evaluating the performance of an edge detector. This work proposes such a framework. The framework is based on using real images, hand-specified ground truth, and a receiver operating characteristic (ROC) curve style of analysis of true positive and false positive edge pixels. The method is illustrated by developing a comparison of six edge detectors. Relative performance rankings seem remarkably stable across different types of imagery. The reputation of the Canny detector as a standard for comparison is reinforced. The Heitger detector may offer a slight performance gain, but at a cost of increased computation.

1: Introduction

Despite that edge detection is one of the most-studied problems in the field of computer vision, the field does not have a standard method of objectively and accurately evaluating the performance of an edge detector. The current prevailing method of showing a few images and visually comparing subsections of resulting edge images lacks the rigor necessary to find fine-level performance differences between edge detectors. In addition, this method does not help to evaluate some aspects of an edge detector, such as how certain parameters affect performance. This lack of objective performance evaluation has resulted in an absence of clear advancement in the "state of the art" of edge detection, or a clear understanding of the relative behavior of different detectors.

This work proposes a method to objectively and accurately evaluate an edge detector on a battery of real images in multiple, distinct categories. The result of this evaluation is a receiver-operating characteristic (ROC) curve which can be plotted against those of other detectors to determine relative performance. This method allows for the comparison of a significant number of edge images across a variety of edge content using a formalized procedure.

2: Related Work

Previous work on methods to evaluate edge images can be split into four distinct groups.

2.1: Human Evaluation

The most recent example of this type of work is that of Heath *et al.* [8]. This study compares the performance of five different edge detectors using a set of 28 real images. The study rates an edge image's content relative to the goal of object recognition, as determined by multiple subjects who rate each individual image.

Human perceptual abilities can likely make more out of an edge image than any current computer vision system. Also, due to the complexities of the experimental design involving ratings by human observers, only a relatively coarse sampling of the parameter space was possible for each edge detector.

2.2: Theory-Based Evaluation

Theoretical approaches, such as one described by Ramesh and Haralick [18], use the mathematical definitions of multiple detectors and an edge in the presence of Gaussian noise or other modelable factors. Using these definitions a model is created from which calculations of false alarm and misdetection rates are made. Approaches such as these do not yet model either the complexity found in real images, or the complexity of modern edge detectors.

2.3: Edge Feature Measurement Evaluation

Another group of methods attempts to quantify edge image quality using measures of various image features, such as smoothness or continuity, without using a ground truth of the original image. Automated methods such as [5][13][15][25][29] are highly fallible on recent edge detectors which can meet their performance criteria automatically, or under undesirable conditions. It was shown in [8] that the Kitchen-Rosenfeld metric scored most highly for the Canny detector when high blur settings were used. This produced images with extremely thin, smooth, and continuous edge chains, but which would have little value in any task. Without the use of some form of ground truth, it does not seem possible to grade edge images accurately.

2.4: Ground-Truth Comparative Evaluation

The final group consists of automated or nearly automated methods that use some form of ground truth. This ground truth may be of the actual edges to be found or not found, or of higher-level features such as lines or object motion.

A branch of works including those by Abdou [1], Fram [6], and Kanungo [12] uses synthetic images with varying noise or simple edge combinations to determine performance. These combinations are not often encountered in real images, and so these results may not transfer to real images. Heath *et al.* [8] also showed how typical synthetic images do not offer a significant challenge to modern edge detectors. When their parameter space is sampled finely enough, all of the detectors examined found almost perfect edge content, even at very high levels of noise.

A second branch of these works attempts to use real images. Bryant and Bouldin [3] put forward the idea of using marked lines for comparison in aerial imagery. Palmer *et al.* [15] described using line-finding methods and evaluating them through ground-truthed lines. Later, Jiang *et al.* [10] developed a method to evaluate edge detection in range images using ground truth. Salotti *et al.* [21] described a method of evaluating using hand made ground truths of images. Recently, Thornton *et al.* [26] performed ground truthing of model board imagery by hand, which was used in a study by Rosin [19] for the purposes of edge saliency and thresholding evaluation.

The work presented here differs from prior work in that a more sophisticated method of ground truthing is used which allows us to exclude ambiguous regions. Also, unlike prior methods, we adaptively sample the parameter space of the edge detectors in a formally defined manner, refining that parameter space equally for each detector. Using this improved ground truth and adaptive sampling, accuracy to true performance is improved dramatically over prior methods.

3: Function of the Framework

In the field of edge detection, there is presently no formalized, objective method of evaluation. The majority of edge detectors are published with only a few parameter settings and a few images examined. These images are compared in a subjective manner with comparison against only a few detectors at single parameter settings, looking at subsections of images.

As a result, there has been very little demonstrated progress in edge detection. Each edge detector published claims to be an improvement over all prior detectors. The primary criteria to decide which edge detectors are published seems to not be how well they perform, but instead how novel their method of finding edges is. Obviously this criterion is not sufficient to measure progress in edge detection. A new method needs to be constructed. We assert that this method must:

(1) Be formally defined such that the experimenter has minimal involvement.

A formally defined method with minimal interaction approaches the objectivity necessary to do accurate comparison. The framework proposed should be refined enough to handle any decisions crucial to the represented performance of a detector.

(2) Be completely repeatable.

A method which returns different results for each evaluation on the same image set introduces variability which obscures the actual performance of an edge detector.

(3) Sample the parameter space of an edge detector to fairly represent its performance.

The observed performance of an edge detector is directly related to the parameter sampling considered. For any detector, increasing the number of parameter samplings will generally produce some increase in observed performance. Any evaluation method must sample the parameter space in a way that (1) samples each parameter at a level proportionate to its importance in determining the performance of the edge detector and (2) samples the parameter space densely enough to reach a level of performance closely approximating the detector's ideal performance. No method previously published recognizes the importance of the above two requirements. Most methods use some fixed number of samplings per detector, which may inadvertently hinder detectors by not adequately sampling the parameter space.

(4) Accurately rate edge detectors in a manner consistent with some foreseeable task.

This is the most difficult requirement to meet. As of yet, few tasks exist which can directly be used to compare with the results of an edge detector evaluation. Instead, we must aim our method in the direction which appears to offer a generalized basis for higher-level tasks.

The framework proposed here attempts to meet all of the above requirements to the greatest extent possible by using the components introduced in the next section.

4: Components of the Framework

4.1: Real Images

An important feature of this framework is its applicability to real, relatively complex images. The images can be from any source and of any subject, providing there is sufficient certainty in the ground truth of the image. For the comparison reported here, four categories of ten images each were selected. The categories were selected to represent different application areas of edge detection.

The first category is "Object images." These images are selected from the study by Heath *et al.* [8]. Each image is approximately 512 by 512 pixels, with 256 levels of grey, and contain some central object in typical surroundings.

Images from the DARPA Fort Hood Aerial Image set were selected for the "Aerial images" category. A set of 10 varied types of scenes was selected from the larger original images, at the original pixel scale. Each selected sub-image is approximately 512 by 512 pixels in dimension, with 256 levels of grey. Images selected include both oblique and vertical views.

A third category of images, "Medical images," contains magnetic resonance images selected from two studies. Five images were selected from the dataset of MR knee cross sections used in the work by Tistarelli [27]. An additional 5 images were selected from the public Texas Brain MR dataset. All 10 images are exactly 256 by 256 pixels, with 256 levels of grey. All of the knee images are T1 spin-echo images. The brain images represent one T1 weighted, one T2 weighted, and three proton-density.

The fourth category of images is made of data from the FERET [17] face recognition evaluation project. Only frontal face images are used, and each image is 256 by 384 pixels with 256 levels of grey. The images were selected to have varying types of people, and also to include varying hair and clothes types. One image was selected to include a subject with eyeglasses.

These various categories of images should reveal any difference in performance between categories that may exist between detectors or over the entire group of detectors. Such distinctions are valuable observations in describing the performance of given edge detectors on bodies of data.

4.2: Edge Detectors

Edge detectors were chosen in part to represent the most modern or best recognized member of a technology of edge detection. Also, in order to avoid problems in implementing the detectors, all but the Canny and Sobel detectors use the original author's implementation. In this work, a total of six edge detectors have been evaluated.

The only restriction upon edge detectors in this framework is that they produce pixel-based edge maps of the original image resolution, and that those edge maps contain only single pixel wide edges (such as those which result from non-maximal suppression). All detectors were set to create PGM format greyscale images where the value 0 represented an edge pixel, with all remaining pixels set to 255.

The Canny [4] and Sobel [23] detectors were chosen as well-recognized historical bench-marks in the development of edge detection. The Canny implementation used here is an identical implementation to that used in [8]. The three parameters are a Gaussian blur width sigma value, the low hysteresis threshold, and the high hysteresis threshold. The Sobel was implemented by the authors of this paper, and supplemented with non-maximal suppression and hysteresis thresholding. The only parameters for the Sobel are the low and high hysteresis thresholds.

The Sarkar-Boyer [22] detector is an "optimal" zero-crossing detector with hysteresis. The Sarkar-Boyer detector has four possible input parameters: Gaussian blur scale, low hysteresis, high hysteresis, and an edge length threshold. Only the first three are used in this experiment, and the last is fixed at the default value.

Representing the Logical/Linear field of detectors is the Heitger [9] detector, which uses a method of "Suppression and Enhancement" to find edges. The detector's parameters are number of search orientations, blur scale, and an edge strength threshold. At first, this detector was evaluated using only the last two parameters. It was also evaluated using all three to see if this improved performance. We used the original author's implementation of the detector, modified slightly to output the edge result in the uniform format.

The detector proposed by Bergholm [2] was chosen to represent the area of scale-space detectors. The implementation used here is identical to that in [8], and comes from the Candela package distributed by KTH. The Bergholm's parameters are starting sigma, stop-ping sigma, and absolute edge strength threshold. All three parameters are used in this comparison.

The final edge detector used is the optimizing topographic detector proposed by Rothwell [20]. The implementation used here is identical to the one in [8] and was rewritten from the C++ implementation in the DARPA IUE to C. The parameters for the Rothwell are: Gaussian blur sigma, an edge neighbor sensitivity threshold "alpha", and an absolute edge strength threshold. All three parameters are used in this comparison.

4.3: Ground Truth

4.3.1 Ideal versus Practical Ground Truth

Complete and correct ground truth for complex real images would be ideal. However, specifying complete ground truth for complex real images is not feasible, either in terms of the effort required or the confidence in the accuracy of the ground truth. Since the idea of what should be an edge in edge images is not fully defined, it is often not possible to mark edge pixels with any certainty in, for example, regions of texture.

Our approach is to use real images and to specify all of the ground truth that can be specified with high confidence. In practice, this tends to cover a large portion of the images used. Regions of the image where it is problematic for a person looking at the image to decide on the presence of an edge are left as "don't-count" regions. While specifying complete ground truth is not feasible, specifying this high-confidence *partial* ground truth is generally significantly less difficult.

4.3.2 Three-Value Label Ground Truthing

With this in mind, we have adopted a three-valued labeling of ground truth for images. One, edges in the image may be marked as instances of *true edges*. If an edge detector reports the presence of an edge in close proximity to a true edge, that edge is counted as a "true positive". Two, regions in the image may be marked as *no-edge regions*. If an edge detector reports an edge within a no-edge region, that edge is counted as spurious (a "false positive"). Three, areas of the image that are not marked as a true edge (i.e., are outside of the proximity parameter to a true edge) or as part of a no-edge region become "don't-count areas." These areas do not contribute towards a count of true positives or false positives.

4.3.3   Merit of Three-Value Ground Truth

At first, there may be some resistance to this use of "partial" ground truth. While this method could, in theory, exclude regions on some images that would influence detector performance for some tasks, there does not appear to be a way to adequately circumvent the ambiguity of the "truth" in that image. Counting the detected edge pixels falling into ambiguous regions as false positives only compounds the potential for error. Conversely, counting edge pixels in those regions as true positives under all circumstances causes a variable number of ground truth pixels for each detector on a given image, resulting often in a skewed and misleading curve.

Two-valued ground truth labeling, such as done for images from the RADIUS project in [19][26], is simply less reliable than the three-valued labeling here. The ground truths for the images from [26] exclude portions which clearly contain an edge, such as windows on the sides of buildings, as shown in Figure 1. An edge detector which finds those features would be penalized versus a detector which did not find them.

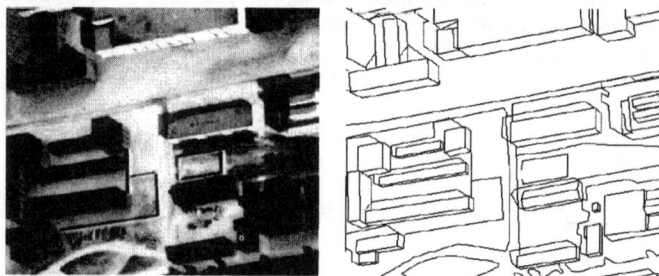

Figure 1. Examples from the RADIUS model board imagery from [26], taken from their web site http://george.ee.washington.edu/.

4.3.4   Ground Truth Methodology

Extensive effort was made to ensure the quality of the ground truths used in this work. As an initial experiment, three subjects with computer vision background ground-truthed two of the object images used in this study (the Airplane and Stapler images). By studying the marks made and combining the ground truths, new superset ground truths were created. The full set of images was then ground truthed by a single individual, with the same or greater level of attentiveness to edge detail as arrived at in the initial experiment. At points throughout the experiment the merit of the ground truths has been re-evaluated, and corrections made when necessary.

Each ground truth was created in a manner similar to that used in the study by Salotti [21]. The image was enlarged such that a magnified portion of the full image was viewed. This magnification level was generally set to three times. Each "section" of the image was ground truthed, using the three-value labeling above. This process requires approximately 3 to 4 hours per 512×512 image.

Unfortunately, ground truths cannot be truly task-independent. When they are created, they must be consciously pushed towards one task or another. The ground truths here are not made such that anything above noise level variation in the grey level is an edge. Instead, for each body of images, the features deemed "vital" or "obvious" are marked.

When no-edge regions are marked, the most obvious regions are those of low intensity changes. However, these regions are not terribly interesting to the researcher. It is vital that edge detectors also not find edges where the intensity changes are above the noise level, but where edges would seem to confuse or otherwise make the task at hand more difficult.

At some point in the future, post processing steps may come about which will be able to interpret the true source of every stray edge pixel. However, for any task used at present these pixels only serve to confuse or hinder any post-processing steps. As such, the ground truths here are created with some degree of specificity to currently existing tasks.

The primary edges focused upon in the object images are object boundaries. Similarly, in aerial images the outlines of buildings and roof/building boundaries, as well as ground markings (roads, etc), were the primary focus. The face and medical image ground truths are significantly more task-dependent. The face images focused on the edges created only by major facial features, such as the eyes, nose, and mouth as well as the facial boundary. This choice was made as edge detection is typically used on these images to find eye positions and to create automatic registration. Medical images focused on major tissue boundaries, as edge detection is typically employed to align MR imagery in an automated fashion.

While "consciously pushing" a ground truth towards a particular state might seem to run counter to the idea of being objective, it has the opposite effect. The act of defining exactly what the ground truth represents makes judgments *less* subjective. Attempting to create a ground truth without defining what is supposed to be an edge or not an edge makes the ground truth highly subjective, and at worst, not representative of any task.

4.4: ROC-style Analysis

Each set of parameter values for each edge detector and image will produce a count of true-edge pixels and false-edge pixels. By sampling broadly enough in the parameter space for an edge detector, and at fine enough intervals, it is possible to produce a representative range of possible tradeoffs in true positives versus false positives. This results in a graphical representation of the possible "true positive / false positive" tradeoffs commonly known as a receiver operating characteristic (ROC) curve.

These items are plotted on two axes, one being the "percent of ground truth missing" and the other being the "ratio of false pixels to possible false pixels". When each parameter's resulting (% missing, % false) pair is plotted, they form a "cloud" of result points in the plot.

Since this method is concerned with the best-case performance of each detector, points are chosen for which there does not exist another point with a lower false pixel ratio and a lower amount of ground truth missing. These points lie along the bottom edge of the "cloud," and are used as the ROC curve points. An example of a cloud, and the ROC curve constructed from that cloud can be found in figure 2.

The "ratio of false pixels to possible false pixels" axis was created to normalize false-pixel counts as the images in this study have a differing number of pixels in no-edge regions. This is particularly the case in images of smaller overall size. To equalize the contribution of these images in aggregate statistics, a ratio is used to express the number of false pixels in each image. This number rarely exceeds 0.30 as the detectors here output "thinned" edges.

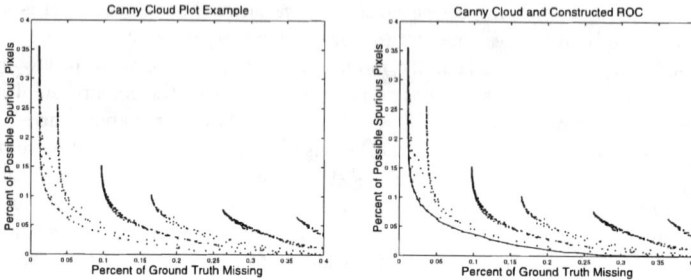

Figure 2. Example of an ROC construction. Though the ideal cloud would be densely populated, some clouds in this study may contain only a couple hundred points. Using the cloud at left, the ROC at right is created.

4.5: Iterative Parameter Set Search

4.5.1   Theory Behind Iterative Sampling

It is important to have an objective method of sampling an edge detector's parameter space. For any edge detector, increasing the number of parameter sets used will result in some degree of improved performance on an ROC curve. Additionally, different detectors attain similar degrees of change in ROC curve performance at different numbers of parameter sets. Therefore, using a constant N number of parameter sets per detector does not guarantee a similar level of performance between detectors at reasonable values of N. Instead, our framework uses an iterative process which expands the parameter set systematically on a per parameter basis until the change in the ROC curve performance is "sufficiently small." As such, each detector should be equally close to its ideal performance.

Not only must the number of sets be selected, but also the range of values for each parameter. That range of values for each parameter must be determined either from the numerical definition of the parameters, or from experimentation. Typically, this involves running the edge detector on a subset of images testing various parameter bound settings until no gain is found in further widening the bounds. Figure 3 lists the parameter bounds used in the experiment here for each edge detector.

4.5.2   Framework Process

For each parameter a detector has, a four value uniform sampling is taken across the bounds defined for each parameter. In the case of a three parameter detector, this results in an

Detector	Parameter 1	Parameter 2	Parameter 3
Canny	sigma = [0.5, 5.0]	low = [0.0, 1.0]	high = [0.0, 1.0]
Sobel	low = [0.0, 1.0]	high = [0.0, 1.0]	
Bergholm	start sigma = [0.5, 5.0]	stop sigma = [0.5, 5.0]	threshold = [0, 60]
Rothwell	sigma = [0.5, 5.0]	threshold = [0.0, 60.0]	alpha = [0.0,1.0]
Sarkar	sigma = [0.5, 5.0]	low = [0.0 , 1.0]	high= [0.0, 1.0]
Heitger (2)	sigma = [0.5, 5.0]	threshold = [0.0, 50.0]	
Heitger (3)	orientations = i [2, 12]	sigma = [0.5, 5.0]	threshold = [0.0, 50.0]

Figure 3. Bounds for each parameter for each detector used in the experiment.

initial $4 \times 4 \times 4$ sampling. From these 64 parameter settings, an initial ROC curve is constructed.

Next, the parameters are checked to see which, if sampled more finely, would produce the best refined ROC curve. A better ROC curve is one which has less area beneath it. A separate tentative refined ROC curve is created using a refined sampling of each parameter. The refined sampling is the original N values in that parameter dimension plus the additional $N - 1$ values that lie between the original values, roughly doubling the number of samples. For our example case of a three parameter detector, in the first stage of refinement, we go from a 64-point sampling of parameter space to a 112-point ($7 \times 4 \times 4$, $4 \times 7 \times 4$, or $4 \times 4 \times 7$) sampling. This becomes the input to another refinement step. The process repeats until the area under the curve has changed by less than 5% between the current and prior iterations. Figure 4 shows the first 4 iterations of this process for one image.

It should be noted that this process is a "greedy" search for the best sampling of parameter space. It does not test all possible parameter branches, and as such is not guaranteed to find the absolutely best ROC result curve. However, we have found this search process works very well, and is necessary to make the search of the parameter space practical. Testing every branch would be an enormously more time consuming process.

4.6: Comparison Tool

In order to compare the ground truth to edge images, a comparison tool is required. In this study, we used a modified implementation of the tool developed by Jiang, et al. [10]. The comparison tool takes the ground truth image and the edge pixel image, and attempts to match each ground truth edge pixel to the closest edge pixel in the edge image, within a defined range. Each ground truth pixel can only be matched to a single edge map pixel, and each match is counted as a "true positive". That edge map pixel is also discarded so it is not matched against multiple ground truth pixels. In that, ground truth and edge map pixels are matched only on a one-to-one basis.

The search range was selected to be three pixels after experimentation and examination of the actual matching results for varying search radii. Image sets which differ dramatically from those used here may require a different setting to produce reliable results.

The remaining, unmatched edge pixels that lie in regions marked as no-edge are counted as the "false positive", or spurious, pixels.

Figure 9 in the appendix shows how the rankings are generally sensible and in accord with overall subjective evaluations.

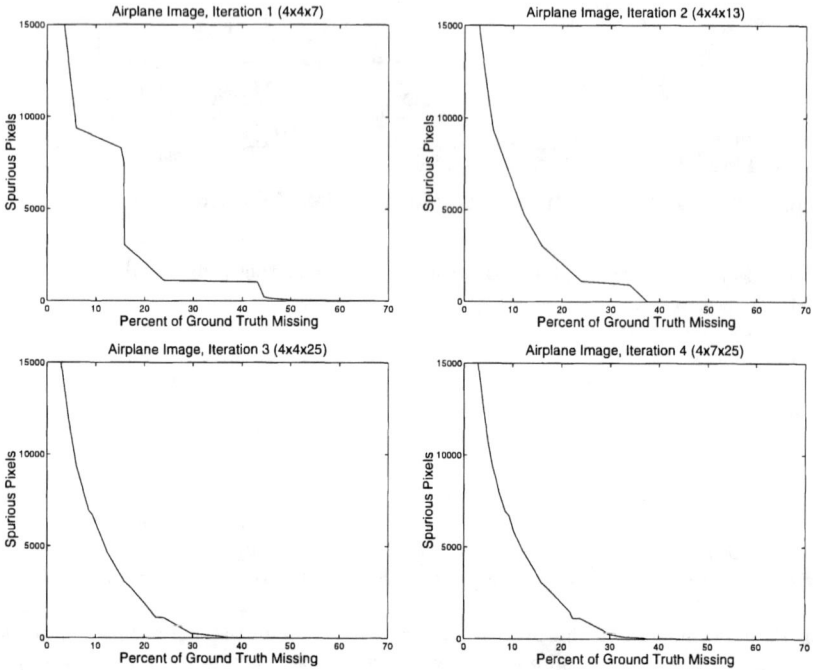

Figure 4. The first four iterations of the Rothwell detector on the Airplane object image. As the sampling becomes finer, the ROC becomes more well defined as points "fill-out" the curve.

5: Methods of Result Examination

The amount of data collected by our method is enormous relative to other methods, and as a result it is difficult to make conclusions over the whole body of results without further processing. Two processes are used to create aggregate result ROC curves over sets of images. These processes look at performance from the two vantage points defined in [8].

5.1: Adapted Parameter Analysis

In the adapted parameter analysis, the parameter space of the edge detector is sampled for each image individually to give the best ROC curve for that image. Then, those ROC curves are combined. Thus the points on the ROC curve for different images may represent different parameter values of the detector, and the points on the aggregate curve do not directly correspond to any particular parameter settings.

The aggregate adapted parameter scenario ROC curve is calculated by summing interpolated values of the *percent of possible spurious pixels* for the curves to be "combined" at fixed intervals of missing ground truth, and then dividing this sum by the number of images being combined. Effectively, parameter values are picked which maximize performance. This method is useful to evaluate an edge detector in a scenario where adaptation of the parameter settings can be done to increase edge image quality. Conversely, this curve does not well characterize the scenario where only a single or very small set of parameter values will be used across a broad set of images.

5.2: Fixed Parameter Analysis

The "fixed parameter method" uses the same parameter set samplings across all images in the aggregate. Instead of refining the parameter set sampling per image, the full set of images is edge detected using the same parameter space sampling. The resulting points on the aggregate ROC curve represent average results for the same parameter setting applied to all images.

As in the standard framework process, this ROC becomes the input to the next refinement step. Since aggregate ROC curves are used in the framework process itself, the ROC at the end of the parameter set refinement needs no further processing and is the fixed-parameter performance curve.

The results presented here only cover the adapted scenario, as the work involved in computing the fixed-parameter results is still in progress.

5.3: Parameter Set Size Analysis

In addition to the ROC curves, each image under the adapted process also gives a unique sampling of its parameters. Similarities between images and image categories in which parameters were sampled with greater fineness can find parameters which are more important for a given edge detector's performance.

The overall number of parameter sets used for images can also be compared against those for other detectors, indicating how much work one detector must do relative to another detector to reach similar levels of performance. This is possible because our framework uses an adaptive sampling process which tunes the parameter space used on each image with equal effort across all detectors.

5.4: Point Cloud Behavior Analysis

There are two ways to approach this method of analysis:

(1) Cloud Formations

After plotting an ROC, very obvious patterns sometimes emerge from the cloud that can indicate which parameter settings may be best, or indicate basic trends in how an edge detector may perform.

(2) Paths Through the ROC Graph

When one parameter is allowed to be variable, while the others are fixed, a path is created through the ROC graph by using a series of parameter values. Paths created may show how a parameter behaves and how to better tune that parameter in a real application.

While neither of these two methods can easily or objectively demonstrate performance differences between detectors, they may give insight into how to go about tuning an edge detector.

6: Experimental Results

The depth of results presented here is to give a general overview of the capabilities of the framework to define the performance characteristics of an edge detector. Evaluation at depth for each edge detector would go beyond the space available.

ROC graphs for 12 images from this study can be found in Figure 13, and aggregate ROC graphs for each category as well as the full set of 40 images can be found in Figure 14.

6.1: Relative Rankings on ROC graphs

6.1.1   Individual ROCs

One of the most remarkable results found is that the differences in ROC rankings between images can be split into three groups of two detectors. This relative ranking exists not only between images in a category, but also between categories.

Over almost every image, either the Canny or Heitger detectors rank first, with the other detector second. The second group is the Rothwell and Bergholm detectors, which are most often ranked third and fourth. The last group is the Sobel and Sarkar detectors, which are most often ranked fifth and sixth.

It should be noted that while the trends found are strong, there are exceptions. One such exception is the Face image labeled "00846fa010a" where the Sarkar detector's ROC curve lies below that of the other detectors until approximately 5% ground truth missing. These cases do not appear frequently enough, however, to alter aggregate rankings.

The three-parameter Heitger shows minimal differences across all images from the two parameter Heitger. Because the addition of this parameter did not significantly improve performance, and the time required to run the detector increases approximately tenfold, the results of the three parameter Heitger are not presented in the ROC curves here.

6.1.2   Aggregate ROCs

Due to the similarities between the ROC curves in individual images, the aggregate ROC graphs largely echo the results found for the individual images. The Canny and Heitger detectors rank first and second, with the Heitger generally having a slight advantage at the lowest values of ground truth missing. The Rothwell and Bergholm detectors rank third and fourth, with the Bergholm failing to perform as well in the Aerial image category. While on the other three categories the performance of the two detectors is similar, the difference in the Aerial category causes the full 40 image aggregate to show the Bergholm's performance behind that of the Rothwell. Finally, the Sarkar and Sobel detectors tend to rank fifth and sixth, their rank interchanging often over the range of ground truth missing.

6.2: Parameter Set Sizes

Figures 5, 6, 7, and 8 show the results found in the experiment here. Due to the number of images and detectors, only statistics of the raw results are shown. These aggregate statistics are the most useful aspect of these results, however.

Figure 5 shows a wide variance between detectors for the number of parameter sets required. The fineness of sampling varies due to the framework's adaptive sampling process.

Average Number of Parameter Sets Per Image

Group	Bergholm	Canny	Rothwell	Sobel	Sarkar	Heitger (2)	Heitger (3)
Aerial	1,106	538	2,154	127	1,025	1,216	12,673
Faces	5,527	2,486	2,805	230	1,339	1,909	18,267
Medical	9,436	2,412	6,086	194	10,493	2,718	17,060
Objects	4,565	645	2,583	129	746	4,549	11,275
Full	4,565	1,520	3,407	170	3,401	1,848	14,819

Figure 5. The average number of parameter sets computed for each image as determined by the framework's adaptive process. Averages for each category as well as over the full set of 40 images are shown.

Average Number of Parameter Sampling Refinements Performed

Group	Bergholm	Canny	Rothwell	Sobel	Sarkar	Heitger (2)	Heitger (3)
Aerial	4.30	3.50	5.20	2.30	4.40	5.60	8.20
Faces	5.80	4.20	5.60	3.10	4.30	5.70	9.00
Medical	6.00	5.10	6.80	2.90	7.10	6.70	8.70
Objects	5.00	3.50	5.00	2.30	3.80	6.00	7.90
Full	5.28	4.08	5.65	2.65	4.90	6.00	8.45

Figure 6. The average number of times the parameter space sampling was doubled (resulting in an increase in fineness of sampling) for each image, presented per category and the over the full set of 40 images.

A detector which requires more parameters than another to reach the equal-effort tuning point would not necessarily be less desirable unless its performance was also poorer. From this figure, it is very clear that the Canny, with the second lowest number of parameter sets used has a clear advantage over all of the detectors in this study. This figure also shows the clear disparity between the three and two parameter Heitger implementations. While the gain in performance due to adding the third parameter is marginal, the size of the parameter setting increases nearly tenfold.

The Heitger detector's analysis is slightly complicated due to the handling of parameters. Figure 5 shows that the Heitger in general requires a similar number of parameter settings as the Canny in the two parameter implementation, but Figure 6 shows that the Heitger requires on average about an extra two steps of refinement. In this case, the Canny has more parameters which individually require less tuning than the Heitger requires for each parameter. Due to the differing number of parameters, the net result of this makes the average number of parameter sets required more or less equal.

The remaining two figures are useful in determining the importance of a given parameter. For instance, in the case of the Canny, Figure 7 shows that the blur sigma and the low hysteresis threshold are less finely sampled than the high hysteresis threshold. This indicates that the performance of the Canny edge detector is most influenced by how well the high hysteresis threshold is sampled. A similar conclusion can be drawn from the Heitger detector where in the both the two and three parameter implementations the threshold is clearly more important than the other parameters.

Average Number of Times Each Parameter's Sampling was Refined

Group	Bergholm			Canny			Rothwell			Sobel	
	stop	start	thres	blur	low	high	blur	thres	alpha	low	high
Aerial	0.40	0.00	3.90	0.50	0.80	2.20	0.10	4.70	0.40	0.10	3.20
Faces	1.20	0.00	4.60	0.00	1.10	3.10	0.00	5.20	0.40	0.10	4.00
Medical	1.80	0.80	3.40	1.50	0.80	2.80	1.40	4.90	0.50	0.50	3.40
Objects	1.20	0.10	3.70	0.40	0.80	2.30	0.00	4.70	0.30	0.00	3.30
Full	1.15	0.23	3.90	0.60	0.88	2.60	0.38	4.88	0.40	0.18	3.48

Group	Sarkar			Heitger (2)		Heitger (3)		
	blur	low	high	blur	thres	ori	blur	thres
Aerial	1.30	0.50	2.60	1.50	5.10	0.60	1.60	6.00
Faces	0.50	1.50	2.30	0.80	5.90	1.60	0.90	6.50
Medical	2.80	1.30	3.00	2.10	5.60	1.00	1.70	6.00
Objects	0.50	0.50	2.80	1.70	5.30	0.20	1.40	6.30
Full	1.28	0.95	2.68	1.53	5.48	0.85	1.40	6.20

Figure 7. The average number of times each parameter's sampling was doubled (resulting in greater fineness), presented for each category and the over the full set of 40 images.

When combined with variance information provided by Figure 8, an even better conclusion can be formed on the importance of parameters. The "ideal" parameter would have both a low sampling requirement, as well as a low variance in that sampling requirement. Parameters which fit this description may also be the most easily set at some fixed value, though a much more in-depth investigation is required to determine what value to pick. Other much more complex factors come into play, such as closeness of a parameter setting to the ROC, in deciding which parameters most fixable.

All of the detectors in this study have one parameter which is sampled significantly more finely than the others. In every case, this parameter is one of the thresholds for that detector. Other parameters such as "number of search orientations" or Gaussian blur window settings do not appear to as strongly influence detector performance.

The parameter set statistics are also useful to make distinctions between categories of images. Despite that ROC performance differences are small between categories, clear distinctions can be found when examining numbers of parameter sets sampled. Across nearly all detectors, the medical images demonstrated they were more "difficult" in some sense because they require an extra refinement step or more. This difference is particularly dramatic for the Sarkar-Boyer detector which requires approximately ten times the parameter sets, or almost three extra refinement steps. Some detectors do not display this tendency, such as the Sobel and three parameter Heitger.

The conclusions that can be drawn from the parameter set statistics must be viewed in comparison with the ROC ranking results. Even though the Sobel requires many fewer parameter settings, and has one parameter which is a strong candidate for being fixed at a single value, it is ranked in the poorest performance group of detectors.

Standard Deviation of the Number of Parameter Doublings

Group	Bergholm			Canny			Rothwell			Sobel	
	stop	start	thres	blur	low	high	blur	thres	alpha	low	high
Aerial	0.52	0.00	0.88	0.71	0.79	1.23	0.32	0.67	0.70	0.32	0.42
Faces	1.62	0.00	0.84	0.00	1.29	1.20	0.00	0.92	0.52	0.32	0.67
Medical	2.04	0.92	1.35	1.18	0.63	1.81	1.17	1.29	0.71	0.71	0.84
Objects	0.42	0.32	1.34	0.70	0.92	1.16	0.00	1.34	0.48	0.00	0.48
Full	1.39	0.58	1.17	0.93	0.91	1.37	0.84	1.07	0.59	0.45	0.68

Group	Sarkar			Heitger (2)		Heitger (3)		
	blur	low	high	blur	thres	ori	blur	thres
Aerial	0.95	0.85	1.17	0.97	0.57	0.97	0.97	0.67
Faces	0.85	1.27	1.64	1.14	0.74	0.84	0.88	0.53
Medical	1.40	1.57	1.56	0.74	0.84	1.05	0.82	0.67
Objects	0.97	0.71	1.23	0.95	0.48	0.63	0.52	0.82
Full	1.40	1.20	1.38	1.04	0.72	1.00	0.84	0.69

Figure 8. The standard deviation of the number of times each parameter's sampling was doubled over the images in each category and over the full set of 40 images. This statistic shows how much a parameter's discovered fineness varies between images in the body examined.

7: Conclusions

We have introduced a framework for experimental performance evaluation of edge detection algorithms. While it is based on a fast, pixel-level examination of the image, the higher level results and interpretations appear to agree with the results of the human survey experiment by Heath *et al.* [8]. The experiments here require approximately six to eight hours of processing per detector, per image on newer computing equipment (Sun Sparc Ultras, or similar performance level machines) and are fully repeatable.

This framework's most obvious potential use is in assisting the creation of improved edge detection methods. With the ability to test new methods quickly and objectively, researchers will be able to focus their attention on areas which can be shown to have advantages.

This framework cannot easily say exactly what caused a particular detector to behave differently than another. However, if this framework was used in connection with the creation of a new detector, additions and modifications could be tested to discover what combination of those actions resulted in a better detector. This process now requires either highly subjective decisions of the designer, or a complex experiment with human subjects as in Heath *et al.* [8]. In either case, only a very small sample of the parameter space can examined. Using the method presented here, both the length of time it takes to evaluate a new detector, and the soundness of the evaluation are vastly improved.

Of the detectors rated, the detectors proposed by Canny and Heitger consistently find better edge images across all levels of ground truth, across all images. The three parameter Heitger performs marginally better than the two parameter Heitger at the cost of greatly increased computation time.

Another interesting observation is that the Sobel detector, while not the best performing, may be suitable for some edge detection tasks due to its relative speed. This is particularly the case in tasks which could tolerate more missing ground truth, since at greater than 15% missing ground truth all detectors perform similarly.

Perhaps the most surprising result, however, is that edge detector performance on the whole does not differ dramatically between different types of images. Though differences do exist, such as the Bergholm detector in the Aerial image category, and the extra parameter setting refinement required by the medical images, the general observations about relative performance of detectors are remarkably similar across image categories.

The work on this framework continues, in several directions. The most obvious direction is the completion of the work in the fixed analysis scenario. Also, work is being done to compare the results this framework reaches as compared with human subject evaluation (as in [8]), as well as a related experiment to determine if human vision object recognition prefers a certain range of percent of ground truth missing. Finally, more edge detectors are being included into the comparison.

Acknowledgments

The authors would like to thank Horst Bunke, Mike Heath, Sudeep Sarkar, Dimitry Goldgof, Min Shin, and Xiayoi Jiang for valuable discussions at different points in this work. This work is supported by NSF grants CDA-9424214, EIA-9729904 and IRI-9731821.

References

[1] Abdou, I. E. and Pratt, W. K. Quantitative design and evaluation of enhancement/thresholding edge detectors, *Proceedings of the IEEE* 67 (5), pp. 753–763, 1979.

[2] Bergholm, F. Edge Focusing, *Trans. PAMI*, vol. 9, 1987, pp. 726-741.

[3] D. J. Bryant and D. W. Bouldin, Evaluation of edge operators using relative and absolute grading, *IEEE Conf. on Pattern Recognition and Image Processing*, pp. 138–145, 1979.

[4] Canny, J. A computational approach to edge detection, *Trans. PAMI*, vol. 8, 1986, pp 679-698.

[5] K. Cho, P. Meer and J. Cabrera, Quantitative evaluation of performance through bootstrapping: edge detection, *International Symposium on Computer Vision*, pp. 491–496, 1995.

[6] J. R. Fram and E. S. Deutsch, On the quantitative evaluation of edge detection schemes and their comparison with human performance, *IEEE Trans. on Computers* 24 (6), 1975.

[7] E. S. Deutsch and J. R. Fram, A quantitative study of the orientation bias of some edge detector schemes, *IEEE Trans. on Computers* 27 (3), pp. 205-213, 1978.

[8] M. Heath, S. Sarkar, T. Sanocki, and K. Bowyer. A robust visual method for assessing the relative performance of edge detection algorithms, *IEEE Trans. PAMI* 19 (12), 1338-1359, December 1997.

[9] Heitger, F. Feature Detection using Suppression and Enhancement, Technical report nr. 163, Image Science Lab, ETH-Zurich, 1995.

[10] Jiang, X.Y., Hoover, A., Jean-Baptiste, G., Goldgof, D., Bowyer, K. and Bunke, H. A methodology for evaluating edge detection techniques for range images, *1995 Asian Conference on Computer Vision*, 1995, pp. 415-419.

[11] X. Y. Jiang, et al, A methodology for evaluating edge detection techniques for range images, *Asian Conference on Computer Vision*, pp. 415–419, 1995.

[12] T. Kanungo, M. Y. Jaisimha, and R. M. Haralick, A methodology for quantitative performance evaluation of detection algorithms, *IEEE Trans. IP* 4, pp. 1667–1674, Dec. 1995.

[13] L. Kitchen and A. Rosenfeld, Edge evaluation using local edge coherence, *IEEE Trans. SMC* 11 (9), pp. 597–605, 1981.

[14] W.H.H.J. Lunscher and M.P. Beddoes, Optimal edge detector evaluation, *IEEE Trans. SMC* 16 (2), pp. 304–312, 1986.

[15] P. L. Palmer, H. Dabis and J. Kittler, A Performance Measure for Boundary Detection Algorithms, *Computer Vision and Image Understanding* 63 (3), pp. 476–494, 1996.

[16] T. Peli and D. Malah, A study of edge detection algorithms, *Computer Graphics and Image Processing* 20, pp. 1–21, 1982.

[17] Phillips, P.J., Moon, H., Patrick, R., Rizvi, S. The FERET Evaluation Methodology for Face-Recognition Algorithms, *Computer Vision and Pattern Recognition*, 1997.

[18] V. Ramesh and R. M. Haralick, Performance characterization of edge detectors, *SPIE Vol. 1708 Applications of AI X* , pp. 252–266, 1992.

[19] Rosin, P. Edges: saliency measures and automatic thresholding, *Machine Vision and Applications 9*, 1997, pp. 139-159.

[20] Rothwell, C., Mundy, J., Hoffman, B., and Nguyen, V. Driving Vision by Topology, *International Symposium on Computer Vision*, Coral Gables, FL, Nov 1995, pp. 395-400.

[21] Salotti, M., Frabrice, B., and Garbay, C. Evaluation of Edge Detectors: Critics and Proposal, *Workshop on Performance Characterization of Vision Algorithms*, 1996. (http://svr-www.eng.cam.ac.uk/Research/Vision/ECCV/)

[22] Sarkar, S. and Boyer K.L. Optimal Infinite Impulse Response Zero Crossing Based Edge Detection, *CVGIP* Vol. 54, No. 2, pp. 224-243.

[23] Sobel, I.E. Camera Models and Machine Perception, Stanford University, 1970, pp. 277-284.

[24] L.J. Spreeuwers and F. van der Heijden, Evaluation of edge detectors using average risk, *International Conference on Pattern Recognition*, the Netherlands, pp. 771-774 of volume 3, 1992.

[25] R. N. Strickland and D. K. Chang, Adaptable edge quality metric, *Optical Engineering* 32 (5), pp. 944–951, 1993.

[26] Thornton, K., Nadadur, D., Ramesh, V.,Lui X., Zhang, X., Bedekar, A., Haralick, R. Groundtruthing The RADIUS Model-Board Imagery, *Image Understanding Workshop*, 1994.

[27] Tistarelli M., Marcenaro G. Using Optical Flow to Analyze the Motion of Human

Body Organs from Bioimages, *Proceedings of the IEEE Workshop on Biomedical Image Analysis*, 1994.

[28] S. Venkatesh and L.J. Kitchen, Edge evaluation using necessary components, *CVGIP: Graphical Models and Image Processing* 54 (1), pp. 23-30, 1992.

[29] Q. Zhu, Efficient evaluations of edge connectivity and width uniformity, *Image and Vision Computing* 14, pp. 21–34, 1996.

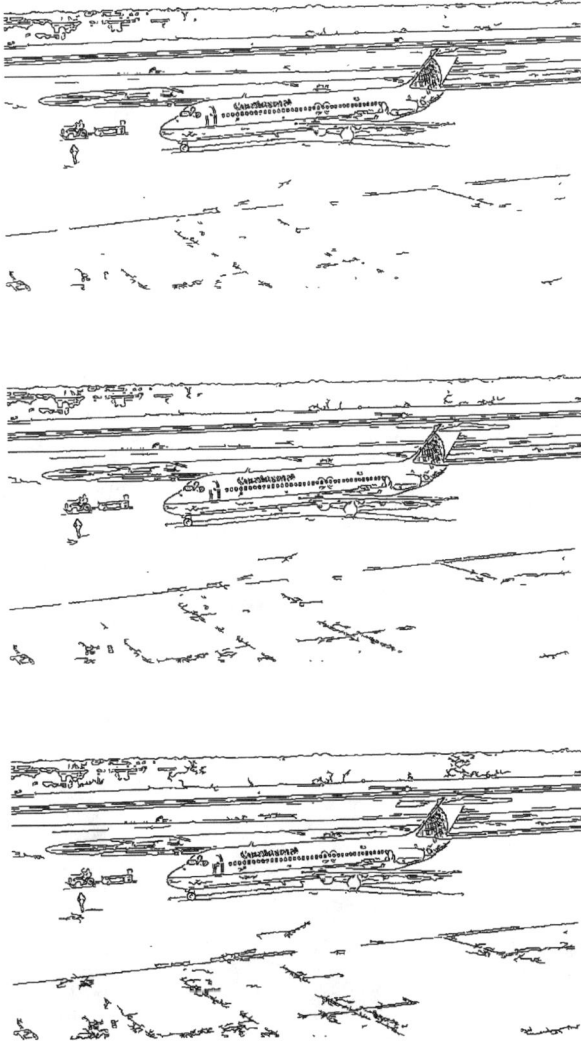

Figure 9. A sample of how the comparison tool ranks images. Going top to bottom, each image, taken from the Canny detector, contains approximately the same number of matched ground truth pixels, but an increasing number of spurious pixels. Despite only minor differences between the images, they range from approximately 1,000 in the top image to over 3,000 spurious pixels in the bottom image.

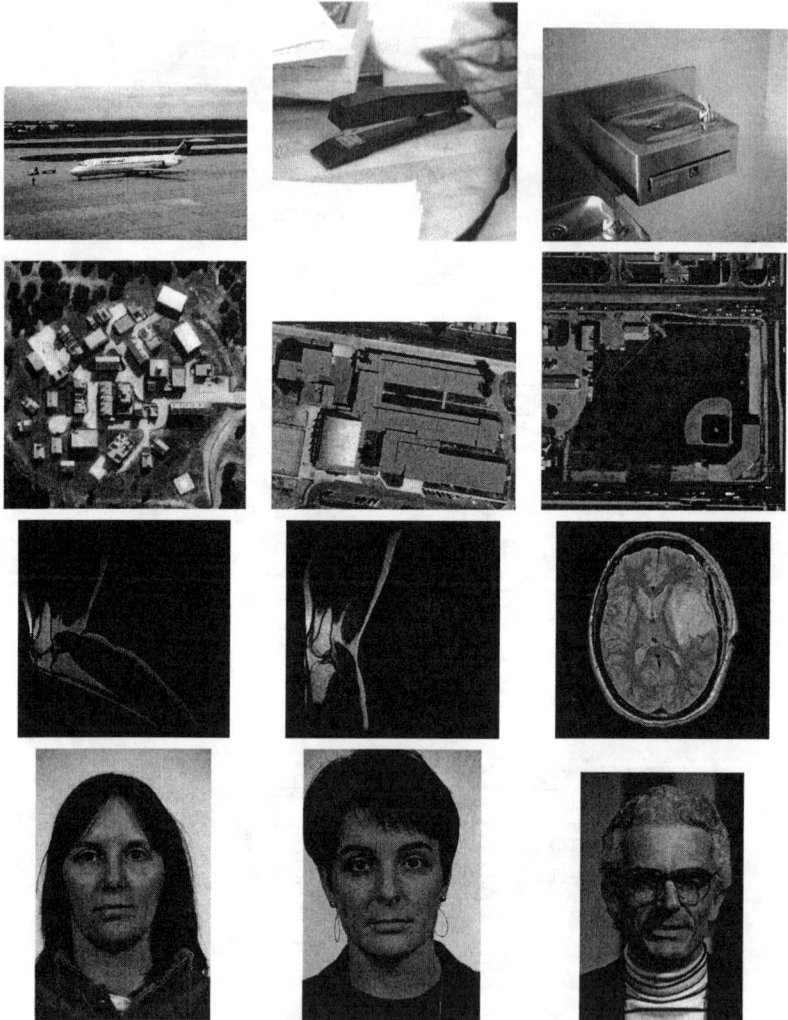

Figure 10. Twelve images selected from the 40 in the study. These images are the ones used to compute the ROC curves presented in this paper. Their ground truths can be found in the next figure.

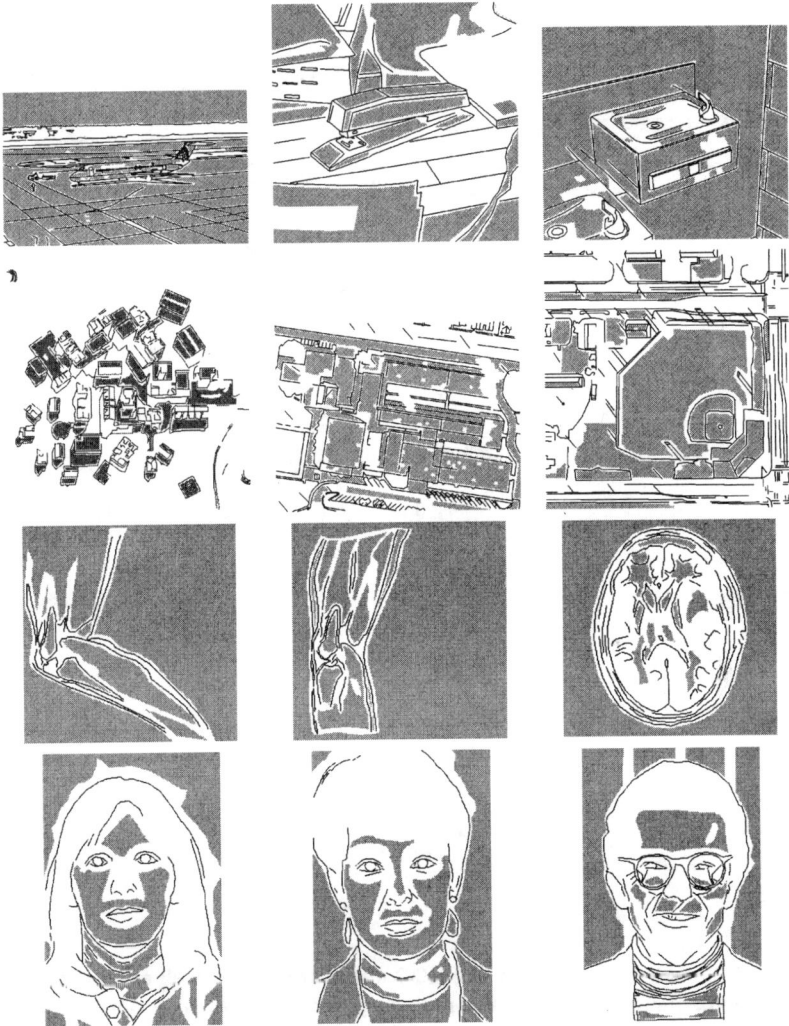

Figure 11. Twelve ground truths for the images in the previous Figure. Black lines in the image are ground truth edges, grey regions are "no-edge" regions, and the white regions are the implicit "don't count" regions.

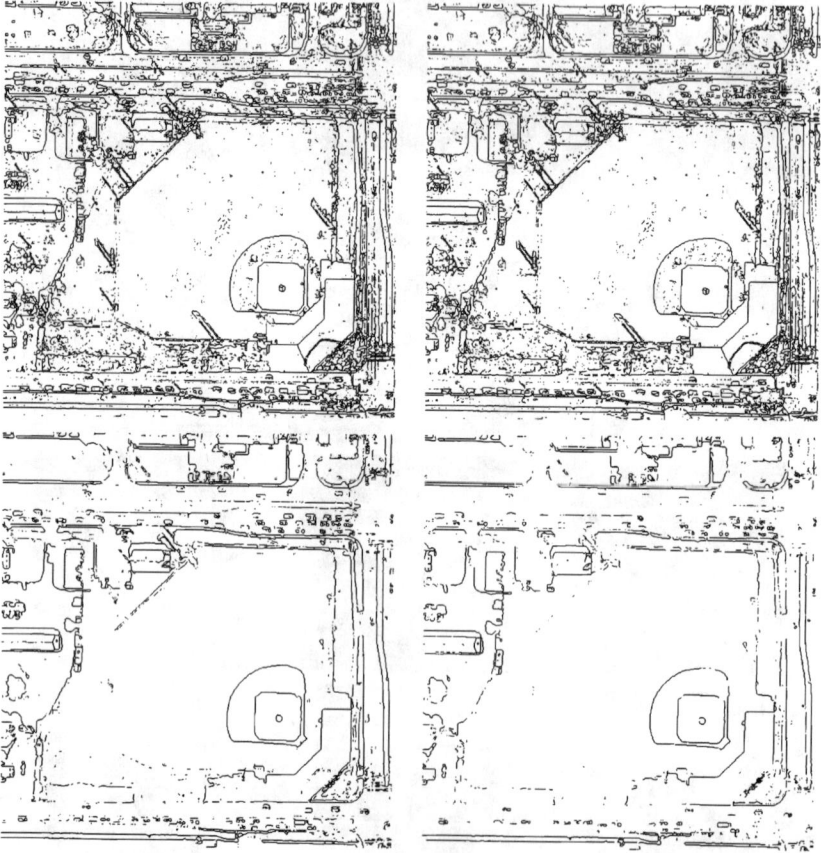

Figure 12. Selected results from the ROC curve of the Rothwell edge detector for various percentages of ground truth missing, for the baseball aerial image. They represent 6%, 10% missing ground truth for the left and right images respectively on the top row, and 30%, and 45% missing ground truth on the bottom row.

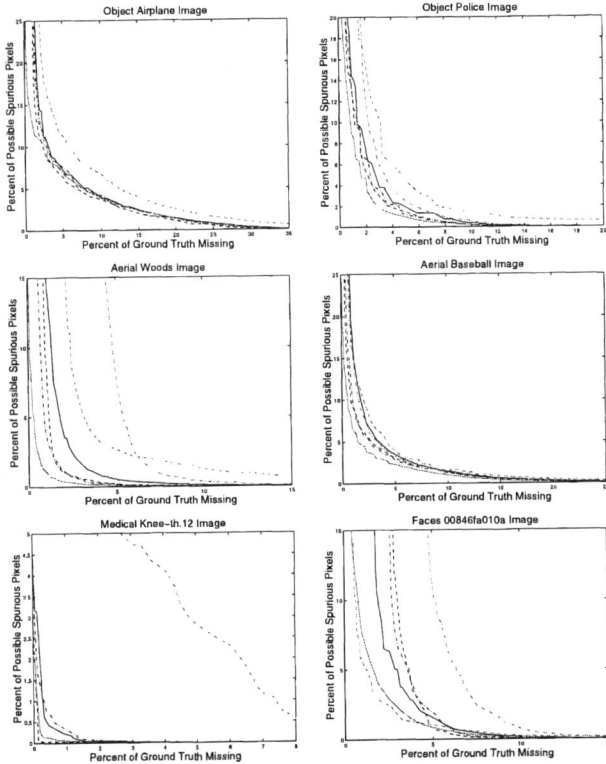

Figure 13. The ROC graphs for 6 images in this study. The labelings are as referred in the table in the figure on the next page. All ROC graphs in this work use this legend. These graphs show some normal cases. However, unusual cases are shown more often than they occur to provide variety. These cases include early asymptote, unusual rankings, or extremely poor rankings.

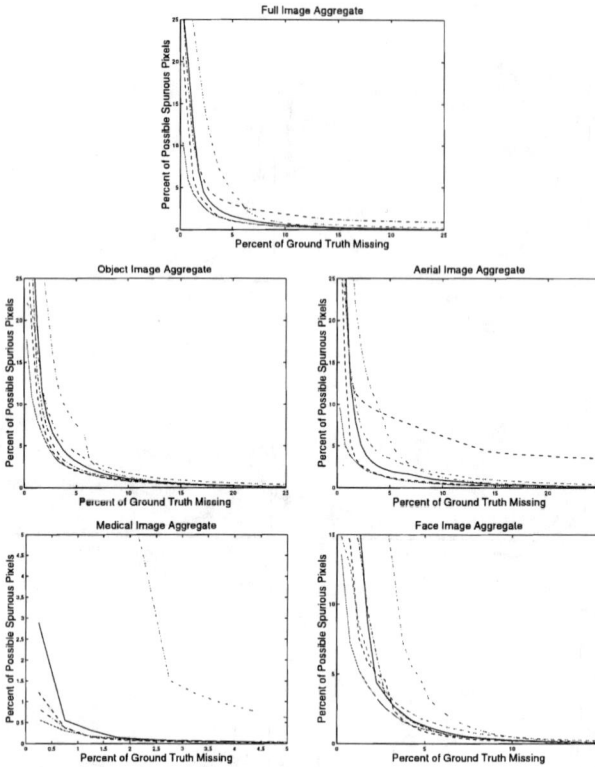

Figure 14. The aggregate ROC curves for the adapted parameter scenario over the full 40 images, as well as over the 10 images in each of the 4 categories. The graphs use the same legend as the figure on the previous page.

An Objective Comparison Methodology of Edge Detection Algorithms Using a Structure from Motion Task

Min C. Shin, Dmitry Goldgof, and Kevin W. Bowyer

Department of Computer Science & Engineering
University of South Florida
Tampa, FL 33620
shin, goldgof or kwb@csee.usf.edu

Abstract

This paper presents a task-oriented evaluation methodology for edge detectors. The performance of an edge detector is measured based on the task of structure from motion. Eighteen real image sequences from 2 different scenes varying in the complexity and scenery types are used for testing. The task-level ground truth for each image sequence is manually specified in terms of the 3D motion and structure. An automated tool computes the accuracy of the motion and structure achieved using the set of edge maps. The parameter sensitivity and speed of algorithms are also analyzed. Four edge detectors are compared. All implementations and data sets are publicly available.

1: Introduction

The lack of an accepted objective empirical methodology of evaluating even the most widely used computer vision algorithms has delayed the further application of computer vision [7]. Without such a methodology, it is difficult for the users to select the best algorithm for their need. Also, researchers who wish to develop better algorithms cannot document the true contribution of their work.

Edge detection has been an active research area for more than 30 years [30]. We have found 23 new algorithms published in just four major journals since 1992. (Refer to Table 1). The necessity of comparing with other works has been realized, in that 19 of the 22 papers compared their detector with one or more other edge detectors visually and/or quantitatively. However, only 9 of the comparisons used any previously published methodology. Four established their own quantitative comparison metrics. Seven others used only visual (qualitative) comparison. That there would be so little serious comparison is surprising considering that the first comparison methodology was published in 1975 [10]. Even more importantly, none of authors have used any real image ground truth to evaluate their algorithms. The quantitative comparisons were drawn from synthetic images. One work has used the edgemap of one edge detector as a ground truth, leading the evaluation to be *relative* to the edge detector used to construct the ground truth [23]. Even though the importance of using real images has been acknowledged, real images simply are not used

Table 1. Summary of Journal-Published Edge Detection Algorithms

source	nature of algorithm	image set	real image ground truth	compared algorithms	visual comparison	quantitative comparison
PR '95 [39]	simple directional edge detection	1 synth 3 real	0	1	Yes	None
PR '95 [41]	multi-channel filtering	3 synth 5 mosaic	0	0	No	None
PR '95 [35]	isotropical 2-D ISEF cascade	1 synth 1 real	0	1	Yes	None
PR '95 [16]	relaxation labeling	1 synth 4 real	0	2	Yes	Built Own [1]
PR '95 [26]	multi-resolution	1 synth 6 real	0	2/5 [2]	Yes	Abdou & Pratt
PR '94 [3]	genetic algorithm optimization	1 synth 1 real	0	2	Yes	Abdou & Pratt
PR '94 [37]	neural network	1 synth 1 real	0	1/5 [3]	Yes	Abdou & Pratt
PR '94 [14]	local structure	1 synth 1 real	0	4	Yes	Haralick
PR '94 [25]	co-occurrence matrix	4 synth 2 real	0	3	Yes	Abdou & Pratt
PR '93 [8]	chi-square test	3 real	0	2	Yes	None
PR '93 [47]	multi-scale	1 real	0	0	No	None
PR '92 [40]	hierarchical structure	2 synth 2 real	0	1	Yes	None
PAMI '97 [22]	facet model & relaxation labeling	3 real	0	2	Yes	Built Own
PAMI '96 [20]	fuzzy reasoning	5 real	0	2	Yes	None
PAMI '96 [44]	expansion matching	1 synth 1 real	0	1	Yes	Abdou & Pratt
PAMI '95 [15]	logical/linear operator	3 real	0	1	Yes	None
PAMI '95 [13]	covariance models	3 real	0	0 [4]	No	None
PAMI '94 [29]	expansion matching	1 synth 1 real	0	1	Yes	Abdou & Pratt
PAMI '93 [11]	refined regularization	2 synth 2 real	0	2	Yes	Haralick
GMIP '96 [24]	computational zero-crossing	1 synth 2 real	0	1	Yes	None
GMIP '93 [23]	edge detection on compressed image	2 real	1 [5]	1	Yes	Abdou & Pratt
SMC '95 [6]	regularized cubic B-spline fitting	2 synth 1 real	0	2	Yes	Built Own

1. on synthetic data only
2. 2 algorithms compared quantitatively
3. 5 algorithms compared quantitatively on synthetic data only
4. no algorithms compared for 2D edge detection
5. edge map of Sobel operator on uncompressed image used as ground truth

GMIP = Graphical Models and Image Processing
PAMI = IEEE Transactions on Pattern Analysis and Machine Intelligence
PR = Pattern Recognition
SMC = IEEE Transactions on Systems, Man, and Cybernetics

Table 2. Summary of Journal-Published Edge Detection Comparison Methods

source	nature of evaluation	images	GT required	# of algorithms compared
Fram & Deutsch '75 [10]	MLE of ratio of true edge signal	1 synth 1 real	Yes	3
Bryant & Bouldin '79 [4]	relative & absolute grading	1 real	Yes/No	3
Abdou & Pratt '79 [1]	probability of false/true detection figure of merit	1 synth 3 real	Yes	4
Shaw '77 [34]	relative edge location & direction test	1 real	Yes	5
Kitchen & Rosenfeld '81 [19]	thinness & continuation of edges	2 synth	No	4
Ramesh & Haralick '92 [28]	probability of false alarm & misdetection	1 synth 2 real	Yes	3
Peli & Malah '82 [27]	contour type, avg squared devisation mean abs value of deviation	2 synth	Yes	5
Lunscher & Beddoes '86 [21]	same as Kitchen & Rosenfeld	2 synth	No	5
Venkatesh & Kitchen '92 [43]	combined error measures with quality based weights	3 synth	Yes	4
Strickland & Chang '93 [38]	6 edge qualities	1 synth	Yes	5
Jiang '95 [18]	pixel by pixel comparison of range images	80 real (range)	Yes	9
Salotti & Bellet '96 [32]	pixel level comparison with edge classification	2 real	Yes	2
Zhu '96 [46]	connectivity & width uniformity check	2 synth 2 real	No	3
Dougherty & Bowyer '97 [9]	pixel by pixel comparison using ROC curve	40 real	Yes	6
Heath *et al* '97 [12]	subjective human object recognition	28 real	No	5

A variation of edge detector is not counted as a separate edge detector compared

during the quantitative evaluation. This is due to the fact that it is extremely difficult to build ground truth for the real images. An additional concern is that the parameter sensitivity of edge detectors is generally not acknowledged. The output of an algorithm changes significantly with the parameter setting. However, none of 23 works in Table 1 states any details about how the parameters of edge detectors compared to were searched. This leaves one to wonder if the result of the edge detectors compared to could have been better if a better parameter setting was used.

Lastly, the purpose of edge detection needs to be remembered. Edge detector is not usually a final result by itself; it is an input for the further processing. Therefore, the true performance of the edge detector lies in how well it prepares the input for the next algorithm.

We propose that a convincing comparison methodology should have the following features.

1. *A comparison method must be objective and quantitative.*

   A subjective evaluation could easily be biased by different factors. The evaluator's knowledge of and experiences with the algorithms could greatly influence the evaluation. The common practice of displaying a set of edge maps side-by-side for visual comparison carries a low significance.

2. *A comparison method must be publicly available and easily applicable.*

   The true value of the research on comparison methodology is not in comparing previously developed algorithms, but the ones which will be developed. Deriving an implementation from the published paper alone can be extremely difficult and time consuming. A comparison methodology (including the implementations and data sets) should be available for any new researchers or users who need to test algorithms for selecting the best one. Also, the method should be easily applicable to a new algorithm. A method which evaluates thoroughly yet needs a lot of adjustments or is extremely time/space consuming might require significantly more effort on the user's part.

3. *A large data set must be carefully designed using real images.*

   Twelve out of 15 edge detection comparison works used *four or less* images (combining real and synthetic) to evaluate edge detectors, while 5 of them have tested solely on synthetic data. (See Table 2). With so few images, it will be difficult to have a wide enough range of difficulty and complexity to thoroughly test algorithms. Synthetic images (even with different types of noises and manipulations) do not result in as great a credibility as using real images.

4. *A comparison method should evaluate the algorithm based on a vision task.*

   A vision task is performed using a set of algorithms. The low-level vision algorithms such as edge detection are mostly used for developing input data for other vision tasks. Though it might sound logical that more true edges and less false edges will enhance the next step using the edge information, it seems impossible to tell how different edge maps would perform on the next task without testing them. The evaluation of one algorithm based on other tasks has been delayed because it is difficult. However, it must be done to meet the true needs of the users.

In this research, the edge detection algorithms have been evaluated based on the task of structure from motion. This is the first attempt to test edge detectors based on another highly researched vision algorithm. Four edge detectors are tested using 18 real image sequences, containing a total of 278 images. Six original sequences were taken from 2 different scenes. Two shorter sequences are derived from each original sequence, resulting in 12 shorter sequences. The comparison tool is developed to objectively evaluate the performance. Once the edge maps are created by an edge detector, the evaluation program will extract and correspond lines, execute the SFM program and evaluate the result using the comparison tool. The data set, comparison tool, and intermediate process implementations are publicly available. The entire comparison process can easily be applied to other edge detectors.

2: Background

2.1: Edge Detectors

Edges are usually located where a significant amount of intensity change occurs. They are believed to contain significant information about the objects in a scene. Such a belief has motivated the computer vision community to continue investing efforts on developing new algorithms.

Four detectors were studied. The Bergholm edge detector applies a concept of **edge**

focusing to find *significant* edges [2]. The image is smoothed using a *coarse* resolution (high smoothing) and the possible edges are located. Then, the neighbors of edges from coarser resolutions are checked in finer resolution. The Canny edge detector is considered as the standard methodology of edge detection [5]. The image is smoothed with a Gaussian filter, and the edge direction and strength is computed. The edges are refined with non-maximal suppression and hysteresis. The Rothwell edge detector is similar to the Canny edge detector except 1) non-maximal suppression is not used since it is claimed that it fails at junction points, and 2) hysteresis is not used due to the belief that the edge strength is not relevant for the higher level image processing tasks [31]. The Sarkar edge detector is an Optimal Zero Crossing Operator (OZCO) [33]. Original implementations of the detectors were obtained where possible: the Bergholm from the Candela package, the Canny from the University of Michigan which was slightly modified by Mike Heath [12], the Rothwell from the ARPA IUE, and the Sarkar-Boyer from Sarkar.

2.2: Structure from Motion

A structure from motion (SFM) algorithm determines the structure (depth information) of a scene and the motion of the camera. For this comparison experiment, the structure and motion from line segments algorithm by Taylor and Kriegman was selected for several reasons [42]. First, the *working* implementation was obtainable from the web site. Second, the work had been extensively tested with synthetic and real data, and this data is also publicly available. Third, the algorithm was already tested against another structure from motion algorithm [45] and was concluded to be more stable under noise [42].

Given n images with m corresponded lines, the SFM algorithm extracts the depth information of the line (3D location of each line in the camera coordinates), and the motion of the camera. It solves the problem in terms of an objective function $\mathcal{O}$ which measures the disparity between the projected 3D line and its corresponding observed 2D line. The algorithm iterates searching for the structure and motion estimate which minimizes $\mathcal{O}$. The algorithm generates an initial random guess of camera positions for each iteration using the initial motion information. It is found that without providing any initial motion information, the algorithm usually managed to converge into a solution but after a far greater number of iterations. In order to speed up the process, a good motion estimate (rotation angle) was provided. A minimum of 3 images and 6 lines is required by the SFM algorithm, and more images or lines are allowed.

3: Framework

The comparison methodology involves four steps : 1) edge detection, 2) intermediate processing, 3) SFM, and 4) evaluation of the result.

3.1: Intermediate Processing

Given the edge maps generated by each edge detector, the intermediate processing involves extracting and corresponding lines. These steps prepare the input for the SFM program. It must be acknowledged that these steps are not perfect nor optimized since the result can vary depending on the sophistication of the algorithm used and/or the parameters selected. The overall goal is to compare the edge detectors, and algorithms that we

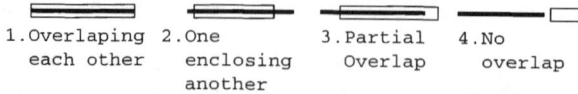

1. Overlapping   2. One   3. Partial   4. No
   each other   enclosing   Overlap   overlap
                     another

Figure 1. Three Possible Orientations of Collinear Lines

believe will give the minimum amount of advantage or disadvantage to any particular edge detectors are adopted.

A simple method of line extraction is implemented. Line segments are represented by two endpoints in pixel coordinates. First, the edge links are created from the edge map by scanning from left to right and top to bottom. A link starts with an edge pixel with only 1 neighbor. The 8-connected neighboring edge pixels are recursively linked until 1) there are no more neighbor edge pixels or 2) there is more than one neighbor, indicating a possible junction or branch. Second, the edge links are divided at a high curvature point with the curvature angle greater than $T_{high\_curv\_angle}$. Third, edge links are further broken to form line edge segments using the Polyline Splitting Technique [17] until the farthest point is closer than $T_{point\_to\_line}$. Then, the line edge segment (a chain of edge pixels) is fitted to a line using the Least Squared Estimation, and the ending edge pixels are projected to the line to the nearest pixel. These two projected points are used to describe a line. Finally, lines which are shorter than $T_{min\_line\_length}$ are eliminated.

The input of the SFM is a set of line correspondences across a sequence of images. Manually matching the lines would provide the most error-free method possible. However, it is not practical to create ground truth by matching lines from 18 image sequences (total of 278 images) for four edge detectors where each edge detector will be tested on minimum of 177 parameters. Also, the manual matching could possibly give advantage or disadvantage to a particular edge detector since the human can actually use the knowledge of the scene to result in better matches. Therefore, an automated line matching program was developed which will give 1) the minimum advantage to any particular edge detector and 2) the minimum mismatches. Note that the SFM program will produce a **wrong** result if any lines are mismatched, so minimizing mismatches is extremely important.

First, the Ground Truth (GT) lines in all images of the sequence (L_{ij} for image i and line j) are manually defined and corresponded. Second, the Machine Estimated (ME) lines (l_{ik} for image i and line k) from the Line Extraction program were corresponded using the GT lines. If a ME line (l_{ik}) matches to a GT line (L_{ij}), l_{ik} is labeled with the index of L_{ij}. Two lines are corresponded if the following two conditions are met.

1. *Collinearity*

   If the sum of the distance between the endpoints of L_{ij} to l_{ik} is less than T_{perp_dist}, they are considered to be collinear.

2. *Overlap*

   Two line segments could be collinear, yet not belong to the same part of the object/image. So, l_{ik} is projected to L_{ij}, resulting in l'_{ik} and L_{ik} is projected to l_{ij}, resulting in L'_{ik}. L_{ij} and l'_{ik} could be oriented in four different ways. (Refer to Figure 1). Obviously, the orientation #1 indicates overlap while #4 indicates non-overlap. Since one GT line segment could be broken down into several ME line segments in one image, if l'_{ik} is completely within L_{ij} (#2), they are corresponded. In case of

Table 3. Properties of Image Sequences

image set name	# of images in sequence	total rotation angle	# of lines	Ave # of corres per line	Ave line length (pixel^2)
LegoHouse1	18	160.00°	122	8.80	80.64
LegoHouse1.A	12	160.00°	122	5.94	80.17
LegoHouse1.B	9	160.00°	122	4.46	81.42
LegoHouse2	19	355.00°	104	6.89	89.44
LegoHouse2.A	13	355.00°	104	4.76	88.41
LegoHouse2.B	10	355.00°	104	3.63	89.19
LegoHouse3	20	190.00°	118	7.58	83.29
LegoHouse3.A	14	190.00°	118	5.31	84.13
LegoHouse3.B	10	180.00°	118	3.83	82.15
WoodBlock1	18	170.00°	29	11.03	132.91
WoodBlock1.A	9	160.00°	29	5.52	133.87
WoodBlock1.B	12	160.00°	29	7.31	133.69
WoodBlock2	28	275.00°	36	15.72	110.85
WoodBlock2.A	14	265.00°	36	7.94	110.56
WoodBlock2.B	17	275.00°	36	9.50	111.86
WoodBlock3	30	285.00°	47	18.17	91.93
WoodBlock3.A	15	270.00°	46	9.24	92.04
WoodBlock3.B	10	260.00°	45	6.18	92.77

# of lines, Ave # of correspondence/line, Ave line length are computed from the manually specified GT lines

the partial overlap (#3), if 1) the intersection of l'_{ik} and L_{ij} is greater than $T_{overlap}$ percent of L_{ij} and 2) the intersection of L'_{ik} and l_{ij} is greater than $T_{overlap}$ percent of l_{ij}, then they are considered overlapping. Note that many GT lines to 1 ME line correspondence is not allowed.

The SFM requires a minimum of 3 correspondences for each line. Lines with less than this T_{min_corr} correspondence are dropped.

3.2: Imagery Design

Some comparison methodologies have concentrated primarily on the evaluation criteria and ignored the importance of the dataset; 13 out of 15 edge detection comparison methodologies have used 4 or less synthetic and real images. (Refer to Table 2). The dataset should be large and thorough in order for the users of the methodology to have confidence in evaluation. Four images (including synthetic) would not seem to be thorough enough.

In this work, a large dataset of 18 sequences (6 original and 12 derived) containing 278 images is carefully designed considering different aspects influencing the edge detectors and the structure from motion task such as following. (Refer to table 3).

1. number of images in a sequence

2. total rotation angle

3. number of lines in the scene

4. average number of correspondences per line

5. average line length

LegoHouse #1 (18 images in sequence - 1st, 9th, and 18th shown)

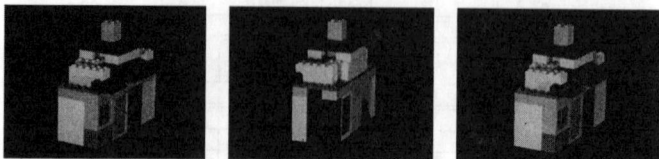

LegoHouse #2 (19 images in sequence - 1st, 9th, and 19th shown)

LegoHouse #3 (20 images in sequence - 1st, 10th, and 20th shown)

WoodBlock #1 (18 images in sequence - 1st, 9th, and 18th shown)

WoodBlock #2 (28 images in sequence - 1st, 14th, and 28th shown)

WoodBlock #3 (30 images in sequence - 1st, 15th, and 30th shown)

Figure 2. Image Scenes (6 original sequences)

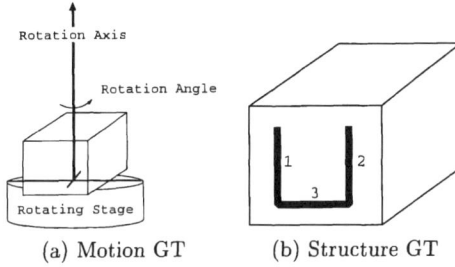

(a) Motion GT (b) Structure GT

Figure 3. Motion & Struct Ground Truth

6. lighting condition of the scene

7. scenery types

8. amount of occlusion, transparency (which results in natural noise)

3.3: Ground Truth

The ground truth is manually defined in terms of 1) the rotation angle between two consecutive images, and the rotation axis, and 2) "structure," measured as the angle between selected pairs of lines.

The object is rotated on a rotation stage according to the predetermined GT rotation angle. (Refer to Figure 3.a). In order to determine the rotation axis of the stage, a cube is placed on the calibrated rotating stage so that the straight edge of the cube is aligned with the rotation axis. The intensity and range images are taken. Two points defining the endpoints of the rotation axis are determined from the intensity image. The 3D location of two points is extracted from the range image, and the vector defined by two points is normalized to represent the rotation axis. The structure GT is defined by a set of two lines of the object and the angle between them. (Refer to Figure 3.b). For instance, line #1 and #3 are a pair making 90°.

3.4: Performance Metrics

The ME result is compared with GT in two areas : motion (rotation axis, rotation angle) and structure. Since the object is rotated along the fixed rotation axis with no translation, the motion of the camera (which SFM produces) is converted to the motion of the object by reversing the sign of the rotation angle while keeping the same rotation axis.

Two measurements for the motion (rotation axis and rotation angle) are combined by the following method. (Refer to Figure 4). First, an arbitrary point at $(1,0,0)$ is set for P_{GT_0} and P_{ME_0}. For each camera orientation j, P_{GT_j} is computed by moving P_{GT_0} with $Angle_{GT_j}$ and $Axis_{GT_j}$ while P_{ME_j} is computed with $Angle_{ME_j}$ and $Axis_{ME_j}$. Then the motion error is computed by $MotionError = \frac{ME_{error}}{GT_{move}} \cdot 100\%$ where ME_{error} is the distance between P_{GT_j} and P_{ME_j}, and GT_{move} is the distance between P_{GT_0} and P_{GT_j}. The percentage error is used since the absolute distance error would hold different significance depending on the amount of the camera movement. The structure error is measured by computing the angle difference between the ME angle and its corresponding GT angle.

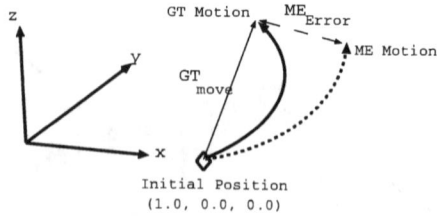

Figure 4. Motion Error Measurement

Table 4. Parameter Space of Edge detectors

edge detector	parameters		
Bergholm	S_{start} (0.5 - 5.0)	S_{end} (0.5 - 5.0)	T (5.0 - 60.0)
Canny	$sigma$ (0.01 - 5.0)	low (0.0, 1.0)	$high$ (0.0 - 1.0)
Rothwell	$sigma$ (0.5 - 4.0)	T (0.0, 60.0)	$alpha$ (0.0 - 1.0)
Sarkar	$sigma$ (0.01 - 5.0)	low (0.0, 1.0)	$high$ (0.0 - 1.0)

3.5: Parameter Training

This final result depends on parameters of the edge detector (3), line extraction (3), and line correspondence (3). Finding the best setting of 9 parameters seemed computationally infeasible. Therefore, the following method was established.

First, good parameter settings for line extraction and line correspondence are found after observing many runs of the experiment: $T_{point\_to\_line}$=5.0, $T_{high\_curv\_angle}$=90.0, $T_{min\_line\_length}$=50.0, T_{perp_dist}=5.0, $T_{overlap}$=80.0, and T_{min_corr}=3. These values are fixed for all experiments.

Second, realizing that the edge detector's performance greatly varies with parameters, an adaptive method of searching for the best parameter values is used. A 5x5x5 initial uniform sampling of parameter space is tested. The area around the best parameter point in this coarse sampling is further subsampled at 3x3x3 (with the previous best at the center). A minimum of 2 subsamplings is executed, resulting in a minimum attempt of 177 different parameter points. Subsampling is continued while there is a minimum of 5% improvement from the previous best. The parameters are trained separately for structure and motion. After the average of 2.67 subsampings and the maximum of 8 subsamplings, the best parameters were found.

4: Results

The results of 9 LegoHouse and 9 WoodBlock sequences are presented. The section is divided into three sections: Train, Test, and Parameter Sensitivity & Speed. The motion performance is obtained with the motion-trained parameters and the structure performance with the structure-trained parameters. The results are categorized into two scene groups (LegoHouse and WoodBlock) and by two metrics (motion and structure) resulting in 4 categories.

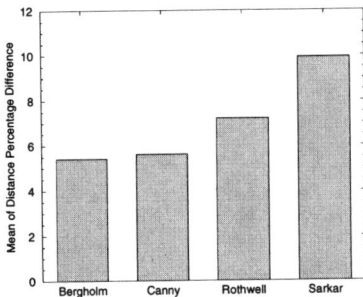

(a) Motion performance on LegoHouse (b) Structure performance on LegoHouse

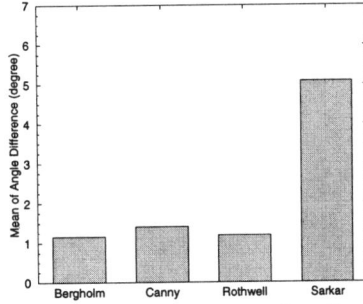

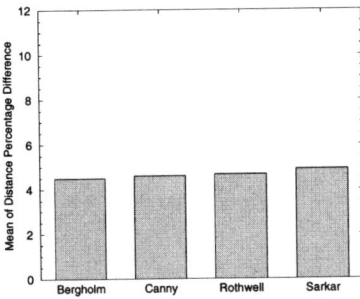

(c) Motion performance on WoodBlock (d) Structure performance on WoodBlock

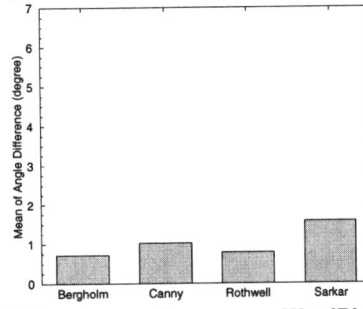

Figure 5. Train Performance

4.1: Train

Edge detectors are trained separately for each image sequence. The performance with trained parameters is the best possible for the given image sequence with the parameter training algorithm described in the previous section.

First, the Bergholm was able to obtain the best train motion and structure performance in both scene types. (Refer to Figure 5). The Canny and the Rothwell performed the second, while the Sarkar performed the worst in all 4 of them. Most of edge detectors performed better with WoodBlock sequences. This could be the result of longer lines with longer correspondences being available in the LegoHouse scenes. (Refer to Table 3). Another interesting observation is that the rankings were identical between 2 scenes for a given metric (motion or structure).

Second, the performance varies greatly with the parameter setting for all edge detectors. The mean of motion and structure performance obtained through all parameter setting attempted during the Train is plotted in Figure 6. Note the difference between the best (Figure 5) and the mean (Figure 6.): the mean motion performance was nearly twice of the best, and the structure performance was nearly 4 times of the best. Epecially, the structure performance showed a great variation among the parameter settings. This confirms the importance of the parameter training for edge detector performance. The Canny had the

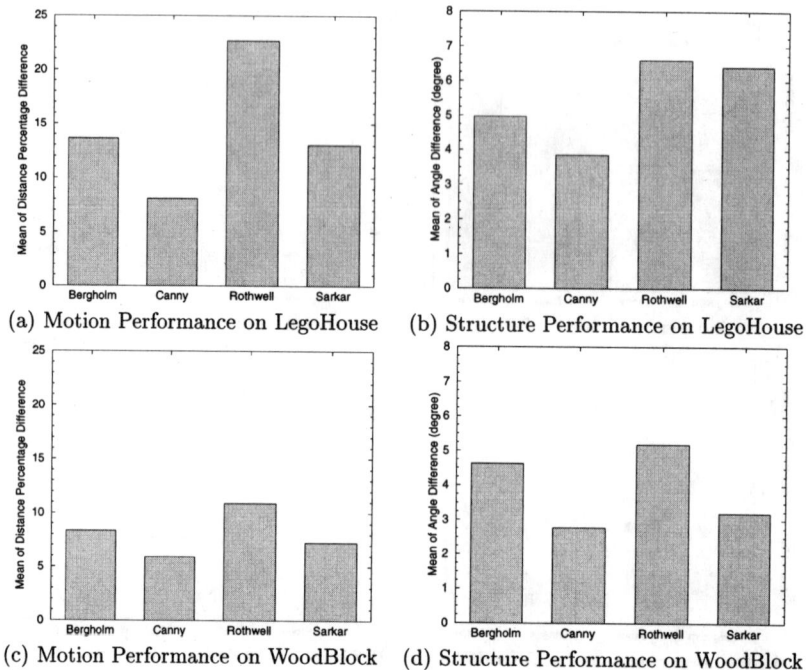

(a) Motion Performance on LegoHouse   (b) Structure Performance on LegoHouse

(c) Motion Performance on WoodBlock   (d) Structure Performance on WoodBlock

Figure 6. Mean of Performance over All Parameter Settings During Train

narrowest motion difference, while the Sarkar had the narrowest structure difference.

4.2: Test

In order to test the performance of the edge detectors on sequences different from those trained on, the parameters trained for one sequence were tested on other sequences within the scene group. In our test data, there are 9 image sequences, where each sequence is tested on all other sequences for motion and structure separately. Therefore, for each edge detector we tested 144 times for each image scene type: 9 sequence × 8 trained parameter settings from other sequences in the group × 2 trainings (for motion and for structure).

It is important to realize that under some instances (settings of edge detector parameters), the edge maps resulted in a set of corresponded lines that the SFM program could *not converge* into any solution after 1750 iterations. In order to take this problem into consideration, the analysis of Test is divided into two sections. First, the convergence rates are compared. Second, the analysis of the test trials where all edge detectors successfully converged into a solution.

First, even though the testing is performed within the same scene group, not all edge detectors converged into good solutions in all testings. The Canny showed the best convergence rate (overall 98.65%), while the Rothwell had the worst (overall 84.72%). (Refer to Figure 7.) Lower convergence rates were achieved from the structure-trained parameters

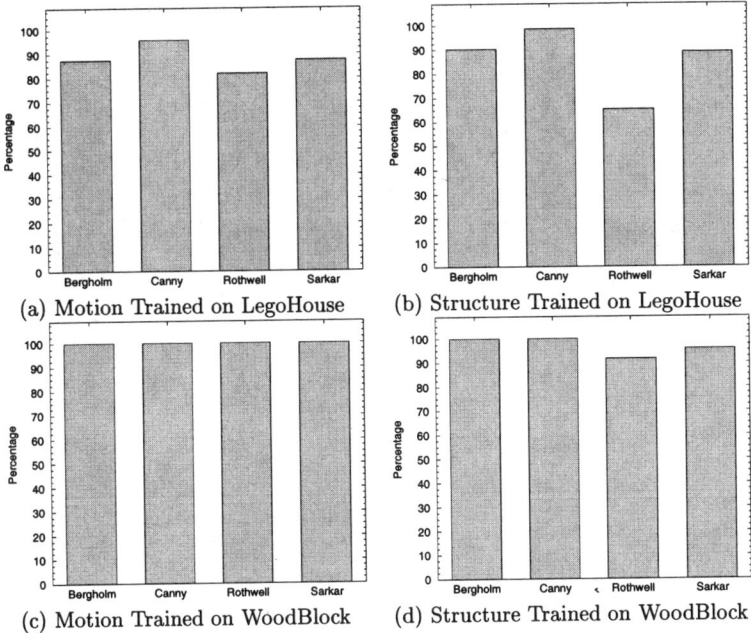

(a) Motion Trained on LegoHouse   (b) Structure Trained on LegoHouse

(c) Motion Trained on WoodBlock   (d) Structure Trained on WoodBlock

Figure 7. Convergence Rate

than for the motion-trained parameters, and also in LegoHouse scenes compared to the WoodBlock scenes.

Second, the test performance is compared. The Canny performed best in all 4 testing categories with all converged sequences. (Refer to Figure 8). The Rothwell performed the worst in 3 categories. The Bergholm, which showed the best train performance, shared the second and the third with the Sarkar. Interestingly, the Sarkar, with the worst trained performance, showed a small degradation in the test. This behavior could have been expected from a relatively small difference between the train performance (Figure 5) and the mean of all attempted parameter settings (Figure 6.)

4.3: Parameter Sensitivity & Speed

The effect of each parameter on the performance is analyzed. First, the standard deviations of the measurements of the converged parameter setting (structure and motion separately) where only one parameter was varying are calculated. Then, the mean of standard deviations is calculated for all image sequences. (Refer to Figure 9 & 10.)

First, the smoothing operators (S_{start}, S_{end}, and $sigma$) were usually the most influential parameters. Second, *low* and *high* of the Sarkar had little effect on either metric. Therefore, for the Sarkar, the parameter tuning *could be* reduced by 2 dimensions within the range of *low* and *high* which will lead to converging results. This finding is extremely valuable, since the parameter tuning is important and time consuming. However, the motion performance

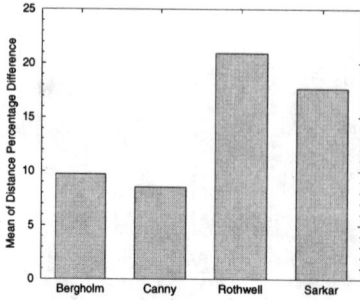

(a) Motion performance on LegoHouse
(48 converged trials out of 72)

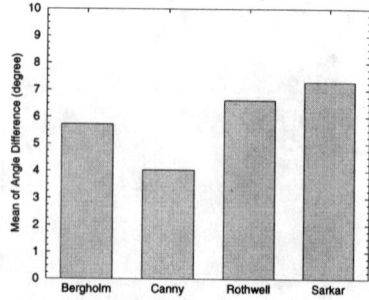

(b) Structure performance on LegoHouse
(40 converged trials out of 72)

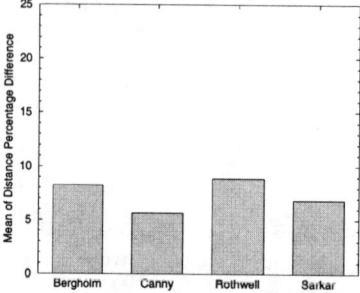

(c) Motion performance on WoodBlock
(72 converged trials out of 72)

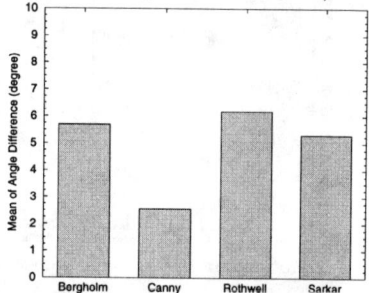

(d) Structure performance on WoodBlock
(63 converged trials out of 72)

Figure 8. Test Performance (All Common Converged Trials)

of the Sarkar in the LegoHouse scene was highly sensitive with the *sigma*. All parameters of the Bergholm and the Rothwell were highly influential while all three parameters of the Canny were usually less influential than other edge detectors' parameters. Also, the ranking of the sensitivity of the parameters (such as the ranking of S_{start}, S_{end}, T of the Bergholm) within each edge detector was similar across the 4 categories.

In order to compute the speed of the edge detectors, one image was run with the initial 5x5x5 initial parameter spaces on a Sun Ultra Sparc Workstation. The average execution times were 40 sec (Bergholm), 8 sec (Canny), 8 sec (Rothwell), and 14 sec (Sarkar). The precise differences in execution time are likely not reliable, as these was no attempt to assure comparable levels of efficiency in the different implementations.

5: Conclusion

First of all, the sensitivity of edge detectors' performance to changes in the parameters is verified. This corresponds with one conclusion of Heath *et al* 's work [12]. The edge detectors showed the average degradation from the best to the mean of the average factor of 2 (motion) and 4 (structure). (Refer to Figure 6 and Figure 5).

(a) on LegoHouse

(b) on WoodBlock

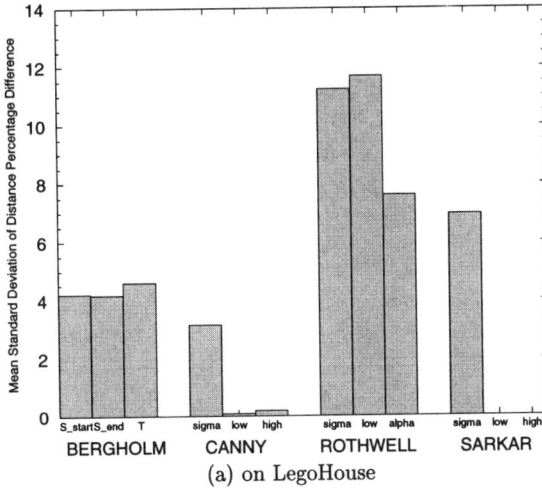

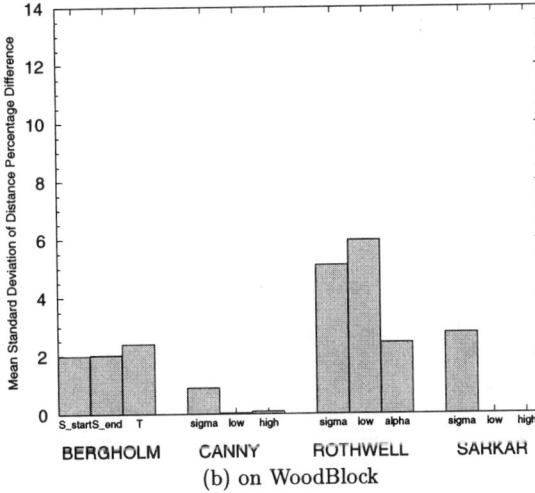

Figure 9. Parameter Sensitivity of the Motion performance

(a) on LegoHouse

(b) on WoodBlock

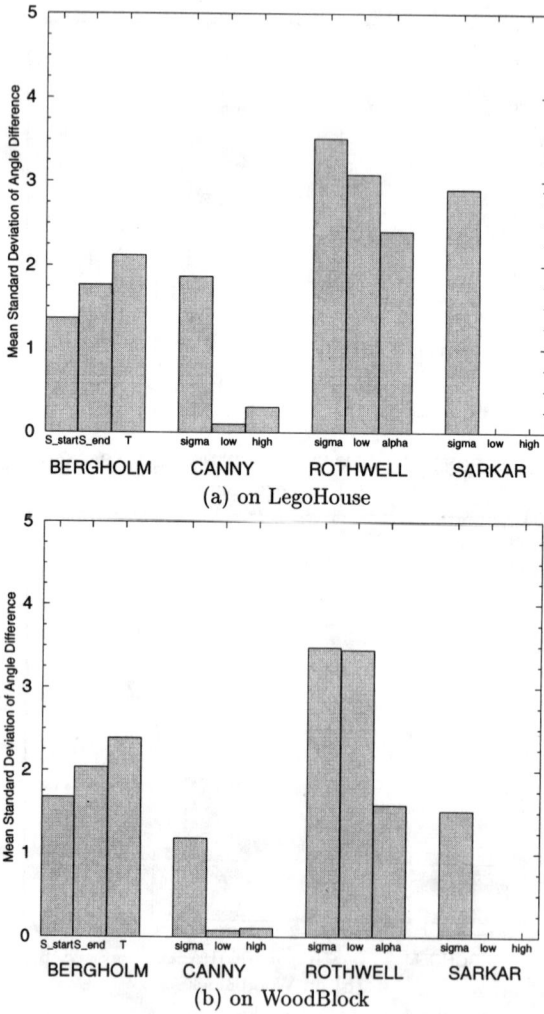

Figure 10. Parameter Sensitivity of the Structure performance

Second, the Bergholm had the best "test-on-training" performance in both metrics (Figure 5), while the Canny and the Rothwell were second. With separate test data, the Canny had the best performance with all image sequences for both motion and structure (Refer to Figure 8). In Heath *et al*'s work, the Canny performed the best when the parameter was adapted for each image and the worst when fixed for all images. Theoretically, it can be concluded that once *the optimal* parameter setting for the image sequences is found, the Bergholm can achieve the best performance, since it was the best performer in the test-on-train. However, in practice, the Canny performed better with any deviation from the training sequence.

Overall, the Canny had the lowest sensitivity to the parameter variations, the best test performance, the fastest speed, and the robustness of highest convergence rate. Thus we concluded that it performs the best for the task of structure from motion. This conclusion is similar to that reached by Heath *et al* [12] in the context of a human visual edge rating experiment, and by Dougherty and Bowyer [9] in the context of ROC curve analysis. Thus it appears that the Canny may be a preferred edge detector for a very broad range of tasks.

There are several desirable directions for extending this work. One is to include additional edge detectors. Another is to include more complex image sequences. A third is to more closely compare the results of this analysis with those of Heath *et al* [12] and Dougherty and Bowyer [9].

6: Acknowledgements

We thank Camillo Tayor and David Kriegman for answering numerous questions regarding the SFM algorithms; Mike Heath, Sean Dougherty and Sudeep Sarkar for many valuable discussion; and the authors of the edge detectors for making the codes of the edge detectors available for the research. The work was supported by the NSF grants CDA-9724422, EIA-9729904, IRI-9619240, IRI-9731821.

References

[1] I.E. Abdou and W.K. Pratt, "Quantitative design and evaluation of enhancement/thresholding edge detctors," *Proceedings of the IEEE*, Vol. 67, pp. 753-763, 1979.

[2] F. Bergholm, "Edge focusing," *IEEE Transactions on Pattern Analysis and Machine Intelligence*, vol. PAMI-9, no. 6, pp. 726, 741, November 1987.

[3] S. Bhandarkar, Y. Zhang, and W. Potter, "An edge detection technique using genetic algorithm-based optimization," *Pattern Recognition*, Vol. 27, pp. 1159-1180, Sept. 1994.

[4] D. J. Bryant and D. W. Bouldin, "Evaluation of edge operators using relative and absolute grading," *IEEE Computer Society Conference on Pattern Recognition and Image Processing*, pg. 138-145, 1979.

[5] J. Canny, "A computational approach to edge detection," *IEEE Transactions on Pattern Analysis and Machine Intelligence*, vol. PAMI-8, no. 6, pp. 679-698, November 1986.

[6] G. Chen and Y. Yang, "Edge detection by regularized cubic B-spline fitting," *IEEE Transactions on Systems, Man, and Cybernetics*, Vol. 25, pp. 636-643, 1995.

[7] H. Christensen and W. Forstner, "Performance characteristics of vision algorithms," *Machine Vision and Applications*, pp. 215-218, 1997.

[8] E. Chuang and D. Sher, "Chi-square test for feature extraction," *Pattern Recognition*, Vol. 26, pp. 1673-1683, 1993.

[9] S. Dougherty and K. W. Bowyer, "A formal framework for the objective evaluation of edge detectors," to be presented at FLAIRS '98. See also the chapter in this volume.

[10] J.R. Fram and E.S. Deutsch, "On the quantitative evaluation of edge detection schemes and their comparison with human performance," *Transactions on Computers*, Vol. 24, pp. 616-628, 1975.

[11] M. Gokmen and C. Li, "Edge detection and surface reconstruction Using Refined Regularization," *IEEE Transactions on Pattern Analysis and Machine Intelligence*, Vol. 15, pp. 492-499, May 1993.

[12] M. Heath, S. Sarkar, T. Sanocki, and K.W. Bowyer, "A Robust Visual Method for Assessing the Relative Performance of Edge Detection Algorithms," *IEEE Trans on PAMI*, 19 (12), 1338-1359, December 1997.

[13] F. Heijden, "Edge and line feature extraction based on covariance models," *IEEE Transactions on Pattern Analysis and Machine Intelligence*, Vol. 17, pp. 16-33, Jan. 1995.

[14] W. Higgins and C. Hsu, "Edge detection using two-dimensional local structure information," *Pattern Recognition*, Vol. 27, pp. 277-294, Feb. 1994.

[15] L. A. Iverson and S. W. Zucker, "Logical/linear operators for image curves," *IEEE Transactions on Pattern Analysis and Machine Intelligence*, Vol. 17, pp. 982-996, Oct. 1995.

[16] S. Iyengar and W. Deng, "An efficient edge detection algorithm using relaxation labeling technique," *Pattern Recognition*, Vol. 28, pp. 519-536, April 1995.

[17] R. Jain, R. Kasturi, and B. G. Schunck, <u>Machine Vision</u>, MIT Press and McGraw-Hill, New York, 1995.

[18] X. Y. Jiang, A. Hoover, G. Jean-Baptiste, D. Goldgof, K. Bowyer, and H. Bunke, "A methodology for evaluating edge detection techniques for range images," *1995 Asian Conference on Computer Vision*, pp. 415-419, 1995.

[19] L. Kitchen and A. Rosenfeld, "Edge evaluation using local edge coherence," *IEEE Transactions on Sysmtes, Man, and Cybernetics*, Vol. 11, pp. 597-605, 1981.

[20] T. Law, H. Itoh, and H. Seki, "Image Filtering, Edge Detection, and Edge tracing Using Fuzzy Reasoning," *IEEE Transactions on Pattern Analysis and Machine Intelligence*, Vol. 18, pp. 481-491, May 1996.

[21] W. Lunscher and M. Beddoes, "Optimal edge detector evaluation," *IEEE Transactions on System, Man, and Cybernetics*, Vol. 16, pp. 304-312, April 1986.

[22] I. Matalas, R. Benjamin, and R. Kitney, "An edge detection technique using the facet model and parameterized relaxation labeling," *IEEE Transactions on Pattern Analysis and Machine Intelligence*, Vol. 19, pp. 328-341, April 1997.

[23] G. McLean, "Codebook edge detection," *CVGIP: Graphical Models and Image Understanding*, Vol. 55, pp. 48-57, Jan. 1993.

[24] R. Mehrotra and S. Zhan, "A computational approach to zero-crossing-based two-dimensional edge detection," *Graphical Models and Image Processing*, Vol. 58, pp. 1-17, Jan. 1996.

[25] D. Park, K. Nam, and R. Park, "Edge detection in noisy images based on the co-occurrence matrix," *Patteren Recognition*, Vol. 27, pp. 765-775, June 1994.

[26] D. Park, K. Nam, and R. Park, "Multiresolution edge detection techniques," *Pattern Recognition*, Vol. 28, pp. 211-229, Feb. 1995.

[27] T. Peli and D. Malah, "A study of edge detection algorithms," *Computer Graphics and Image Processing*, Vol. 20, pp. 1-21, 1982.

[28] V. Ramesh and R. M. Haralick, "Performance characterization of edge detectors," *SPIE Vol. 1708 applications of Artificial Intelligence X: Machine Vision and Robotics*, pp. 252-266, 1992.

[29] K. Rao and J. Ben-Arie, "Optimal edge-detection using expansion matching and resotration," *IEEE Transactions on Pattern Analysis and Machine Intelligence*, Vol. 16, pp. 1169-1182, Dec. 1994.

[30] L. G. Roberts, "Machine perception of three-dimensional solids," *Optical and ElectroOptical Information Processing*, D. A. Berkowitz, L. C. Clapp, C. J. Koester, and A. Vanderburgh Jr., Eds., pp. 159-197, MIT Press, Cambridge, Massachusetts, 1965.

[31] C. A. Rothwell, J.L. Mundy, W. Hoffman, and V. D. Nguyen, "Driving vision by topology," *International Sympsium on Computer Vision, November 1995, pp. 395-400.*

[32] M. Salotti, F. Bellet, and C. Garbay, "Evaluation of edge detectors: Critics and proposal," *Workshop on Performance Characterization of Vision Algorithms*, 1996.

[33] S. Sarkar and K. L. Boyer, "Optimal infinite impulse response zero crossing based edge detectors," *Computer Vision, Graphics, and Image Processing: Image Understanding*, vol. 54, pp. 224-243, September 1991.

[34] G. B. Shaw, "Local and regional edge detectors: Some comparison," *Computer Graphics and Image Processing*, Vol. 9, pp. 135-149, 1979.

[35] Jun Shen, "Multi-edge detection by isotropical 2-D ISEF cascade," *Pattern Recognition*, Vol. 28, pp. 1871-1885, 1995.

[36] M. Shin, D. Goldgof, and K.W. Bowyer, "An Objective Comparison Methodology of Edge Detection Algorithms for Structure from Motion Task," Technical Report, Dept. of Computer Science & Engineering, University of South Florida, Tampa, FL, January 1998. Short version to appear in *Proceedings of IEEE Conference on Computer Vision and Pattern Recognition, 1998*.

[37] V. Srinivasan, P. Bhatia, and S. Ong, "Edge detection using a neural network," *Pattern Recognition*, Vol. 27, pp. 1653-1662, Dec. 1994.

[38] R. N. Strickland and D. K. Chang, "Adaptable edge quality metric," *Optical Engineering*, Vol. 32, pp. 944-951, 1993.

[39] Paul Tadrous, "A simple and sensitive method for directional edge detection in noisy images," *Pattern Recognition*, Vol. 28, pp. 1575-1586, 1995.

[40] C. Tan and S. Loh, "Efficient edge detection using hierarchical structures," *Pattern Recognition*, Vol. 26, pp. 127-135, Jan. 1993.

[41] T. N. Tan, "Texture edge detection by modelling visual cortical channels," *Pattern Recognition*, Vol. 28, pp. 1283-1298, 1995.

[42] C. Taylor and D. Kriegman, "Structure and Motion from Line Segments in Multiple Images," *IEEE Transactions on Pattern Analysis and Machine Intelligence*, Vol. 17, pp. 1021-1032, Nov. 1995.

[43] S. Venkatesh and L. Kitchen, "Edge evalution using necessary components," *CVGIP: Grahpical Models and Image Understanding*, Vol. 54, pp. 23-30, Jan. 1992.

[44] Z. Wang, K. Rao, and J. Ben-Arie, "Optimal ramp edge detection using expansion matching," *IEEE Transactions on Pattern Analysis and Machine Intelligence*, Vol. 18, pp. 1092-1097, Nov. 1996.

[45] J. Weng, Y. Liu, T. Huang, and N. Ahuja, "Structure from Line Correspondences: A Robust Linear Algorithm and Uniqueness Theorems," *Proceedings of IEEE Conference on Computer Vision and Pattern Recognition*, pp. 387-392, 1988.

[46] Q. Zhu, "Efficient evaluations of edge connectivity and width uniformity," *Image and Vision Computing*, pp. 21-34, 1996.

[47] D. Ziou and S. Tabbone, "A multi-scale edge detector," *Pattern Recognition*, Vol. 26, pp. 1305-1314, Sept. 1993.

Author Index

IEEE Computer Society Publications

The world-renowned Computer Society publishes, promotes, and distributes a wide variety of authoritative computer science and engineering texts. These books are available in two formats: 100 percent original material by authors preeminent in their field who focus on relevant topics and cutting-edge research, and reprint collections consisting of carefully selected groups of previously published papers with accompanying original introductory and explanatory text.

Submission of proposals: For guidelines and information on Computer Society books, send e-mail to cs.books@computer.org or write to the Project Editor, IEEE Computer Society, P.O. Box 3014, 10662 Los Vaqueros Circle, Los Alamitos, CA 90720-1314. Telephone +1 714-821-8380. FAX +1 714-761-1784.

IEEE Computer Society Proceedings

The Computer Society also produces and actively promotes the proceedings of more than 130 acclaimed international conferences each year in multimedia formats that include hard and softcover books, CD-ROMs, videos, and on-line publications.

For information on Computer Society proceedings, send e-mail to cs.books@computer.org or write to Proceedings, IEEE Computer Society, P.O. Box 3014, 10662 Los Vaqueros Circle, Los Alamitos, CA 90720-1314. Telephone +1 714-821-8380. FAX +1 714-761-1784.

Additional information regarding the Computer Society, conferences and proceedings, CD-ROMs, videos, and books can also be accessed from our web site at http://computer.org/cspress